Symmetry and the Standard Model

Matthew Robinson

Symmetry and the Standard Model

Mathematics and Particle Physics

Matthew Robinson

ISBN 978-1-4899-9777-7 ISBN 978-1-4419-8267-4 (eBook)
DOI 10.1007/978-1-4419-8267-4
Springer New York Dordrecht Heidelberg London

Softcover re-print of the Hardcover 1st edition 2011

Printed on acid-free paper

Springer is part of Springer Science+Business Media (www.springer.com)

To our teachers

Celeste Brockington
Joe Stieve
Virginia Cooke
Bruce Hrivnak
Chris Hughes
Scotty Johnson
Donald Koetke
Rick Kuhlman
Robert Manweiler
Stan Zygmunt

for showing us that anything that
can be explained can be explained clearly

Preface

Motivation for This Series

First of all, we want to point out that this book is by no means meant to compete with or take the place of any of the standard quantum field theory, particle physics, or mathematics texts currently available. There are too many outstanding choices to try to add yet another to the list. Our goal is, simply put, to teach physicists the math that is used in particle physics.

The origin of this goal is the plight of upper-level undergraduate and first/second year graduate students in physics, especially those in theory. Generally, after four years of standard undergraduate coursework and two years of standard graduate coursework, the road to understanding a modern research paper in particle theory is a long, hard hike. And as the physics becomes more and more advanced, the necessary math becomes more sophisticated at an overwhelming rate. At least, it is overwhelming for those of us who don't understanding everything immediately.

To make matters worse, the way physicists and mathematicians think about nearly everything in math and physics can be (and usually is) vastly different. The way a mathematician approaches differential equations, Lie groups, or fiber bundles is typically unlike the way a physicist approaches them. The language used is often very different, and the things that are important are almost always different. When physics students realize that they need a better understanding of how one does analysis on manifolds (e.g. at a graduate level), reading a graduate-level book about analysis on manifolds (by a mathematician) is very often frustrating and unhelpful. This shouldn't be taken to reflect poorly on our friends in the math department or on their pedagogical abilities. It is simply indicative of the wide gulf between two different disciplines.

Nevertheless, despite being different disciplines, the language of physics is mathematics. If you want to understand the inner workings of nature, you have to understand analysis on manifolds, as well as countless other topics, at a graduate level at least. But because physicists are not used to thinking in the way mathematicians think,[1] they will make much more progress when things are explained in a way that is "friendly to a physicist", at least at first. For example, if you show a physicist the formal definition of an ideal or of cohomology (when they've never encountered those ideas before) they will usually find it very difficult to intuit what they actually are. However, if you say something like "an ideal is essentially all the multiples of something, like all the multiples of 7 on the real line" or "cohomology is essentially a way of measuring how many holes are in something", and slowly build up the formal definition from there, progress will be much faster. The downside to this approach is that any mathematicians standing nearby will likely get very annoyed because "friendly to a physicist" usually translates to "lacking rigor".

While mathematicians are correct in pointing out that we often (usually) lack rigor in how we think about their ideas, it is still good for us to do it. For example, after understanding that cohomology is essentially a way of measuring how many holes there are in something, the physicist will have a much easier time parsing through the formal definition. If there are a few non-trivial (but still simple) examples scattered along the way, there is a good chance that the physicist will develop a very good understanding of the "real" mathematical details.

However, with only a few exceptions, there are math books and there are physics books. When physicists write physics books they generally try to go as far as they can with as little advanced math as they can, or they assume that the reader is already familiar with the underlying math. And when mathematicians write math books, they either don't care about the physical applications or they mention them only briefly and maintain abstract mathematical formalism with some physics vocabulary. While neither of these situations is the fault of the authors, it can often be to the detriment of helping eager physics students get any real intuition for what lines drawn around a hole have to do with magnetic fields.

So that is the context of this series of books. We're trying to teach math in a way that is ~~lacking rigor~~ friendly to physicists. The goal is that after reading this, one of the many excellent introductory texts on relativistic quantum mechanics, quantum field theory, or particle physics will be much more accessible.

Outline of This Series

This is the first in a series of books intended to teach math to physicists. The current plans are for at least four volumes. Each of the first four volumes will discuss a variety of mathematical topics, but each will have a particular emphasis. Furthermore, a substantial portion of each will discuss, in detail, how specific mathematical ideas are used in particle physics.

The first volume will emphasize algebra, primarily group theory. In the first part we will discuss at length the nature of group theory and the major related

[1]The converse is typically very true as well.

ideas, with a special emphasis on Lie groups. The second part will then use these ideas to build a modern formulation of quantum field theory and the tools that are used in particle physics. In keeping with the theme, the formulations and tools will be approached from a heavily algebraic perspective. Finally, the first volume will discuss the structure of the standard model (again, focusing on the algebraic structure) and the attempts to extend and generalize it. As a comment, this does not mean that this volume is *solely* about algebra. We will talk about and use a variety of mathematics (i.e. we'll use analysis, geometry, statistics, etc.) – we'll just be primarily using algebra.

The second volume will emphasize geometry and topology in a fairly classical way. The first part will discuss differential geometry and algebraic topology, and the second part will combine these ideas to discuss more fundamental formulations of classical field theory and electrodynamics, and gravitation.

The third volume will once again emphasize geometry and topology, but in a more modern context – namely through fibre bundles. The major mathematical goal will be to build up a primer on global analysis (mathematical relationships between locally defined geometric data and globally defined topological data). The physical application in the second part of this volume will be a fairly comprehensive and robust overview of gauge theory, and a reformulation of the standard model in modern terms.

Finally, the fourth volume will emphasize real and complex analysis. The physical application will then be the study of particle interactions (a topic that is glaringly, but deliberately, absent from the first volume). This will include detailed discussions of renormalization, scattering amplitudes, decay rates, and all of the other topics that generally make up the major bulk of introductory quantum field theory and particle physics courses.

So, over the first four volumes, we will cover algebra, geometry, topology, and real and complex analysis – four of the major areas in mathematics. Furthermore, we will have discussed how all of this math ties in to classical field theory, quantum field theory, general relativity, gauge theory, non-perturbative quantum field theory, particle interactions, and renormalization. In other words, the first four volumes are intended to be a fairly comprehensive introduction to modern physics.

However, we do wish to reiterate that while we are hoping to be as comprehensive as possible, we only mean in scope, not in depth. Once again, our goal is *not* to replace any of the standard physics or math texts currently available. Rather, our goal is that these volumes will act as either a primer for those texts (so that after reading these books, you will find those to be highly approachable), or as supplemental references (to assist you when an idea is not clear).

As a final comment before moving on, we do have long term plans for more volumes. We would likely break the theme of having one mathematical emphasis and simply build on the math from the first four volumes as necessary. The tentative plan for the topics in the later volumes is:

- Volume V – Supersymmetry and Supergravity
- Volume VI – Conformal Field Theory and Introductory String Theory

- Volume VII – D-Branes and M-Theory
- Volume VIII – Algebraic Geometry and Advanced Topics in String Theory
- Volume IX – Cosmology and Astrophysics
- Volume X – String Cosmology

As a warning, this series is a "spare-time project" for all of us and the timeline for these volumes will, unfortunately, be quite drawn out.

Outline of This Volume

As we said, the emphasis of this book is algebra, and the physical application is the (algebraic) structure of the standard model. However, because this book is the first in the series, there is quite a bit of additional material that is not entirely vital to the logical flow of this volume or the series as a whole.

Chapter 1 (which is vital to the overall flow) is a primer in the classical prerequisites. In short, it is undergraduate physics using graduate notation. Consequently it is much more cursory than the subsequent chapters. The major idea is to review:

- The variational calculus formulation of classical mechanics
- Special relativity
- Classical field theory (primarily electromagnetism)

Chapter 2 is meant to serve as a reminder that despite all of the math, we are still trying to describe something physical at the end of the day. This chapter is therefore an overview of how experimentalists think about particle physics. We talk about the history of particle physics and how particles and interactions are organized. The content of this chapter is not vital to the overall flow of the book, but is vital to any self-respecting theorist and should be read carefully.

Chapter 3 (which is vital) is meant to serve as an introduction to group theory. There are three major sections:

- Basic group theory. This section focuses on finite, discrete groups because many group theoretic ideas are easier to intuit in this setting. We won't use finite discrete groups much later in the book (or series), but the ideas illustrated by them will be constantly used.
- Basic Lie group theory.
- A specific Lie group: the Lorentz group of special relativity.

Chapter 4 (which is vital) then begins with the real physical content of this book. Using the algebraic machinery developed previously, we discuss the three major types of fields: spin-0 fields (also called scalar fields or Klein-Gordon fields), spin-$1/2$ fields (also called spinor fields or Dirac fields), and spin-1 fields (also called vector fields). Then we discuss gauge theory, the algebraic framework that seems to describe all of particle physics. Next is quantization, then symmetry breaking and non-Abelian gauge theory, and finally we look at the standard model itself.

Finally, Chap. 5 (which is not vital, but is highly recommended) is a survey of several of the extensions and generalizations of the standard model, including $SU(5)$ and $SO(10)$, supersymmetry, and approaches to quantum gravity. As a warning, this chapter is meant to be a vast, mountaintop overview of several ideas. It is not meant to be a thorough introduction to anything, and you will likely find that a lot of it leaves you wanting more. We will be coming back to almost every topic in Chap. 5 in later volumes in much greater detail. We encourage you to view it as a way of getting familiar with the generic ideas and vocabulary, not as a way of gaining deep understanding.

Acknowledgments

I would first of all like to thank Laura Serna for her thorough editing of the entire book.

Next I'd like to thank Dr. Tibra Ali of Perimeter Scholars International. He has taught me more physics and math in the past 8 years than I can begin to recall. And perhaps more importantly, he more than anyone else taught me to seek intuition through rigor. I know few people who are as deliberate about "deep understanding".

I'd also like to thank the faculty of Baylor University for their endless support throughout the process of writing this book, most notably Dr. Gregory Benesh and Dr. Anzhong Wang.

Dr. Gerald Cleaver wishes to acknowledge and thank V.H. Satheeshkumar for useful discussions on quantum gravity, especially with regard to Hořava-Lifshitz theory.

And finally, I'd like to thank those who helped edit various portions of the book, including Markus Hunziker, Ben McKeown, and Joe Orville.

Contents

Contributing Authors

Despite the single name on the cover, this book was by no means a solo effort. I've been fortunate enough to work with several other outstanding physicists/writers who are far more qualified than I am to write on many (or perhaps all) topics, and their contributions have made this book much better than I ever could have made it.

Chapter 2 was written by Dr. Jay Dittmann and Dr. Karen Bland of Baylor University. Dr. Dittmann is a professor at Baylor and he leads the Experimental High Energy Physics group's research on the CDF experiment at Fermilab, where he and Dr. Bland measure the properties of fundamental particles in proton-antiproton collisions.

Section 4.8 was written by Dr. Mario Serna of the United States Air Force Academy and myself.

Chapter 5 was written by Dr. Gerald Cleaver and Dr. Serna. Dr. Cleaver is a professor at Baylor University and the head of the Early Universe Cosmology and Strings group of Baylor's Center for Astrophysics, Space Physics, and Engineering Research (CASPER).

As for the rest of the book, I had tremendous help from all of the other authors. From editing to many (extremely) lengthy discussions on what needed to be added, changed, taken out, improved, and removed, the entire book is genuinely a collaborative effort. This is the reason for the use of first person plural throughout the book.

Chapter 1
Review of Classical Physics

As we said above, this chapter will be somewhat cursory. We assume that if you're reading this you are already at least somewhat familiar with these ideas, and we therefore won't spend much time detailing them. The following chapters will include much, much more explanation of the ideas contained in them.

The flip side of this is that you can think of this chapter as a good test of whether or not you have the prerequisites for this book. If you find that you can get through this chapter (even if you're limping a little at the end) then you'll probably be fine for the rest of the book. If, on the other hand, you can't follow what we're doing here at all, it may be best to look through some of the texts mentioned in the further reading section at the end of this chapter before diving in here.

1.1 Hamilton's Principle

Just about everything in physics begins with a **Lagrangian**, which is defined (the first time you see it) as the kinetic energy minus the potential energy,[1]

$$L = T - V, \tag{1.1}$$

where $T = T(q, \dot{q})$ and $V = V(q)$, and q is a general position coordinate (like x, y, r, θ, etc.). The dot in the term $\dot{q}$ represents a time derivative, so $\dot{q}$ represents velocity. Then the **Action** is defined as the integral of the Lagrangian from an initial time to a final time,

$$S = \int_{t_i}^{t_f} dt L(q, \dot{q}). \tag{1.2}$$

[1] We'll discuss a possible reason for this definition in a few pages.

M. Robinson, *Symmetry and the Standard Model: Mathematics and Particle Physics*, DOI 10.1007/978-1-4419-8267-4_1, © Springer Science+Business Media, LLC 2011

It is important to realize that S is a "functional" of the particle's world-line in $(q, \dot{q})$ space, not a function. This means that it depends on the entire path $(q, \dot{q})$ rather than a given point on the path (hence the integral). The only fixed points on the path are $q(t_i)$, $q(t_f)$, $\dot{q}(t_i)$, and $\dot{q}(t_f)$. The rest of the path is generally unconstrained, and the value of S depends on the entire path.

An analogy that may help is a room full of sand. Imagine that you have to walk from one corner of the room to another corner of the room in a certain amount of time, and at each step you have to pick up some amount of sand and drop some other amount of sand. And let's say that how much sand you have to pick up is a function of not only where you are in the room, but also how fast you're moving. And let's say that how much sand you drop is a function only of where you are in the room. If we then call the amount of sand you pick up $p = p(q, \dot{q})$, and the amount of sand you drop $d = d(q)$, then at each point you are gaining $l = p(q, \dot{q}) - d(q)$ units of sand.[2] So, after traveling some path across the room, you will have a total of

$$s = \int_{t_i}^{t_f} dt\, l(q, \dot{q}) \tag{1.3}$$

units of sand. This total amount of sand is a function of the path you took to cross the room as well as how fast you moved along that path. This poses an interesting mathematical problem. Let's say we want to move through the room collecting as little sand as possible. How could we go about finding the $(q, \dot{q})$ path that would do this?

This problem, it turns out, leads to one of the most fundamental ideas in physics: **Hamilton's Principle**. It says that as a particle moves through space, it "picks up" some kinetic energy $T(q, \dot{q})$ and "drops" some potential energy $V(q)$, and that nature will always choose a path that extremizes the amount of energy the particle picks up along the path.[3] In more formal language, Hamilton's principle says that nature will always choose a path in $(q, \dot{q})$ space that extremizes the functional S. Nature always chooses the most efficient way to get from one place to another.

But there is a practical problem with finding this "most efficient" path. Because S is a functional (and not a function), which depends on the entire path in $(q, \dot{q})$ space rather than a point, it cannot be extremized in the "Calculus I" sense of merely setting the derivative equal to 0. We'll need to do something slightly different.

To motivate the solution, let's look more closely at the "Calculus I" way of finding an extremum point. As we said above, we set the derivative equal to 0. In other words, x_0 is an extremal point of a function $f(x)$ if

$$\left.\frac{df(x)}{dx}\right|_{x_0} = f'(x_0) = 0 \tag{1.4}$$

[2]For the sake of the analogy let's say that you can have a "negative" amount of sand.

[3]Note that potential energy can be positive or negative.

(the term in the middle is a slight abuse of notation, but its meaning should be clear). Then, looking at the general Taylor expansion of $f(x)$ around some arbitrary point $x = x_0$, we have

$$f(x) = f(x_0) + f'(x_0)(x - x_0) + \frac{1}{2!}f''(x_0)(x - x_0)^2 + \cdots . \tag{1.5}$$

Therefore, x_0 is an extremum of $f(x)$, the first-order term $(f'(x_0)(x - x_0))$ will be zero.

Put another way, equation (1.5) allows us to approximate the value of a function at one point in terms of information about the function at another point. Let's say we know everything about $f(x)$ at $x = x_1$ but we want to approximate $f(x)$ at $x = x_2$. If x_1 is reasonably close to x_2 we can keep only first-order terms and (1.5) gives

$$f(x_2) = f(x_1) + f'(x_1)(x_2 - x_1). \tag{1.6}$$

But as we just noted, if x_1 is an extremal point of $f(x)$, then we have

$$f(x_2) = f(x_1) \tag{1.7}$$

for x_2 sufficiently close to x_1. In this sense, we can think of the first-order term

$$\delta f = f'(x_1)(x_2 - x_1) \tag{1.8}$$

as a perturbation from $f(x_1)$ when moving a distance $(x_2 - x_1)$. So

$$f(x_2) = f(x_1) + \delta f. \tag{1.9}$$

And as we just saw, if x_1 is an extremal point, we have

$$\delta f = 0, \tag{1.10}$$

where δf is the first-order shift away from the original point x_1. In other words, a point is an extremum point if and only if a point very close to it has no first-order correction.

So, looking back at our functional case, while this general idea will work, there is a complication because we can't do a Taylor expansion around a point – we have to expand around an entire $(q, \dot{q})$ path. Doing this (and the results that follow) is the underlying idea in **Variational Calculus**, which we will use in a great deal of this book. Our approach will be to start with the functional action S along some $(q, \dot{q})$ path, move slightly away from this path, to the path $(q + \delta q, \dot{q} + \delta \dot{q})$, expand this to get the first-order term δS, and then set this equal to 0 to see what constraints this places on $(q, \dot{q})$.

Consider some arbitrary path $(q_0, \dot{q}_0)$ with action

$$S_0 = \int_{t_i}^{t_f} dt\, L(q_0, \dot{q}_0). \tag{1.11}$$

We can shift slightly away from this path, and the resulting action will be

$$S_\delta = \int_{t_i}^{t_f} dt\, L(q_0 + \delta q, \dot{q}_0 + \delta \dot{q}). \tag{1.12}$$

Taylor expanding this to first order gives

$$\begin{aligned} S_\delta &= \int_{t_i}^{t_f} dt\, L(q_0, \dot{q}_0) + \int_{t_i}^{t_f} dt \left(\delta q \frac{\partial L}{\partial q} + \delta \dot{q} \frac{\partial L}{\partial \dot{q}} \right) \\ &= \int_{t_i}^{t_f} dt\, L(q_0, \dot{q}_0) + \int_{t_i}^{t_f} dt \left(\delta q \frac{\partial L}{\partial q} + \frac{\partial L}{\partial \dot{q}} \frac{d}{dt} \delta q \right). \end{aligned} \tag{1.13}$$

Integrating the second term in the parentheses by parts (and taking the variation of δq and $\delta \dot{q}$ to be 0 at t_i and t_f – these are the fixed points of the path because we require it to start and stop at particular places), we have

$$\begin{aligned} S_\delta &= \int_{t_i}^{t_f} dt\, L(q_0, \dot{q}_0) + \int_{t_i}^{t_f} dt \delta q \left(\frac{\partial L}{\partial q} - \frac{d}{dt} \frac{\partial L}{\partial \dot{q}} \right) \\ &= S_0 + \delta S. \end{aligned} \tag{1.14}$$

So, our constraint on $(q, \dot{q})$ to be an extremal path (setting the first-order term equal to 0) is

$$\delta S = \int_{t_i}^{t_f} dt \delta q \left(\frac{\partial L}{\partial q} - \frac{d}{dt} \frac{\partial L}{\partial \dot{q}} \right) = 0. \tag{1.15}$$

The only way to guarantee this for an arbitrary variation δq from the path $(q, \dot{q})$ is to require

$$\frac{d}{dt} \frac{\partial L}{\partial \dot{q}} - \frac{\partial L}{\partial q} = 0. \tag{1.16}$$

This equation is called the **Euler-Lagrange** equation, and it produces the equations of motion of the particle.

The generalization to multiple coordinates q_i $(i = 1, \ldots, n)$ is straightforward:

$$\frac{d}{dt} \frac{\partial L}{\partial \dot{q}_i} - \frac{\partial L}{\partial q_i} = 0. \tag{1.17}$$

If, for example, we have motion in two dimensions (x and y) with Lagrangian L, we would have two equations of motion

$$\begin{aligned} \frac{d}{dt}\frac{\partial L}{\partial \dot{x}} - \frac{\partial L}{\partial x} &= 0, \\ \frac{d}{dt}\frac{\partial L}{\partial \dot{y}} - \frac{\partial L}{\partial y} &= 0. \end{aligned} \tag{1.18}$$

A simple mathematical example of this is to find the extremal path between two points in a plane. The Lagrangian in this case can be simply taken as the length travelled, ds:

$$S_{length} = \int_a^b ds, \tag{1.19}$$

where a and b are the initial and final x values for the path. The (infinitesimal form of the) Pythagorean formula allows us to rewrite ds using $ds^2 = dx^2 + dy^2$. So, we have

$$\begin{aligned} S_{length} &= \int_a^b ds \\ &= \int_a^b \sqrt{dx^2 + dy^2} \\ &= \int_a^b \sqrt{1 + \left(\frac{dy}{dx}\right)^2} dx \\ &= \int_a^b \sqrt{1 + (y')^2} dx. \end{aligned} \tag{1.20}$$

So the Lagrangian is $L = \sqrt{1 + (y')^2}$, where x takes the place of t and y takes the place of q. The Euler-Lagrange equation (1.17) here gives:

$$\begin{aligned} \frac{\partial L}{\partial y} &= 0 \\ \frac{d}{dx}\frac{\partial L}{\partial y'} &= \frac{d}{dx}\frac{\partial}{\partial y'}\left(\sqrt{1 + (y')^2}\right) \\ &= \frac{d}{dx}\left(\frac{y'}{\sqrt{1 + (y')^2}}\right). \end{aligned} \tag{1.21}$$

So

$$\frac{d}{dx}\left(\frac{y'}{\sqrt{1+(y')^2}}\right) = 0,$$
$$\Longrightarrow \left(\frac{y'}{\sqrt{1+(y')^2}}\right) = A, \tag{1.22}$$

where A is some constant. Rearranging this gives

$$(y')^2 = A^2\big(1+(y')^2\big) \Longrightarrow y' = \sqrt{\frac{A^2}{1-A^2}} = M, \tag{1.23}$$

where M is yet another constant. Integrating this is straightforward:

$$y' = \frac{dy}{dx} = M \Longrightarrow y(x) = Mx + B, \tag{1.24}$$

where B is a final constant. This is the equation of a straight line, and we have therefore proven that the shortest distance between two points in the plane is a line. While this isn't a terribly profound realization, the fact that our variational calculus approach has allowed us to prove this is certainly remarkable. And while we will look at many more physical examples later, suffice it now to say that different Lagrangians result in different $(q, \dot{q})$ paths just as a different choice for the Pythagorean theorem (say, the length formula on a sphere, which is very different than on the plane) result in different "shortest paths".[4]

1.2 Noether's Theorem

When taking introductory physics, students often notice that a lot of the tools they are using are "conservation" laws: conservation of energy, conservation of momentum, conservation of mass, conservation of charge, etc. This is a very important observation, and it turns out that these conservation laws are actually specific manifestations of very deep aspects of the mathematical structure of physics. The idea of all of the conservation laws is that no matter how the system changes, there is something (energy, momentum, charge, etc.) that stays the same.

[4]As a preview, this is in some sense the fundamental idea behind general relativity. The form the Pythagorean theorem (which we'll see in a few pages is also called the "metric") takes depends on the geometry of the space you're in – i.e. it is different on the flat plane than on the curved sphere. Mathematically the fundamental equation of general relativity, Einstein's field equation, is a relationship between energy and the metric. So just as the "most efficient path" on a sphere is different than in a plane, when energy (i.e. mass) changes the metric via Einstein's field equation, matter follows different paths. We call this effect 'gravity'.

It turns out that this idea flows very naturally from the Lagrangian structure we discussed in the previous section. Given a Lagrangian, we can find a special collection of mathematical transformations on the Lagrangian (see below) that correspond to the physical conservation laws mentioned above.

To see this, consider a Lagrangian $L = L(q, \dot{q})$, and then make an infinitesimal transformation away from the original path:

$$q \to q + \epsilon \delta q, \tag{1.25}$$

where ϵ is some infinitesimal constant ($\epsilon \ll 1$) included for later convenience. This transformation will give

$$L(q, \dot{q}) \to L(q + \epsilon \delta q, \dot{q} + \epsilon \delta \dot{q}) = L(q, \dot{q}) + \epsilon \delta q \frac{\partial L}{\partial q} + \epsilon \delta \dot{q} \frac{\partial L}{\partial \dot{q}}. \tag{1.26}$$

If the Euler-Lagrange equations of motion are satisfied, so that $\frac{\partial L}{\partial q} = \frac{d}{dt}\frac{\partial L}{\partial \dot{q}}$, then under $q \to q + \epsilon \delta q$,

$$L \to L + \epsilon \delta q \frac{\partial L}{\partial q} + \epsilon \delta \dot{q} \frac{\partial L}{\partial \dot{q}} = L + \epsilon \delta q \frac{d}{dt}\frac{\partial L}{\partial \dot{q}} + \epsilon \frac{\partial L}{\partial \dot{q}} \frac{d}{dt} \delta q = L + \frac{d}{dt}\left(\frac{\partial L}{\partial \dot{q}} \epsilon \delta q\right). \tag{1.27}$$

So under $q \to q + \epsilon \delta q$ we have the first-order change $\delta L = \frac{d}{dt}\left(\frac{\partial L}{\partial \dot{q}} \epsilon \delta q\right)$. We define the **Noether Current**, j, as

$$j = \frac{\partial L}{\partial \dot{q}} \delta q. \tag{1.28}$$

Now, if we can find some transformation δq that leaves the action invariant, or in other words such that $\delta S = 0$, then

$$\delta L = \frac{d}{dt}\left(\frac{\partial L}{\partial \dot{q}} \epsilon \delta q\right) = \frac{dj}{dt} = 0, \tag{1.29}$$

and so the current j is a constant in time. In other words, j is *conserved.*

As a familiar example, consider a projectile described by the Lagrangian

$$L = \frac{1}{2} m (\dot{x}^2 + \dot{y}^2) - mgy. \tag{1.30}$$

This will be unchanged under the transformation

$$x \to x + \epsilon, \tag{1.31}$$

where ϵ is any constant (here, $\delta q = 1$ in the above notation), because

$$x \to x + \epsilon \Rightarrow \dot{x} \to \dot{x}. \tag{1.32}$$

So,

$$j = \frac{\partial L}{\partial \dot{q}} \delta q = m\dot{x} \tag{1.33}$$

is conserved. We recognize $m\dot{x}$ as the momentum in the x-direction, which we expect to be conserved by conservation of momentum from Physics I.

For the sake of illustration, notice that we can find the equations of motion directly from this using the Euler-Lagrange equations (1.17):

$$\begin{aligned} \frac{d}{dt}\frac{\partial L}{\partial \dot{x}} - \frac{\partial L}{\partial x} &= 0, \\ \Longrightarrow \frac{d}{dt}(m\dot{x}) - 0 &= 0, \\ m\ddot{x} = F_x &= 0. \end{aligned} \tag{1.34}$$

and

$$\begin{aligned} \frac{d}{dt}\frac{\partial L}{\partial \dot{y}} - \frac{\partial L}{\partial y} &= 0, \\ \Longrightarrow \frac{d}{dt}(m\dot{y}) + mg &= 0, \\ m\ddot{y} = F_y &= -mg. \end{aligned} \tag{1.35}$$

So, using only the action, we have "derived" the conservation of momentum and we have derived Newton's Laws, from which we can derive the equations of motion for the particle:

$$\begin{aligned} m\ddot{x} = 0 &\Longrightarrow \dot{x} = v_{0,x} \\ &\Longrightarrow x = v_{0,x}t + x_0. \end{aligned} \tag{1.36}$$

and

$$\begin{aligned} m\ddot{y} = -mg &\Longrightarrow \dot{y} = v_{0,y} - gt \\ &\Longrightarrow y = y_0 + v_{0,y}t - \frac{1}{2}gt^2. \end{aligned} \tag{1.37}$$

You should recognize these as the standard kinematical equations for a projectile.

So to summarize the primary point of this section, **Noether's Theorem** says that whenever there is a continuous symmetry in the action for a physical object there is a corresponding conserved quantity. Conservation of linear momentum comes from an action that is invariant under continuous linear translations. Conservation of angular momentum comes from an action that is invariant under continuous rotations (write out, for example, the action for a planet in orbit and see if you can derive the conservation of angular momentum). Things like conservation of charge are a little more complicated, but we will see them later in this book.

While this is an enormously powerful idea in physics, it is actually fairly simple. Consider the projectile again. It should be clear that the Lagrangian should be invariant under translations in x. The path the ball moves in certainly shouldn't depend on where I'm standing across the surface of the earth (my x position). And if the motion is unaffected by where I am standing, then there's no reason its behavior "in the x direction" should ever change – whatever it starts off doing is what it should keep doing.[5] And therefore its x-motion, or momentum in the x direction, remains the same. However, this argument fails with the y component because of the gravitational field; the path the ball follows does depend on the y direction, and therefore the y momentum is not conserved.

The same thing holds for angular momentum. The behavior of a planet in orbit is unaffected by rotations around the planet, and therefore the Lagrangian is unaffected by these rotations. And, consequently, the momentum around the planet (or the angular momentum) is constant. All conservation laws are really saying is that if nothing changes in a certain direction, motion in that direction won't change either.

1.3 Conservation of Energy

In the last section we saw that the conservation of *momentum* comes from the invariance of the action under translations in *space*. In this sense momentum and space have a special relationship. Now, we'll see that energy and time share a similar relationship – namely conservation of energy comes from the invariance of an action under translations in *time*.

Consider the rate of change of the Lagrangian with respect to time. There is no reason to expect the Lagrangian to be constant with respect to time and therefore $\frac{dL}{dt} \neq 0$. However, there is a very important property we can get from $\frac{dL}{dt}$ if we assume the equations of motion of the particle are satisfied. Starting with the general total derivative of L with respect to time, we have

$$\frac{dL}{dt} = \frac{d}{dt}L(q,\dot{q}) = \frac{\partial L}{\partial q}\frac{dq}{dt} + \frac{\partial L}{\partial \dot{q}}\frac{d\dot{q}}{dt} + \frac{\partial L}{\partial t}. \tag{1.38}$$

---

[5]In terms of mechanics this is simply Newton's First Law. There is a very profound geometric generalization of this involving geodesics and curved spacetime manifolds that we'll look at in the next book in this series.

But because L does not depend explicitly on time $\left(\frac{\partial L}{\partial t} = 0\right)$ this can be rewritten as

$$\frac{dL}{dt} = \frac{\partial L}{\partial q}\dot{q} + \frac{\partial L}{\partial \dot{q}}\ddot{q} = \left(\frac{d}{dt}\frac{\partial L}{\partial \dot{q}}\right)\dot{q} + \frac{\partial L}{\partial \dot{q}}\ddot{q} = \frac{d}{dt}\left(\frac{\partial L}{\partial \dot{q}}\dot{q}\right) \tag{1.39}$$

where we have used the Euler-Lagrange equation to get the second equality. So, taking the far left and far right sides of this equation we have $\frac{dL}{dt} = \frac{d}{dt}\left(\frac{\partial L}{\partial \dot{q}}\dot{q}\right)$, or

$$\frac{d}{dt}\left(\frac{\partial L}{\partial \dot{q}}\dot{q} - L\right) = 0. \tag{1.40}$$

For a general non-relativistic system, $L = T - V$, so $\frac{\partial L}{\partial \dot{q}} = \frac{\partial T}{\partial \dot{q}}$ because V is a function of q only, and normally

$$T \propto \dot{q}^2 \quad \Rightarrow \quad \frac{\partial L}{\partial \dot{q}}\dot{q} = 2T. \tag{1.41}$$

So,

$$\frac{\partial L}{\partial \dot{q}}\dot{q} - L = 2T - (T - V) = T + V = E \tag{1.42}$$

is the total energy of the system. And, by (1.40) this is

$$\frac{dE}{dt} = 0, \tag{1.43}$$

the total energy is conserved. We call this total energy $T + V = H$, the **Hamiltonian**, or total energy function, for the system.

Finally, note that the first term in the parentheses in equation (1.40) is the same as in the definition of a general conserved Noether current as in (1.28). This leads us to actually *define* the momentum of the particle in terms of this expression,

$$\frac{\partial L}{\partial \dot{q}} = p. \tag{1.44}$$

Then, the relationship between the Lagrangian and the Hamiltonian is the Legendre transformation[6]

$$p\dot{q} - L = H. \tag{1.45}$$

[6]Don't worry about the name "Legendre Transformation". We include it merely for your own reference.

1.4 Special Relativity

Now that we've established the underlying mathematical framework we'll be using (variational calculus, Lagrangians, and Noether currents), we'll begin with an actual discussion of physics. The first of these ideas is Einstein's theory of special relativity. We'll proceed in a somewhat unusual way, not following the route most textbooks follow, but rather taking an approach that is a bit more mathematical, and should prepare us better for what is to come. Bear with us for a few pages while we build up the following ideas.

1.4.1 Dot Products and Metrics

We'll start with something that should already be familiar: dot products. Working in two dimensions for simplicity, the dot product of two vectors, say $\mathbf{v}$ and $\mathbf{w}$, is

$$\mathbf{v} \cdot \mathbf{w} = v_x w_x + v_y w_y. \tag{1.46}$$

But if we want to write this out carefully, sticking with the usual matrix multiplication rules, there is an important thing that is often overlooked – the multiplication is actually done between a *row* vector and a *column* vector:

$$\mathbf{v} \cdot \mathbf{w} = \mathbf{v}^T \mathbf{w} = \begin{pmatrix} v_x & v_y \end{pmatrix} \begin{pmatrix} w_x \\ w_y \end{pmatrix} = v_x w_x + v_y w_y = \mathbf{w}^T \mathbf{v} = \mathbf{w} \cdot \mathbf{v}. \tag{1.47}$$

So if we're going to be very careful about dot products we need to be specific about when we mean a row vector and when we mean a column vector.[7] We therefore adopt the notational convention as shown below:

$$\begin{aligned} \mathbf{v} = v^i &= \begin{pmatrix} v_x \\ v_y \end{pmatrix}, \\ \mathbf{v}^T = v_i &= \begin{pmatrix} v_x & v_y \end{pmatrix}. \end{aligned} \tag{1.48}$$

A superscript indicates a column vector and a subscript indicates a row vector.[8] Then, adopting the convention that anytime a superscript index is the same as a

[7] It may seem at this point that we're splitting hairs, but it will be more clear why this distinction is important in a few pages. And it will be *much* clearer in the next book.

[8] If you have studied any differential geometry or general relativity you should be aware that there are very deep geometrical things happening with these raised and lowered indices. If you haven't studied those things that's fine; it won't be necessary for this book and we'll discuss it in great length in the next book. For now, just be aware of when we mean row vectors and when we mean column vectors and how they multiply.

subscript index they are summed, we can write the dot product as:

$$\mathbf{v} \cdot \mathbf{w} = \mathbf{w} \cdot \mathbf{v} = v_i w^i = v^i w_i = w_i v^i = w^i v_i = v_x w_x + v_y w_y. \qquad (1.49)$$

A repeated index like this is called a "dummy" index.[9]

This new convention makes writing dot products very easy, but it takes a very important idea for granted. We have implicitly assumed that, given the components of a column vector v^i, we can "turn it into" a row vector by simply taking a transpose. This is because we are used to thinking in terms of Euclidian space where this is in fact the case. However, it doesn't need to be. What we want is a more general way of turning the components of a column vector into a row vector so that we can take dot products. We'll stick with linear transformations (for reasons that will be clear later), so the most general way we can do this is to say that the row vector components are given in terms of some matrix M_{ij}, which is called the **Metric** matrix.[10] So, for a column vector v^i, we have the row vector components given by

$$v_i = M_{ij} v^j \qquad (1.50)$$

(remember that the repeated index on the right-hand side is summed, whereas the index that is not repeated is not summed). In the Euclidian space we're used to, all we're doing is taking a transpose, so we simply have $M_{ij} = \delta_{ij}$, the Kronecker delta function, and so

$$v_i = \delta_{ij} v^j = v^i \qquad (1.51)$$

as usual. However, let's say we want something more elaborate.[11] Let's take the "rule" for turning a column vector into a row vector (so that we can take dot products) to be given by

$$M_{ij} = \begin{pmatrix} -1 & 1 \\ 1 & 0 \end{pmatrix}. \qquad (1.52)$$

Now we have (switching to numbers for indices instead of x and y)

$$\begin{aligned} v_1 &= M_{1j} v^j = M_{11} v^1 + M_{12} v^2 = -v^1 + v^2, \\ v_2 &= M_{2j} v^j = M_{21} v^1 + M_{22} v^2 = v^1. \end{aligned} \qquad (1.53)$$

[9]This convention for summing over repeated indices is often called the "Einstein Summation Convention". Usually the only time it is called this is when it is first introduced in textbooks.

[10]As you may be aware, metrics are an extremely, extremely deep mathematical structure. They will take center stage for much of the second book in this series. For now, you can think of them as merely defining dot products.

[11]For now don't worry about why we'd want to do this; we'll motivate it later.

And now the dot product of v^i and w^i is

$$v_i w^i = M_{ij} v^i w^j = (-v^1 + v^2)w^1 + v^1 w^2. \tag{1.54}$$

This may seem like a very strange way to think about dot products, but it is actually what you have been doing every time you take a dot product in introductory physics courses – you just always take $M_{ij} = \delta_{ij}$. But there is no reason that a dot product *must* be defined this way; any rule for turning a column vector into a row vector is as good as any other rule.

So, given two column vectors v^i and w^i, the matrix M_{ij} allows us to take a dot product easily; it is simply

$$\mathbf{v} \cdot \mathbf{w} = M_{ij} v^i w^j. \tag{1.55}$$

But there are a few properties of M_{ij} that we want to notice. First of all, it should be a symmetric matrix because $\mathbf{v} \cdot \mathbf{w} = \mathbf{w} \cdot \mathbf{v}$:

$$\mathbf{v} \cdot \mathbf{w} = M_{ij} v^i w^j = M_{ji} v^i w^j = M_{ji} w^j v^i = \mathbf{w} \cdot \mathbf{v}. \tag{1.56}$$

(The first equality is the definition of a dot product, the second assumes M_{ij} is symmetric, the third exploits the fact that we can relabel dummy indices however we want, and the fourth is again the definition of a dot product).

Second, if we can turn v^i into v_i, it makes sense that we should be able to invert this and turn v_i into v^i. This requires that M_{ij} be invertible. So denote the inverse of M_{ij} as $M^{ij} = (M_{ij})^{-1}$. Then, we have

$$v^i = M^{ij} v_j = M^{ij} M_{jk} v^k \quad \Longleftrightarrow \quad M^{ij} M_{jk} = \delta^i_k. \tag{1.57}$$

The conventions we've specified in this section will be in use throughout the rest of this book and the books to follow. The important idea is that the dot product between vectors (which are assumed to be column vectors by default) actually involves turning one of them into a row vector so we can use normal matrix multiplication rules. And we therefore need some way of getting the row vector components from the column vector components, which is provided by the metric. The Euclidian case, $M_{ij} = \delta_{ij}$ is the simplest and most familiar, but as we will see shortly this is not the only case nature seems to make use of.

1.4.2 The Theory of Special Relativity

Now that we have some new notation and a new way of thinking about dot products, we are ready to introduce special relativity. The first idea is to combine space and time into a single geometric structure called *spacetime*. This is done simply by adding an extra dimension to a vector. Whereas a spatial vector was $v^i = \begin{pmatrix} x & y \end{pmatrix}^T$ (the T superscript indicates the matrix transpose), we now refer to the space*time* vector v^μ as

$$v^\mu = \begin{pmatrix} v^0 \\ v^1 \\ v^2 \end{pmatrix} = \begin{pmatrix} ct \\ x \\ y \end{pmatrix} \tag{1.58}$$

(we're still ignoring the third spatial dimension for simplicity – the generalization is straightforward). The value c is a constant with dimensions $\frac{length}{time}$ (to make the units match) – specifically, it is the speed of light. We can choose our units so that $c = 1$ however, so from now on we will simply write $ct = t$. Also, we're initiating a convention that will stick with us for the remainder of this book. Whenever we use a Greek index ($\mu, \nu, \rho, \lambda, \sigma$, etc.) we will mean an index that runs over all space and all time. When we use only a Latin index (i, j, k, l, etc.) we will mean an index that runs only over the spatial components. So, for example, we have

$$\begin{aligned} v^\mu &= (t, x, y, z)^T \\ v^i &= (x, y, z)^T . \end{aligned} \tag{1.59}$$

We will also use Latin indices for other types of components (non-spacetime); the context should make it clear which we mean in those cases. In this section we will always be using either spacetime or space indices.

The real content of special relativity is in defining the metric. One may guess that the correct generalization from 2-dimensional space to 3-dimensional spacetime is to make our metric the 3-dimensional Kronecker delta matrix δ_{ij}. However, it turns out that the spacetime we actually live in is a bit different. Rather than the metric

$$\delta_{ij} = \begin{pmatrix} 1 & 0 & 0 \\ 0 & 1 & 0 \\ 0 & 0 & 1 \end{pmatrix} \tag{1.60}$$

we instead take the metric to be

$$\eta_{\mu\nu} = \begin{pmatrix} -1 & 0 & 0 \\ 0 & 1 & 0 \\ 0 & 0 & 1 \end{pmatrix} . \tag{1.61}$$

It turns out that this apparently subtle change accounts for almost the entire content of special relativity. The metric $\eta_{\mu\nu}$ is called the **Minkowski** metric, and the resulting dot product (of, say, a vector with itself) is

$$\mathbf{v} \cdot \mathbf{v} = \eta_{\mu\nu} v^\mu v^\nu = -t^2 + x^2 + y^2 . \tag{1.62}$$

The underlying hypothesis of special relativity is that two observers who each measure an event in different inertial (non-accelerating) reference frames will generally measure different values for its spatial and temporal components. However, according to the theory, all inertial reference frames are physically equivalent. Mathematically this is requiring that the inner products be the same. In other words if one observer sees t, x, and y, and the other sees t', x', and y' which may in general not be equal, we will still have

$$-t^2 + x^2 + y^2 = -t'^2 + x'^2 + y'^2. \tag{1.63}$$

This requirement places very strict constraints on how the two observers' reference frames can be related. It turns out that they are related according to the following transformations. The first is a simple rotation of spacetime; for example, around the z axis (mixing x and y):

$$\begin{aligned} t' &= t, \\ x' &= x \cos\theta + y \sin\theta, \\ y' &= -x \sin\theta + y \cos\theta. \end{aligned} \tag{1.64}$$

A transformation of this type is called a **Euler Transformation**. The second, more complicated type of transformation is one that mixes a space and time dimension. For example, if we assume that the second (primed) frame is moving only in the x direction with respect to the first (unprimed) frame (and that their x and y axes are aligned), then the transformation is

$$\begin{aligned} ct' &= \gamma(ct - \beta x), \\ x' &= \gamma(x - \beta ct), \\ y' &= y. \end{aligned} \tag{1.65}$$

(where we have included the c's for completeness) where $\beta = \frac{v}{c}$ and $\gamma = \frac{1}{\sqrt{1-\beta^2}}$. These transformations are called **Lorentz Boosts**. The two different types of transformations (1.64) and (1.65), along with the rotations and boosts involving the other spacetime dimensions, are the complete set of all continuous transformations that leave the dot product (1.62) unchanged. Together, Euler rotations and Lorentz boosts are called **Lorentz Transformations**.

Notice that when an object is moving very slowly, $\beta = \frac{v}{c}$ is small and can be taken to be zero. This makes γ equal to 1, and the transformations (1.65) reduce to the standard Galilean transformation rules for frames moving relative to each other:

$$\begin{aligned} ct' &= ct, \\ x' &= x - vt, \\ y' &= y. \end{aligned} \tag{1.66}$$

We could go on to discuss the significant number of physical implications of these transformation laws, including time dilation, length contraction, the relationship between energy and mass, etc. But because all of this can be found any almost any introductory text (cf. the further reading section at the end of this chapter), and because we won't need it for this book, we merely encourage you to read it on your own to whatever extent you are interested.

1.4.3 Lorentz Transformations Revisited

Before moving on, we will stop and look at the structure of Lorentz transformations in a bit more detail.

As we briefly mentioned at the beginning of Sect. 1.2, one of the major themes we'll be working with is the idea of something that doesn't change when other things do change. The major example of this above was a Noether current. We've now seen another one – transformations that preserve a dot product. The two examples we saw above were Euler rotations and Lorentz boosts. We'll now formalize this type of thinking.

Consider some transformation matrix $R = R^{\mu}_{\nu}$ which acts[12] on a vector v^{μ} as

$$v^{\mu} \longrightarrow v'^{\mu} = R^{\mu}_{\nu} v^{\nu}. \tag{1.67}$$

We are interested in the set of all transformations R that don't change the dot product. In other words, we want to find the most general matrix R such that

$$\begin{aligned} M_{\mu\nu} v^{\mu} w^{\nu} &= \mathbf{v} \cdot \mathbf{w} = \mathbf{v}' \cdot \mathbf{w}' = M_{\mu\nu} v'^{\mu} w'^{\nu} = M_{\mu\nu} (R^{\mu}_{\rho} v^{\rho})(R^{\nu}_{\sigma} w^{\sigma}) \\ &= (M_{\mu\nu} R^{\mu}_{\rho} R^{\nu}_{\sigma}) v^{\rho} w^{\sigma}. \end{aligned} \tag{1.68}$$

This will only hold if we impose

$$M_{\mu\nu} R^{\mu}_{\rho} R^{\nu}_{\sigma} = M_{\rho\sigma}. \tag{1.69}$$

This will depend heavily on the metric. A Euclidian metric will result in very different dot product preserving transformations than a Minkowski metric, or some other more exotic metric like (1.52).

So let's look at the specific (and easy) example: $M_{ij} = \delta_{ij}$, the usual Euclidian case. We could go through the trouble of deriving it explicitly, but we know that for this metric the transformations that preserve the dot product are rotations of the vectors around the origin. Or in other words, transformations of the form (1.64):

$$R = \begin{pmatrix} \cos\theta & \sin\theta \\ -\sin\theta & \cos\theta \end{pmatrix}. \tag{1.70}$$

[12] As usual we'll assume that R is an invertible transformation.

You can convince yourself that this form satisfied equation (1.69) for $M_{ij} = \delta_{ij}$, for all θ. This is a straightforward result – the dot product we are preserving can be written (for a single vector dotted with itself)

$$x^2 + y^2 = const. \tag{1.71}$$

which is the equation of a circle – exactly what (1.70) represents.

Now let's consider the Minkowski metric.[13] What kind of transformations preserve this dot product? Looking again at (1.69) we want transformations Λ that satisfy

$$\eta_{\mu\nu}\Lambda^{\mu}_{\rho}\Lambda^{\nu}_{\lambda} = \eta_{\rho\lambda}. \tag{1.72}$$

We know that a subset of these transformations will be rotations in space as before:

$$\Lambda^{\mu}_{\nu} = \begin{pmatrix} 1 & 0 & 0 \\ 0 & \cos\theta & \sin\theta \\ 0 & -\sin\theta & \cos\theta \end{pmatrix}. \tag{1.73}$$

This is because the part of the metric that affects transformations between the two spatial dimensions is simply the lower right (2×2) δ_{ij} subset of η_{ij}.

But what about "rotations" mixing a space and time dimension? Consider a rotation between the time dimension and the x spatial dimension. We could write out the equations (1.69) explicitly and solve for the constraints on Λ^{μ}_{ν}, but instead we'll simply look at the dot product. The dot product we are trying to preserve here is (cf. equation (1.71) above)

$$-c^2t^2 + x^2 = const. \tag{1.74}$$

You can (should) recognize this as the equation for a hyperbola, and therefore from the general hyperbolic trig[14] relationship $\cosh^2\phi - \sinh^2\phi = 1$ for angle ϕ, we have

$$\Lambda^{\mu}_{\nu} = \begin{pmatrix} \cosh\phi & -\sinh\phi & 0 \\ -\sinh\phi & \cosh\phi & 0 \\ 0 & 0 & 1 \end{pmatrix}. \tag{1.75}$$

[13]We already looked at the correct form above, but now we rederive them with a bit more mathematical insight.

[14]If you're not familiar with hyperbolic trig, a quick read should provide what you need. In all honesty, the Wikipedia page on "Hyperbolic Function" is enough.

So, if we want to preserve the dot product (which is the fundamental assumption of special relativity), we are allowed transformations that mix spatial dimensions (i.e. (1.73)) and transformations that mix space and time (i.e. (1.75)).

An important question to ask now is "what is the meaning of the angle ϕ?" We have an intuitive understanding of θ, an angle mixing spatial dimensions, but the meaning of ϕ is less obvious. To begin with, define

$$\gamma = \cosh\phi \tag{1.76}$$

and

$$\beta = \tanh\phi. \tag{1.77}$$

(Using γ and β is deliberate). So, $\sinh\phi = \beta\gamma$, and Λ becomes

$$\Lambda^\mu_\nu = \begin{pmatrix} \gamma & -\beta\gamma & 0 \\ -\beta\gamma & \gamma & 0 \\ 0 & 0 & 1 \end{pmatrix}. \tag{1.78}$$

Then, the identity $\cosh^2\phi - \sinh^2\phi = 1$ gives

$$\gamma^2 - \beta^2\gamma^2 = 1 \implies \gamma = \frac{1}{\sqrt{1-\beta^2}}. \tag{1.79}$$

So now we need to understand what β is. First of all it is dimensionless and therefore we can guess that it's a ratio. Second, notice that when there is no transformation we have $\beta = 0$ and therefore $\gamma = 1$. And the $\gamma = 1$ case corresponds to $\phi = 0$. By equation (1.74) and the general definition of the hyperbolic trig functions this corresponds to a vector with only a temporal component and no spatial component. Then, both (1.77) and (1.79) tell us that β can only approach 1 but never reach it, which from (1.74) corresponds to $-c^2t^2 + x^2$ approaching 0, or ct approaching x. And $ct = x$ is equivalent to $v = \frac{x}{t} = c$. In other words, as β approaches 1 the velocity of the object in question approaches c, but it can never exceed or even reach c.

So, β is a dimensionless value that is 0 when there is no transformation and can at most approach 1, which corresponds to velocity being c. The proper choice for β therefore seems to be[15]

$$\beta = \frac{v}{c}, \tag{1.80}$$

[15]Admittedly we already knew the answer from the previous section, making this a bit easier to see.

and the general Λ that leaves the dot product unchanged is a combination of spatial rotations and rotations mixing space and time,

$$\Lambda^{\mu}_{\nu} = \begin{pmatrix} \frac{1}{\sqrt{1-\frac{v^2}{c^2}}} & -\frac{\frac{v}{c}}{\sqrt{1-\frac{v^2}{c^2}}} & 0 \\ -\frac{\frac{v}{c}}{\sqrt{1-\frac{v^2}{c^2}}} & \frac{1}{\sqrt{1-\frac{v^2}{c^2}}} & 0 \\ 0 & 0 & 1 \end{pmatrix} . \tag{1.81}$$

We call these space/time mixing transformations **Lorentz Boosts** – it is exactly what we had before in equations (1.65).

So, we have seen two types of transformations: Lorentz boosts and spatial rotations. Taken together they are the full set of transformations Λ that satisfy (1.72). The set of all Lorentz boosts and spatial rotations are called **Lorentz Transformations**, and they are the primary subject of special relativity. The important thing to take away is that *any* two inertial reference frames differ only by rotations. The simpler case is a spatial rotation, in which case one observer is simply rotated relative to another. The more complicated case of one observer moving with some velocity relative to another is actually *still* merely a rotation, but one that mixes space and time. It is the minus sign in the metric ($\eta_{\mu\nu}$ instead of δ_{ij}) that makes rotations involving time appear so radically different than rotations involving only space.

We should note that our exposition of these results is not at all how they proceeded historically, and it is not at all how Einstein originally thought about them. Our interest here has been to show the deeply geometrical nature of special relativity.

1.4.4 *Special Relativity and Lagrangians*

We make a brief note before moving on. While it is unlikely that any physics student would doubt the usefulness of Hamilton's principle and the Lagrangian formulation of classical mechanics as far as yielding results, students may be left with the feeling that the definition of the Lagrangian (1.1) doesn't seem well motivated. We use it because it works, but more curious students may find that unsatisfactory. And while Lagrangians were introduced over a century before special relativity, it turns out that special relativity provides a nice geometrical interpretation of the Lagrangian.

Consider the general idea of Hamilton's principle: nature chooses a path that extremizes the action. Physically we can loosely think of this as meaning that nature is lazy – it chooses the path that minimizes the energy that needs to be expended in traversing it. We have just seen that a fundamental vector in special relativity is the spacetime 4-vector

$$x^{\mu} = (t, x, y, z)^T . \tag{1.82}$$

Another important vector is the energy-momentum 4-vector corresponding to the quantities that are conserved by symmetries in the components of x^μ (cf. Sects. 1.2 and 1.3)[16]:

$$p_\mu = (E, p_x, p_y, p_z). \tag{1.83}$$

The energy and momentum of a particle are functions of the particle's location in and velocity through spacetime. So, if we want to consider a path the particle travels in spacetime and then take a variation of that path to find the extremum in "energy-momentum space", we would set up our action as

$$\int_{path} p_\mu dx^\mu, \tag{1.84}$$

where we are using $\eta_{\mu\nu}$ to define the dot product as in equation (1.62). This integral represents the total energy-momentum along a particular path in spacetime.

To see better what this integral means, we can expand it out[17]:

$$\begin{aligned}\int p_\mu dx^\mu &= \int (-p^0 dx^0 + p_1 dx^1 + p_2 dx^2 + p_3 dx^3) \\ &= \int (-E dt + \mathbf{p} \cdot d\mathbf{x}) \\ &= \int (\mathbf{p} \cdot d\mathbf{x} - H dt)\end{aligned} \tag{1.85}$$

(we are using bold for spatial vectors and indices for spacetime vectors) where we have replaced the energy function with the Hamiltonian (we are considering a non-relativistic limit so there is no difficulty with this).

Now we can rewrite this as

$$\int (\mathbf{p} \cdot d\mathbf{x} - H dt) = \int (\mathbf{p} \cdot \dot{\mathbf{x}} - H) dt \tag{1.86}$$

and comparison with (1.45) reveals this to be

$$\int (\mathbf{p} \cdot \dot{\mathbf{x}} - H) dt = \int L dt \tag{1.87}$$

which is the same as (1.2). So, we have that

$$\int L dt = \int p_\mu dx^\mu. \tag{1.88}$$

We can think of the Lagrangian in the action as being a sort of non-relativistic limit of an action defined by integrating energy-momentum over spacetime.

---

[16] We use lower indices for reasons that will be clear later.

[17] Doing this makes it not relativistically invariant, but for the moment we are interested in non-relativistic mechanics so this is acceptable.

1.4.5 Relativistic Energy-Momentum Relationship

As we said, we're not including a discussion of topics in special relativity like length and time contraction, but there is one very important relationship we want to derive because it is used later in the book – namely we want to write out the total energy of a particle. We know from the work energy theory in classical physics (actually Physics I physics) that the change in energy ΔK is equal to the integral of the force times the displacement, or

$$\Delta K = \int \mathbf{F} \cdot d\mathbf{x}. \tag{1.89}$$

But we also know from introductory physics that force is simply the time derivative of momentum,

$$\mathbf{F} = \frac{d\mathbf{p}}{dt}. \tag{1.90}$$

Substituting this in to the ΔK expression and assuming (for simplicity) that the motion is along the x axis, this becomes

$$\Delta K = \int \frac{dp}{dt} dx. \tag{1.91}$$

But we can simplify this even further as follows:

$$\frac{dp}{dt} dx = dp \frac{dx}{dt} = v dp, \tag{1.92}$$

where $v = \frac{dx}{dt}$ is simply the classical expression of the velocity. So the change in energy is

$$\Delta K = \int v dp. \tag{1.93}$$

Now, we also know from the previous section that momentum is part of the energy-momentum 4-vector (1.83). So, from the relativistic expression for momentum (taking $c = 1$ as usual),[18]

$$p = \frac{mv}{\sqrt{1 - v^2}}, \tag{1.94}$$

[18] Which we would have derived had we talked about those standard topics in special relativity. If you're not familiar with where this comes from, you probably aren't ready for this book, but if you're committed enough you can read about it in almost any introductory physics book ever written that has a chapter on special relativity.

we can solve for v to get

$$v = \frac{p}{\sqrt{m^2 + p^2}}. \tag{1.95}$$

Plugging this into our work-energy integral, we have

$$\Delta K = \int v dp = \int \frac{p\, dp}{\sqrt{m^2 + p^2}}. \tag{1.96}$$

Evaluating this integral is simply a matter of doing the right substitutions. We take $u = m^2 + p^2$, so we get

$$\Delta K = \frac{1}{2} \int u^{-\frac{1}{2}} du = \sqrt{u} = \sqrt{m^2 + p^2}. \tag{1.97}$$

If we take this to correspond to a particle starting at rest, so that ΔK is the total energy E, we have the relationship

$$E^2 = m^2 + p^2. \tag{1.98}$$

We will see that this relationship plays the role of a fundamental relativistic constraint.

Before moving on, (simply because it's interesting) we'll point out that if we restore the c's to this expression it is

$$E^2 = (mc^2)^2 + (pc)^2. \tag{1.99}$$

In the case of $p = 0$ this gives the famous equation

$$E = mc^2. \tag{1.100}$$

1.4.6 Physically Allowable Transformations

The Lorentz group is defined by (1.72) – the set of all transformations Λ^μ_ν that satisfy that constraint are the set of all Lorentz transformations. We've already discussed the six standard transformations (three rotations and three boosts).

But let's consider the determinant of (1.72). This gives

$$\det\left(\eta_{\mu\nu} \Lambda^\mu_\rho \Lambda^\nu_\sigma\right) = \det\left(\eta_{\rho\sigma}\right) \implies \eta\Lambda^2 = \eta \implies \Lambda = \pm 1. \tag{1.101}$$

(where $\Lambda = \det(\Lambda^\mu_\nu)$ and $\eta = \det(\eta_{\mu\nu})$). So, the determinant of a Lorentz transformation is either 1 or -1. You can go back and show that all of the six transformations discussed so far all have determinant 1. So what about the determinant -1 transformations? Consider, for example, the transformations

$$(\Lambda_P)^\mu_\nu = \begin{pmatrix} 1 & 0 & 0 & 0 \\ 0 & -1 & 0 & 0 \\ 0 & 0 & -1 & 0 \\ 0 & 0 & 0 & -1 \end{pmatrix}. \tag{1.102}$$

It has determinant -1, but you can check that it satisfies (1.72). Its effect on an arbitrary vector will be

$$\Lambda_P \cdot (t, \mathbf{x})^T = (t, -\mathbf{x})^T. \tag{1.103}$$

It is easy to show that it is not possible to carry out the transformation $(t, \mathbf{x})^T \rightarrow (t, -\mathbf{x})^T$ with the 6 standard transformations already discussed. Notice also that this transformation changes the "handedness" of the system.[19] If (x, y, z) is right-handed, then $(-x, -y, -z)$ is left-handed. For this reason we call Λ_P a **Parity** transformation.

We call transformations that have determinant $+1$ the **Proper** Lorentz transformations. There are, of course, other transformations that preserve (1.72) and have determinant -1 besides parity. And because they preserve (1.72) they are legitimate Lorentz transformations, but because you can't perform them with any combination of the proper transformations, such transformations are called **Improper**.

So, we can classify Lorentz transformations into two categories: the proper transformations and the improper transformations. Only the proper transformations are physical in the sense that they describe possible reference frames. However, as we will see, for a given physical theory to be relativistically invariant it must be invariant under *all* valid Lorentz transformations (anything that satisfies (1.72)).

There is another way we can classify Lorentz transformations. Consider the 0, 0 component in equation (1.72),

$$\eta_{00} = \Lambda^\mu_{\ 0} \Lambda^\nu_{\ 0} \eta_{\mu\nu} \Longrightarrow -1 = -\Lambda^0_{\ 0}\Lambda^0_{\ 0} + \sum_{i=1}^{3} \Lambda^i_{\ 0}\Lambda^i_{\ 0},$$

$$\Longrightarrow (\Lambda^0_{\ 0})^2 = 1 + \sum_{i=1}^{3} (\Lambda^i_{\ 0})^2. \tag{1.104}$$

The term on the right-hand side of the last line will *always* be greater than or equal to 1, and therefore $(\Lambda^0_{\ 0})^2 \geq 1$, which implies either

$$\Lambda^0_{\ 0} \geq 1,$$
$$\Lambda^0_{\ 0} \leq -1. \tag{1.105}$$

[19]This fact will prove to enormously important later in the book.

It is again easy to show that the $0, 0$ components of the 6 proper Lorentz transformations all satisfy $\Lambda^0_0 \geq 1$. We call any Lorentz transformation satisfying $\Lambda^0_0 \geq 1$ **Orthochronous**, and any transformation satisfying $\Lambda^0_0 \leq -1$ **Non-Orthochronous**. As with improper transformations it is not possible to perform a non-orthochronous transformation as any combination of orthochronous transformations and therefore these transformations aren't physical. However, they satisfy (1.72) and therefore a relativistically invariant theory must account for them.

The simplest example of a non-orthochronous transformation is the **Time Reversal** transformation

$$(\Lambda_T)^\mu_\nu = \begin{pmatrix} -1 & 0 & 0 & 0 \\ 0 & 1 & 0 & 0 \\ 0 & 0 & 1 & 0 \\ 0 & 0 & 0 & 1 \end{pmatrix}, \tag{1.106}$$

which takes $(t, \mathbf{x})^T \to (-t, \mathbf{x})^T$.

So, Lorentz transformations are anything that satisfies (1.72), but only the proper and orthochronous transformations are "physical". The improper and non-orthochronous transformations, however, are still required in a relativistically invariant theory.

1.5 Classical Fields

When deriving the Euler-Lagrange equations, we started with an action S which was an integral over time only ($S = \int dtL$). If we are eventually interested in a relativistically acceptable theory, this is obviously no good because it treats time and space differently (the action is an integral over time but not over space).

So, let's consider an action defined not in terms of the Lagrangian, but of the "Lagrangian per unit volume", or the **Lagrangian Density** $\mathcal{L}$. The Lagrangian will naturally be the integral of $\mathcal{L}$ over all space,

$$L = \int d^n x \mathcal{L}. \tag{1.107}$$

The integral is in n-dimensions, so $d^n x$ means $dx^1 dx^2 dx^2 \cdots dx^n$.

Now, the action will be

$$S = \int dtL = \int dtd^n x \mathcal{L}. \tag{1.108}$$

In the normal 1 + 3-dimensional Minkowski spacetime we live in, this will be

$$S = \int dtd^3x\mathcal{L} = \int d^4x\mathcal{L}. \tag{1.109}$$

Before, L depended not on t, but on the path $q(t)$, $\dot{q}(t)$. In a similar sense, $\mathcal{L}$ will not depend on $\mathbf{x}$ and t, but on what we will refer to as **Fields**, $\phi(\mathbf{x}, t) = \phi(x^\mu)$, which exist in spacetime.

Following a similar argument as the one leading to (1.17), we get the relativistic field generalization

$$\partial_\mu\left(\frac{\partial\mathcal{L}}{\partial(\partial_\mu\phi_i)}\right) - \frac{\partial\mathcal{L}}{\partial\phi_i} = 0, \tag{1.110}$$

for multiple fields ϕ_i $(i = 1, \ldots, n)$.

Noether's Theorem says that, for $\phi \to \phi + \epsilon\delta\phi$, we have a current

$$j^\mu = \frac{\partial\mathcal{L}}{\partial(\partial_\mu\phi)}\delta\phi, \tag{1.111}$$

and if $\phi \to \phi + \epsilon\delta\phi$ leaves $\delta\mathcal{L} = 0$, then

$$\partial_\mu j^\mu = 0 \Rightarrow -\frac{\partial j^0}{\partial t} + \nabla\cdot\mathbf{j} = 0, \tag{1.112}$$

where j^0 is the **Charge Density**, and $\mathbf{j}$ is the **Current Density**. The total charge will naturally be the integral over the charge density,

$$Q = \int_{all\ space} d^3x\ j^0. \tag{1.113}$$

Finally, we also have a **Hamiltonian Density** and momentum

$$\mathcal{H} = \frac{\partial\mathcal{L}}{\partial\dot{\phi}}\dot{\phi} - \mathcal{L}, \tag{1.114}$$

$$\Pi = \frac{\partial\mathcal{L}}{\partial\dot{\phi}}. \tag{1.115}$$

For the remainder of this book, we will ultimately be seeking a relativistic field theory, and therefore we will never make use of Lagrangians. We will always use Lagrangian densities. We will always use the notation $\mathcal{L}$ instead of L, but we will refer to the Lagrangian densities simply as Lagrangians. We drop the word "densities" for brevity, and because there will never be ambiguity.

Because we'll be working with field Lagrangians in much, much more detail for the remainder of this book (and books later in the series), we'll hold off further comments until then.

1.6 Classical Electrodynamics

We choose our units so that $c = \mu_0 = \epsilon_0 = 1$. The magnitude of the force between two charges q_1 and q_2 is $F = \frac{q_1 q_2}{4\pi r^2}$. In these units, Maxwell's equations are

$$\nabla \cdot \mathbf{E} = \rho, \tag{1.116}$$

$$\nabla \times \mathbf{B} - \frac{\partial \mathbf{E}}{\partial t} = \mathbf{J}, \tag{1.117}$$

$$\nabla \cdot \mathbf{B} = 0, \tag{1.118}$$

$$\nabla \times \mathbf{E} + \frac{\partial \mathbf{B}}{\partial t} = 0. \tag{1.119}$$

If we define the **Potential** 4-vector

$$A^\mu = (\phi, \mathbf{A})^T, \tag{1.120}$$

then we can define

$$\mathbf{B} = \nabla \times \mathbf{A} \tag{1.121}$$

and

$$\mathbf{E} = -\nabla\phi - \frac{\partial \mathbf{A}}{\partial t}. \tag{1.122}$$

Writing **B** and **E** this way will automatically solve the homogenous Maxwell equations, (1.118) and (1.119).[20]

Then, we define the totally antisymmetric **Electromagnetic Field Strength Tensor** $F^{\mu\nu}$ as[21]

$$F^{\mu\nu} = \partial^\mu A^\nu - \partial^\nu A^\mu = \begin{pmatrix} 0 & -E_x & -E_y & -E_z \\ E_x & 0 & -B_z & B_y \\ E_y & B_z & 0 & -B_x \\ E_z & -B_y & B_x & 0 \end{pmatrix}. \tag{1.123}$$

We define the 4-vector current as $J^\mu = (\rho, \mathbf{J})^T$. It is straightforward, though tedious, to show that

[20] You are strongly encouraged to work out these details to convince yourself of this.

[21] For the purposes of this book, you can simply think of a "tensor" as an object with indices. For example, a vector is a tensor with a single index, a matrix is a tensor with two indices, etc. The actual definition is more involved, but we won't get to it until the next book.

$$\partial^\lambda F^{\mu\nu} + \partial^\nu F^{\lambda\mu} + \partial^\mu F^{\nu\lambda} = 0 \Rightarrow \nabla \cdot \mathbf{B} = 0 \quad \text{and} \quad \nabla \times \mathbf{E} + \frac{\partial \mathbf{B}}{\partial t} = 0, \tag{1.124}$$

$$\partial_\mu F^{\mu\nu} = J^\nu \Rightarrow \nabla \cdot \mathbf{E} = \rho \quad \text{and} \quad \nabla \times \mathbf{B} - \frac{\partial \mathbf{E}}{\partial t} = \mathbf{J}. \tag{1.125}$$

So, all of classical electrodynamics can be formulated in terms of an equation (that is really just an identity) and a single source equation.

1.7 Classical Electrodynamics Lagrangian

Bringing together the ideas of the previous sections, we now want to construct a Lagrangian density $\mathcal{L}$ which will, via Hamilton's Principle, produce Maxwell's equations.

First, we know that $\mathcal{L}$ must be a scalar (no uncontracted indices). From our intuition with "Physics I" type Lagrangians, we know that kinetic terms are quadratic in the derivatives of the fundamental coordinates (i.e. $\frac{1}{2}m\dot{x}^2 = \frac{1}{2}m(\frac{dx}{dt}) \cdot (\frac{dx}{dt})$). The natural choice is to take A^μ as the fundamental field. It turns out that the correct choice is

$$\mathcal{L}_{EM} = -\frac{1}{4}F_{\mu\nu}F^{\mu\nu} - J^\mu A_\mu \tag{1.126}$$

(note that the F^2 term is quadratic in $\partial^\mu A^\nu$). So,

$$S = \int d^4x \Big[-\frac{1}{4}F_{\mu\nu}F^{\mu\nu} - J^\mu A_\mu \Big]. \tag{1.127}$$

Taking the variation of (1.127) with respect to A^μ,

$$\begin{aligned} \delta S &= \int d^4x \Big[-\frac{1}{4}F_{\mu\nu}\delta F^{\mu\nu} - \frac{1}{4}\delta F_{\mu\nu}F^{\mu\nu} - J^\mu \delta A_\mu \Big] \\ &= \int d^4x \Big[-\frac{1}{2}F_{\mu\nu}\delta F^{\mu\nu} - J^\mu \delta A_\mu \Big] \\ &= \int d^4x \Big[-\frac{1}{2}F_{\mu\nu}(\partial^\mu \delta A^\nu - \partial^\nu \delta A^\mu) - J^\mu \delta A_\mu \Big] \\ &= \int d^4x \Big[-F_{\mu\nu}\partial^\mu \delta A^\nu - J^\mu \delta A_\mu \Big]. \end{aligned} \tag{1.128}$$

Integrating the first term by parts, and choosing boundary conditions so that δA vanishes at the boundaries,

$$\begin{aligned} \delta S &= \int d^4x \Big[\partial_\mu F^{\mu\nu}\delta A_\nu - J^\nu \delta A_\nu \Big] \\ &= \int d^4x \Big[\partial_\mu F^{\mu\nu} - J^\nu \Big]\delta A_\nu. \end{aligned} \tag{1.129}$$

So, to have $\delta S = 0$, we must have $\partial_\mu F^{\mu\nu} = J^\nu$, and if this is written out one component at a time, it will give exactly the inhomogeneous Maxwell equations (1.116) and (1.117). And as we already pointed out, the homogenous Maxwell equations become identities when written in terms of A^μ.

As a brief note, the way we have chosen to write equation (1.126), in terms of a "potential" A_μ, and the somewhat mysterious antisymmetric "field strength" $F_{\mu\nu}$, is indicative of an extremely deep and very general mathematical structure that goes well beyond classical electrodynamics. We will see this structure unfold as we proceed through this book. We just want to mention now that this is not merely a clever way of writing electric and magnetic fields, but a specific example of a general theory.

1.8 Gauge Transformations

Gauge Transformations are usually discussed toward the end of an undergraduate course on E&M. Students are typically told that they are extremely important, but the reason why is not obvious. We will briefly introduce them here, and while their significance may still not be transparent, they are almost the entire content of this series of books and of particle physics in general.

Given some specific potential A^μ, we can find the field strength action as in (1.127). However, A^μ does not uniquely specify the action. We can take any arbitrary function $\chi(x^\mu)$, and the action will be invariant under the transformation

$$A^\mu \to A'^\mu = A^\mu + \partial^\mu \chi \tag{1.130}$$

or

$$A^\mu \to A'^\mu = (\phi - \frac{\partial \chi}{\partial t}, \mathbf{A} + \nabla \chi). \tag{1.131}$$

Under this transformation, we have

$$\begin{aligned} F'^{\mu\nu} &= \partial^\mu A'^\nu - \partial^\nu A'^\mu = \partial^\mu (A^\nu + \partial^\nu \chi) - \partial^\nu (A^\mu + \partial^\mu \chi) \\ &= \partial^\mu A^\nu - \partial^\nu A^\mu + \partial^\mu \partial^\nu \chi - \partial^\mu \partial^\nu \chi \\ &= F^{\mu\nu}. \end{aligned} \tag{1.132}$$

So, $F'^{\mu\nu} = F^{\mu\nu}$.

Furthermore, $J^\mu A_\mu \to J^\mu A_\mu + J^\mu \partial_\mu \chi$. Integrating the second term by parts with the usual boundary conditions,

$$\int d^4x J^\mu \partial_\mu \chi = - \int d^4x (\partial_\mu J^\mu) \chi. \tag{1.133}$$

But, according to Maxwell's equations, $\partial_\mu J^\mu = \partial_\mu \partial_\nu F^{\mu\nu} = 0$ because $F^{\mu\nu}$ is totally antisymmetric. So, both $F^{\mu\nu}$ and $J^\mu \partial_\mu \chi$ are invariant under (1.130), and therefore the action of S is invariant under (1.130).

While the importance of gauge transformations may not be obvious at this point, it will become perhaps the most important idea in particle physics. As a note before moving on, recall previously when we mentioned the idea of "what doesn't change when something else changes" when talking about Lorentz transformations. A gauge transformation is exactly this (in a different context): the fundamental fields are changed by χ, but the equations which govern the physics are unchanged.

We will eventually see why gauge transformations are so important (in Chap. 4 and especially in Sect. 4.5), but there is quite a bit of math and physics we need first. We'll start that climb in Chap. 3, but before doing so we'll take a few pages to do what any theorist should do (but often doesn't): understand how all of this relates to the real world.

1.9 References and Further Reading

The material in this section can be found in nearly any introductory text on Classical Mechanics, Classical Electrodynamics, and Relativity. The primary sources for this section are [18, 31, 33].

For further reading, we recommend [22, 24, 49, 51, 62, 81, 101].

Chapter 2
A Preview of Particle Physics: The Experimentalist's Perspective

The upcoming chapters in this book, as well as the books to come later in this series, are very, very mathematical. The current structure of elementary particle theory is in many ways barely distinguishable from pure math. As one works through the intricate mathematical formalism, it can be easy to lose sight of how it all relates to what we observe in nature. Therefore, before going any farther, we'll take a few pages to summarize the current state of affairs, and how we got here, from an experimentalist's point of view.

The culmination of theoretical particle physics is called the Standard Model of Particle Physics, or usually just the **Standard Model**. For three and a half decades, the standard model has provided an excellent description of the inner workings of nature. Over the years, carefully designed experiments have probed numerous properties of the physical universe, and every experimental measurement has affirmed the standard model.

Starting in Chap. 3, we'll build up a mathematical foundation for the standard model in a way that will seem obvious – as if it simply has to be that way. Of course, we'll be providing this foundation based on our modern-day knowledge, with the benefit of many decades of hard work and tireless dedication. The twentieth century was marked by a continual exchange of ideas and measurements. New ideas stimulated the design and execution of clever new experiments, and experimental measurements either supported an idea or relegated it to the sidelines. No one dreamed up the $SU(3) \times SU(2) \times U(1)$ gauge theory[1] back in the 1950s when a slew of new particles started popping up in laboratories. It was the careful categorization of those particles, the baryons and mesons, that led to the quark model of nature. In this way, theory and experiment have worked hand in hand, each extending the reach of the other, to bring us where we are today.

Elementary particle physics in the twenty-first century continues to be a healthy and animated exchange between theorists and experimentalists. Bright young graduate students pursuing research in particle physics are usually channeled towards

[1] We'll discuss what this means in plenty of detail later.

M. Robinson, *Symmetry and the Standard Model: Mathematics and Particle Physics*, DOI 10.1007/978-1-4419-8267-4_2, © Springer Science+Business Media, LLC 2011

either theory or experiment. It is unfortunate, perhaps, that the skill set needed to succeed as an experimentalist has become quite different than the skill set needed to succeed as a theorist, and vice versa. While a theorist might study non-Cartan generators, local symmetry breaking, and Kähler geometry, an experimentalist gains expertise in things like data acquisition systems, field programmable gate arrays, ROOT, and the nuances of polymorphism and inheritance. The collective expertise needed to design, commission, test, and successfully run a modern-day particle physics experiment is immense. The day-to-day activities of an experimentalist rarely involve calculating matrix elements of Feynman diagrams.

So, what has happened is that theorists see the standard model as mathematical formalism, where "particles" aren't so much real matter – instead they're fields like Z_μ. At the other extreme, experimentalists think of Z^0 bosons as invisible particles that are "seen" only when they decay to something like a e^+e^- or $\mu^+\mu^-$ pair, mere remnants that we can identify in particle detectors. We authors have remarked to each other that theorists and experimentalists sometimes speak about the standard model in such different ways that it seems like we are not even talking about the same thing at all, and we are reminded of the well-known story of the blind men describing the elephant.

In the next few chapters, you'll learn about a series of mathematical tricks for various types of "fields." We'll talk about "massless scalars with a $U(1)$ charge," and about things "in a $j = \frac{1}{2}$ representation of $SU(2)$." While the primary purpose of this text is indeed to provide the mathematical tools with which particle physics is performed, we are physicists, not mathematicians. It is therefore apt that we reunite theory and experiment and proceed with a "nature-based" preview of elementary particle physics.

2.1 The Ultimate "Atoms"

Since the time of the ancient Greeks, physicists have been progressing toward a simple, elegant, all-encompassing model that attempts to explain the workings of the universe. Humankind's curiosity about the nature of nature can be traced back to the fifth century BC, when a Greek named Empedocles combined the ideas of several others before him to say that all structures of the world were made up of earth, air, fire, and water, and that there are two divine powers, Love and Strife, which govern the way these four elements combine and behave. More scientifically, he was saying that matter is made up of smaller substances that interact with each other through attraction and repulsion. Democritus, a contemporary of Empedocles, dared to propose that all matter is composed of invisible, indestructible particles called *atoms* from the Greek $\alpha\tau\omega\mu\omega\sigma$, meaning "uncuttable" or "indivisible."

Over the centuries, the yearning to identify the elementary constituents of matter has brought us from earth, air, fire, and water to a microworld over a million-billion times smaller than the book in your hands. The basic questions posed by Democritus over 2,400 years ago continue to drive the field of elementary particle physics today.

Are there fundamental, indivisible particles and if so, what are they? How do they behave? How do they group together to form the matter that we see? How do they interact with each other? Today, using the most sophisticated particle probes on earth, we think we might have finally discovered the ultimate $\alpha\tau\omega\mu\omega\sigma$. We call them *quarks* and *leptons*.

2.2 Quarks and Leptons

The twentieth century was a marvelous one for particle physics. It all began in 1897 when J.J. Thompson discovered the first truly elementary particle: the *electron*.[2] With this observation came the realization that the atoms of the nineteenth century – like hydrogen, oxygen, and lead – were not in fact the most basic building blocks of matter. In 1911, Ernest Rutherford and his associates bombarded thin gold foils with α-particles and found that some of them were deflected by huge angles, indicating the presence of a small yet massive kernel inside the atom: the atomic nucleus. The ensuing years revealed that the nucleus consisted of even smaller components, the *proton* and *neutron*, collectively referred to as *nucleons*. Physicists realized that every element in the periodic table could be constructed of a single atomic nucleus with a distinct number of protons and neutrons, surrounded by a cloud of electrons. And with that, modern elementary particle physics was born.

The notion that protons and neutrons were elementary particles was shattered in the late 1950s and 1960s by a population explosion of newly observed particles. With the construction of large particle accelerators, experiments produced hundreds of "elementary" particles, called *hadrons*, with properties very similar to the nucleons. Underlying symmetries in the masses, charges, and intrinsic angular momenta (spins) of the hadrons pointed to an even deeper order within the chaos. In 1963, Murray Gell-Mann and George Zweig independently proposed a scheme in which hadrons are composed of yet smaller particles, called *quarks*.[3] Some hadrons, like the proton and neutron, consist of three quarks. Experimental evidence for the proton's substructure was eventually established in 1968 by a team at the Stanford Linear Accelerator Center (SLAC). In an experiment not so different than Rutherford's, a high-energy beam of electrons was aimed at a small vat of liquid hydrogen. The resulting scattering pattern revealed that the proton is not elementary at all.

---

[2]It is truly elementary, as far as we currently know.

[3]Although quark may sound inherently like a scientific term, its origin is surprisingly from literature. For the name of this type of particle, Murray Gell-Mann came up not with the word first, but with the sound (which he described as "kwork", the sound a duck makes). Soon thereafter, Gell-Mann came across the phrase "Three quarks for Muster Mark" in *Finnegans Wake* by James Joyce. Gell-Mann immediately latched on to quark as the spelling – which seemed very appropo since he was theorizing that hadrons were composed of three different types of elementary particles. Zweig sought (unsuccessfully) to attach the name *aces* to the particles, in connection with his expectation of the discovery of a fourth such particle.

The original quark model of Gell-Mann and Zweig required only three *flavors* of quarks – the *up* (u), *down* (d), and *strange* (s) – to explain the proliferation of new hadrons. Nucleons are comprised of combinations of up and down quarks. Strange quarks explained the existence of odd, short-lived particles in cosmic rays. Each flavor of quark also has an associated *antiquark*, a corresponding particle with an identical mass but opposite electric charge.

Since the early 1970s, three more quarks have been discovered, bringing the total to six. For reasons we'll see shortly, they are often grouped in pairs, or *doublets*, as shown here:

$$\begin{pmatrix} u \\ d \end{pmatrix} \begin{pmatrix} c \\ s \end{pmatrix} \begin{pmatrix} t \\ b \end{pmatrix}$$

The *charm* (c) quark was discovered in 1974 in the form of the *J/psi* (J/ψ) meson, a bound charm-anticharm pair, by two independent teams led by Samuel Ting at Brookhaven National Laboratory and Burton Richter at SLAC. In 1977, Leon Lederman and colleagues at the Fermi National Accelerator Laboratory (Fermilab) found the analogue of the J/ψ for *bottom* (b) quarks, which was named the *upsilon* (Υ). By this point, one more quark was obviously needed to pair with the b quark and fill the gaping hole in the third doublet. The last of the quarks, the *top* (t), was discovered in 1995 in high-energy proton-antiproton collisions by the Collider Detector at Fermilab (CDF) and DZero (DØ) collaborations. Far more massive than anyone expected – more than 186 times the protons mass! – the top quark's fleeting existence prevents it from joining with other quarks to create hadrons.

The arrangement of the six quark flavors in three *generations* of doublets, as shown above, reflects their intrinsic properties. The up and down quarks are the lightest of all and therefore the most stable. As a result, they make up ordinary matter. The proton, with a total electric charge of $+1$, contains two up quarks, each with charge $+2/3$, and a down quark with a charge $-1/3$. The udd configuration of the neutron gives it a net charge of zero. The second and third generations are just heavier duplicates of the first, with quarks that are produced only in high-energy interactions.

The theoretical and experimental advances that led to the quark model also predicted the existence of *leptons*, a second set of six elementary particles, together with their corresponding antiparticles. Like the quarks, the leptons can be arranged in three generations of doublets:

$$\begin{pmatrix} e^- \\ \nu_e \end{pmatrix} \begin{pmatrix} \mu^- \\ \nu_\mu \end{pmatrix} \begin{pmatrix} \tau^- \\ \nu_\tau \end{pmatrix}$$

Of the three charged leptons, the lightest is the familiar electron. The *muon* (μ), a heavy replica of the electron, was first observed in 1938 in cosmic rays by Carl David Anderson. The heaviest known lepton, the *tau* (τ), was discovered decades later in 1975 by Martin Perl with colleagues at SLAC. Unlike the electron, the muon and tau are unstable and exist for only fractions of a second before decaying to less massive particles.

Each of the three charged leptons is complemented by a neutral partner, the *neutrino* (ν). Wolfgang Pauli originally proposed the idea of a neutrino in 1930 as the mysterious, unobserved particle that carried energy from nuclear β-decay. Neutrinos weren't actually "seen" until twenty-six years later, when Clyde Cowan and Fred Reines observed the interactions of electron antineutrinos with protons in a huge instrumented tank of water. Then, in 1961, a group led by Melvin Schwartz, Leon Lederman, and Jack Steinberger developed a neutrino beam at Brookhaven National Laboratory which resulted in the discovery of the second species of neutrino: the muon neutrino. The tau neutrino was ultimately discovered at Fermilab in 2000. Neutrinos, particles with a tiny mass, interact with matter only via the weak interaction. They interact so weakly, in fact, that a single neutrino can pass unscathed through millions of miles of solid steel!

The table below summarizes several details relating to the elementary quarks and leptons[4]:

Particle name	Symbol	Charge ($\lvert e \rvert$)	Mass (MeV/c^2)	Spin
Quarks				
Up	u	+2/3	1.7–3.3	1/2
Down	d	−1/3	4.1–5.8	1/2
Charm	c	+2/3	1180–1340	1/2
Strange	s	−1/3	80–130	1/2
Top	t	+2/3	$\approx$ 172000	1/2
Bottom	b	−1/3	4130–4370	1/2
Leptons				
Electron	e	−1	0.51100	1/2
Electron neutrino	ν_e	0	≈ 0	1/2
Muon	μ	−1	105.66	1/2
Muon neutrino	ν_μ	0	< 0.19	1/2
Tau	τ	−1	1776.8	1/2
Tau neutrino	ν_τ	0	< 18.2	1/2

2.3 The Fundamental Interactions

At the most intuitive level, a *force* is any kind of push or pull on an object. You experience forces every day. To push open a door, for example, your hand exerts a contact force on the door. The force of friction ultimately stops a book that slides across a table. Every "Physics I" student has drawn a free-body diagram with the gravitational pull pointing down and the so-called normal force pointing up.

[4]For complete, up-to-date information, see `http://pdg.lbl.gov`.

If all matter can be described in terms of a few fundamental building blocks, can we also categorize the everyday forces in terms of a few fundamental forces? We believe the answer is yes. Physicists have identified four known *interactions* that appear to underlie all of the phenomena we observe in nature. They are *gravitation*, *electromagnetism*, the *weak interaction*, and the *strong interaction*. The term interactions has become a common way to describe them, for they are the ways that the simplest particles in the universe interact with each other. They are fundamental in that all other forces, even the everyday forces, can be described in terms of them.

2.3.1 Gravitation

Everyone is intimately familiar with gravity. One of Sir Isaac Newton's many discoveries was that the mysterious force that pulls common objects down toward the earth's center is the same force that holds the moon in place in its orbit. As the timeless story goes, Newton observed an apple fall from a tree, and with one brilliant revelation, he unified gravity on the earth and gravity in the heavens.

Newton's Law of Universal Gravitation states that every particle in the universe attracts every other particle with a force that is proportional to the product of the masses of the particles and inversely proportional to the square of the distance between them. The magnitude of the force can be written as

$$F = G\frac{m_1 m_2}{r^2}, \tag{2.1}$$

where G, the *gravitational constant*, has an accepted value equal to $6.67 \times 10^{-11}\ \mathrm{N \cdot m^2/kg^2}$. Despite the way it feels, gravity is by far the weakest of the four interactions. To calculate the force of gravity that you experience on Earth's surface, we must first multiply your mass by the mass of the *entire* earth in the equation above. It is the huge mass of the earth that compensates for the tiny value of G.

Although gravitation is the oldest of the known interactions, it is in many respects the least well understood.The gravitational constant G is among the least well-measured physical constants, and scientists are not completely certain that the above equation remains accurate for distances less than a fraction of a millimeter.[5] The general theory of relativity, published by Albert Einstein in 1915, beautifully predicts the behavior of objects in the celestial realm of stars and planets. Unfortunately, it completely breaks down in the tiny realm of atoms and nuclei. Einstein spent the last decades of his life in a desperate attempt to unify gravity and electromagnetism but came up empty-handed.

---

[5]The reader may be surprised to learn that the $1/r^2$ form of Newtonian gravity was not actually verified for distances below 1 cm until very recently, as discussed in Sect. 5.4.1.

Despite its broad success in describing physical phenomena in the microscopic realm, the standard model is an incomplete theory because it fails to describe gravitation. Physicists continue to work towards a theory that describes all four fundamental interactions, with *string theory* currently showing the most promise.

2.3.2 Electromagnetism

Most of the forces described at the beginning of this section – like contact forces, friction, and the normal force – are actually manifestations of the interaction called electromagnetism. The force we experience when we push or pull ordinary material objects, such as doorknobs, comes from the intermolecular forces between the individual molecules in our bodies and those in the objects. These result from the forces involved in interactions between atoms, which in turn can be traced to electromagnetism acting on the electrically charged protons and electrons inside the atoms. When you rub your hands together, the charged particles near the surface of your hands experience the electromagnetic interaction, giving rise to friction. Electromagnetism, acting between the soles of our shoes and the floor, is responsible for the upward normal force which keeps us all from falling through solid ground toward the center of the earth.

The term "electromagnetism" hints at the interesting history surrounding this interaction. Originally, electricity and magnetism were considered two separate forces. Then, while preparing for a lecture in 1820, Hans Christian Ørsted made a surprising discovery. As he was setting up his materials, he observed the deflection of a compass needle from magnetic north when the electric current from a nearby battery was switched on and off. This deflection convinced him that magnetic fields radiate from all sides of a wire carrying an electric current, and it confirmed a direct relationship between electricity and magnetism.

In 1873, Scottish theoretical physicist and mathematician James Clerk Maxwell published a set of equations that relate both electric and magnetic forces to their sources: charges and currents. *Maxwell's equations* not only brilliantly intertwined the two forces into one unified force, but they also explained the origin of electromagnetic radiation, which includes x-rays, radio waves, visible light, and more. His famous set of equations, reviewed in Sect. 1.6, demonstrated that electricity, magnetism, and light are all manifestations of the same physical phenomenon: the electromagnetic field. His electromagnetic theory successfully synthesized previously unrelated experimental observations and equations of electricity and magnetism into a consistent theory. Maxwell's groundbreaking work has been dubbed the "second great unification in physics" after the first one achieved by Sir Isaac Newton.

2.3.3 The Strong Interaction

In the 1920s, only two of the four fundamental interactions, gravitation and electromagnetism, were known. Gravity controls the motion of the heavenly bodies and keeps our feet on the ground. Electromagnetism dominates all atomic interactions and is ultimately responsible for all that we see and feel. Believing these to be the only two forces, Theodor Kaluza and Oskar Klein developed a theory of unified gravity and electromagnetism.[6] Although it won the support of Albert Einstein, Kaluza-Klein theory faded from importance during the next decade, as the true nature of the nucleus became more and more clear. Studies of the atom during the 1930s led to the realization that gravity and electromagnetism were not the only two forces.

Imagine an atomic nucleus. What holds the constituent protons and neutrons together? Gravity is far too weak, and the electromagnetic interaction would push protons apart because of their like charge – not hold them together. In the 1930s, physicists had to admit that another force was needed to overcome the electromagnetic repulsion and allow nuclei to remain stable.

Enter the strong interaction. By far the strongest of the four fundamental interactions, the strong interaction not only binds protons and neutrons into atomic nuclei, it also unites quarks into composite particles – the hadrons. The range of the strong interaction is extremely small; it acts over a mere 10^{-15} m, a millionth of a billionth of a meter, the size of a proton. As we'll see shortly, the nature of the strong interaction is completely unlike the other interactions.

2.3.4 The Weak Interaction

The weak interaction is rather unfamiliar in our day-to-day existence. It is responsible for certain types of radioactive decay; for example, it permits a proton to turn into a neutron and vice versa. Aptly named, the strength of the weak interaction is 100 billion times less than the strength of the electromagnetic interaction! Like the strong force, it only acts over very short distances – about 10^{-18} m, a billionth of a billionth of a meter.

The weak interaction is the only one that can cause a quark to change its flavor. For instance, it can transform an up quark into a down quark. That is, in fact, *precisely* what happens in the case of β-decay. If one of the two up quarks in a proton changes to a down quark, we're left with the udd structure of the neutron. In the process, electric charge is conserved by the emission of a positively charged *positron* (the antimatter partner of the electron), and lepton number is conserved by the emission of a neutrino.

[6] Kaluza-Klein theory, of which an extended version is a natural result of string theory, is reviewed in Sect. 5.4.1 of Chap. 5 and will be discussed in depth in a later text in this series.

2.3.5 *Summary*

A table summarizing all four fundamental interactions is shown here. The approximate relative strengths have been normalized to unity for the strong interaction. We say "approximate" because we will learn later[7] that the strength of a force depends on the length scale being considered.

Interactions	Acts On	Strength	Range
Strong	Hadrons	1	10^{-15} m
Electromagnetism	Electric Charges	10^{-2}	∞ $(1/r^2)$
Weak	Leptons and Hadrons	10^{-5}	10^{-18} m
Gravitation	Mass	10^{-39}	∞ $(1/r^2)$

2.4 Categorizing Particles

In the middle of the twentieth century, before quarks were discovered, elementary particle physicists were shocked by the sudden population explosion of new particles discovered in the laboratory. Things seemed far too disorganized. How could all of these particles be elementary? Over time, the properties of these particles were measured and eventually it became apparent that they were not elementary at all, but *composite*: comprised of two or more other particles. Just as early biologists sorted living organisms by their appearance and defining features, physicists classified particles based on their measured properties such as mass, electric charge, and intrinsic angular momentum. The identification of common characteristics within the "zoo" of new particles ultimately led to the quark model of nature. Moreover, the names of the categories of particles have become a part of the daily vocabulary of experimental particle physics.

2.4.1 *Fermions and Bosons*

Every type of particle, elementary or composite, has an intrinsic angular momentum, or quantum mechanical spin. A particle with a half-integer spin (1/2, 3/2, 5/2, ...), in units of Planck's constant $\hbar$, is a *fermion*. A particle with an integer spin (0, 1, 2, ...) is a *boson*. The spin, in addition to being the particle's intrinsic angular momentum, governs the statistics of a set of such particles, so fermions and bosons may also be

[7] Much later – in a later book, in fact.

defined according to the statistics they obey. Fermions obey Fermi-Dirac statistics, and also the Pauli exclusion principle, which says that no two identical fermions can be found in the same quantum state at the same time. Bosons, on the other hand, obey Bose-Einstein statistics, which means that any number of the same type of particle can be in the same state at the same time.

The elementary particles that make up matter, the quarks and leptons, all have spin 1/2 and are thus fermions. As we will see, there are also elementary particles that govern the fundamental interactions of the standard model – the *photon*, W and Z bosons, and the *gluons* – which have spin 1, and are thus bosons. The Higgs boson, as yet undiscovered, is predicted to have spin 0.

2.4.2 Baryons and Mesons

One way to distinguish the various elementary fermions is by whether or not they interact via the strong interaction: quarks interact via the strong interaction, while leptons do not. Hadrons are composite particles constructed of quarks bound together by the strong interaction. They can be either fermions or bosons, depending on the number of quarks that comprise them. Three bound quarks (or three bound antiquarks) form spin-1/2 or spin-3/2 hadrons, which are called *baryons*. The baryons are made of "normal" matter quarks and their antimatter counterparts are made of the corresponding antiquarks. The most well known examples of baryons are protons and neutrons. *Mesons* are spin-0 or spin-1 hadrons consisting of a quark and antiquark, though not necessarily of the same flavor. Examples include the π^+ $(u\bar{d})$ (a positively-charged *pion*) and the K^- $(\bar{u}s)$ (a negatively-charged *kaon*). Because of their values of spin, all baryons are fermions and all mesons are bosons.

One of the reasons for the plethora of particles discovered in the past century is the numerous possible combinations of different quark flavors one can put into a three-quark baryon or two-quark meson. Additionally, each of these combinations can be in one of multiple quantum mechanical states. For example, a ρ^+ meson has the same combination of quarks as a π^+, but the ρ^+ is a spin-1 particle whereas the π^+ is a spin-0 particle.

2.4.3 Visualizing the Particle Hierarchy

Newcomers to elementary particle physics quickly notice that particle names typically end in "-on," like "proton" and "pion." However, as we have just seen, even categories of particles have names that end in "-on," like "fermion" and "lepton." A muon is a lepton and a fermion. A pion is a meson, hadron, and boson. This tangled taxonomy is enough to bewilder the brightest minds. Names have even been assigned to particles that have yet to be observed, such as the *preon*: the hypothetical subcomponents of quarks and leptons.

To help make sense of it all, it is useful to visualize the various categories of particles in a Venn diagram, as shown here. A few of the specific particles mentioned throughout this chapter are included as examples.

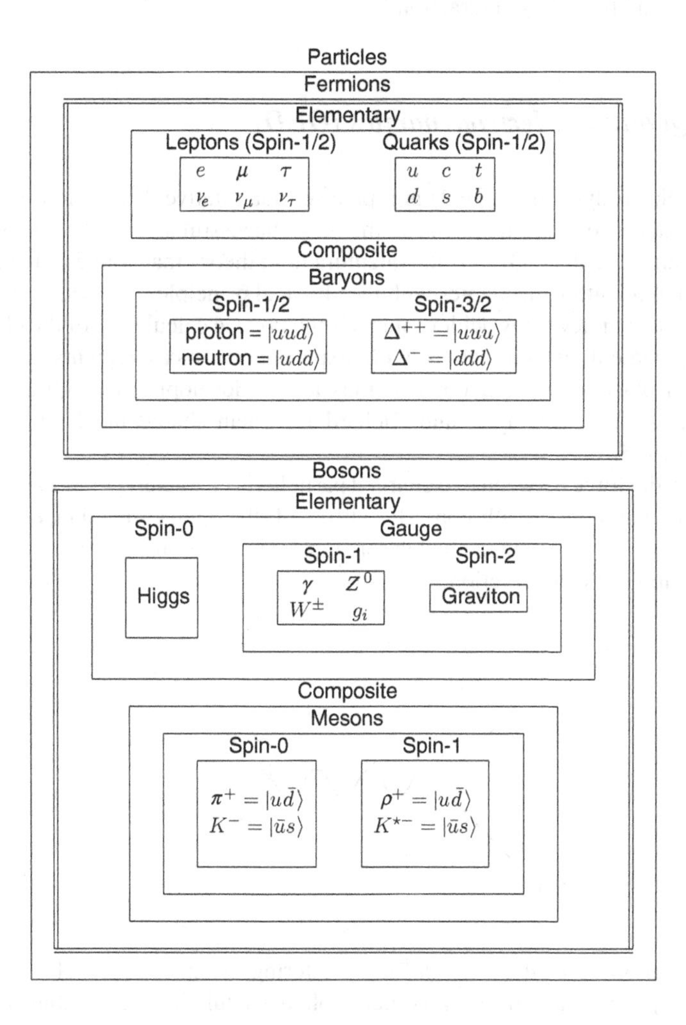

2.5 Relativistic Quantum Field Theories of the Standard Model

Most theories in modern particle physics, including the Standard Model of Particle Physics, are formulated as *relativistic quantum field theories.* Quantum field theory is widely considered to be the only correct approach for combining quantum mechanics and special relativity. In perturbative quantum field theory, the forces between particles are mediated by other particles, the *gauge bosons*. In the following sections, we'll see how the gauge bosons are intricately connected to the three fundamental interactions of the standard model: electromagnetism, the weak interaction, and the strong interaction.[8]

2.5.1 Quantum Electrodynamics (QED)

Quantum electrodynamics (QED) is a precise, quantitative description of electromagnetic interactions. Arguably one of the most successful theoretical achievements of the twentieth century, QED is the quantum field theory that connects the modern formalism of quantum mechanics with the classical principles of electromagnetism. One of its many noteworthy achievements is the precise calculation of the electron's magnetic moment, which agrees with experimental measurements to at least 10 decimal places. For their contributions to the development of QED, Sinitiro Tomonaga, Julian Schwinger, and Richard Feynman shared the Nobel Prize in Physics in 1965.

In QED, the force between two charged particles is characterized by the exchange of a field quantum, the photon. By virtue of the gauge invariance of QED, electric charge is conserved in all electromagnetic interactions. Below is a graphical representation of this interaction:

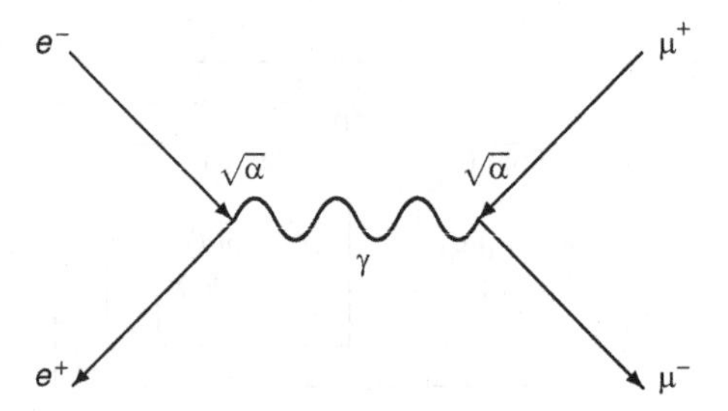

This diagram of $e^+e^- \longrightarrow \mu^+\mu^-$ scattering is an example of a Feynman diagram. Feynman diagrams play a crucial role in calculating measurable quantities such as *cross-sections*, which tell us about the probability for particular interactions to occur, and *decay rates*, which tell us how quickly particular particles decay. Every line and vertex of the diagram is associated with a mathematical term in the QED calculation. For example, each vertex contributes a factor proportional to $\sqrt{\alpha}$ to the amplitude $\mathcal{M}$, where $\alpha = e^2/4\pi$ represents the strength of the electromagnetic

[8] There is currently no complete quantum theory of the remaining fundamental interaction, gravitation, but many of the proposed theories postulate the existence of a spin-2 particle that mediates gravity, the *graviton*.

coupling between photons and charged particles. (Here, e is the magnitude of the electron's charge.) The mathematical evaluation of the above diagram yields a cross-section proportional to $|\mathcal{M}|^2$ (and therefore α^2):

$$\sigma(e^+e^- \longrightarrow \mu^+\mu^-) = \frac{4\pi\alpha^2}{3s}, \tag{2.2}$$

where $\sqrt{s}$ is the center of mass energy of the e^+e^- collision.[9]

An interesting physical ramification of QED is the spontaneous production of *virtual* electron-positron pairs due to the uncertainty inherent in quantum mechanics. Because of this uncertainty, energy conservation can be violated for a very short time period, $\Delta t < \hbar/\Delta E$, where ΔE is the "borrowed" energy. This has important implications for the nature of the electromagnetic interaction. An electron in QED can spontaneously emit a virtual photon, which in turn can produce a virtual e^+e^- pair, and so on, until a single "bare" electron is surrounded by a cloud of virtual electrons and positrons. The diagram[10] below illustrates this.

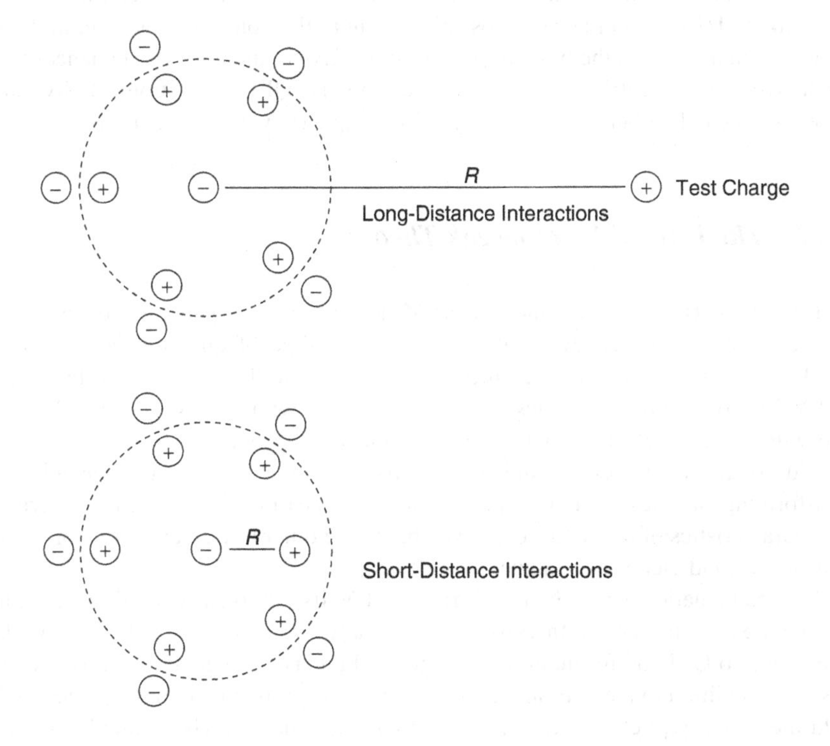

[9]As usual, we've chosen units where $\hbar = c = 1$.

[10]The diagram was adapted from Fig. 1.6 in *Quarks and Leptons* by F. Halzen and A.D. Martin, 1984.

Because opposite charges attract, the positrons will be located preferentially closer to the electron. If one measures the charge of the electron from a location outside of the e^+e^- cloud, the bare charge is reduced by the intervening positrons. This is referred to as *charge screening*. As one moves closer to the electron, penetrating the cloud of nearby positrons, the observed charge of the electron increases.

Since the strength of the electromagnetic coupling α is proportional to the square of the electric charge (e^2), the effect of charge screening is to reduce the coupling strength for long distance (low energy) interactions. Thus, α depends on the energy scale associated with the interaction. The value of α decreases asymptotically with energy to a constant value of $\approx 1/137$. Historically, this quantity is known as the *fine structure constant*.

Equation (2.2) gives the leading-order approximation to the exact $e^+e^- \longrightarrow \mu^+\mu^-$ scattering cross-section. A full QED calculation requires summing an infinite series of diagrams with additional vertices and internal loops! Generally, as more photons are added to the diagrams, the number of vertices (and hence the order of α) increases and the calculations become quite cumbersome. Fortunately, the small value ($\alpha \approx 1/137 \ll 1$) makes it possible to ignore the contributions from higher-order diagrams. This is the basis of *perturbation theory*, and it greatly enhances the predictive power of QED. In most cases, very precise QED predictions of physical observables can be obtained using only a few simple (low order) diagrams.

2.5.2 The Unified Electroweak Theory

In 1954, C.N. (Frank) Yang and Robert Mills formulated a generalized principle of gauge invariance that eventually led to a new type of quantum field theory. Unlike QED, with a single force-mediating photon, the theory proposed by Yang and Mills required three massless gauge bosons: one with positive electric charge, one with negative electric charge, and one electrically neutral. The introduction of additional gauge bosons implied the existence of a force that is capable of transforming particles from one type to another. At the time, this seemed to describe the characteristics of the weak force, which, among other things, converted protons to neutrons (and vice versa) in nuclear β decay.

The mathematical groundwork of Yang and Mills led to substantial theoretical developments in the 1960s. In 1961, Sheldon Glashow irreversibly linked the weak interaction to QED by formulating a gauge field theory with three massless vector bosons in addition to the photon. There was only one problem: no massless charged field-mediating particles had ever been observed in nature. The conundrum was solved by the identification of *spontaneous symmetry breaking* by Jeffrey Goldstone and Peter Higgs. The *Higgs mechanism* was applied to Glashow's theory by Steven Weinberg in 1967 and Abdus Salam in 1968, thereby giving the gauge bosons mass, described further in Sect. 2.6. The result was a self-consistent unified electroweak theory that predicted one massless particle (the photon) and three new massive particles: the W^+, W^-, and Z^0 bosons.[11] The discovery of the W and Z bosons at the European Center for Nuclear Research (CERN) sixteen years later confirmed the theoretical predictions and marked a tremendous advance for the standard model. At sufficiently high energies, the difference between the electromagnetic and weak interactions becomes negligible and the two act together as a single, unified electroweak interaction.

The first measurements of the W and Z boson masses by the UA1 and UA2 collaborations in 1983 were based on a handful of events from $p\bar{p}$ collisions at the CERN SPS collider. To the surprise of many, these mediators of the electroweak force turned out to be over 85 times more massive than the proton! The masses of the W and Z bosons were about $80\,\mathrm{GeV\,c^{-2}}$ and $91\,\mathrm{GeV\,c^{-2}}$, respectively. These huge masses of the W and Z bosons mean that they are extremely short-lived, which explains the relatively small interaction strength of the weak interaction.

In the electroweak theory, the masses of the W and Z bosons are intricately connected with two gauge coupling constants, g and g', via the Weinberg angle θ_W:

$$\tan\theta_W = \frac{g'}{g}, \qquad \cos\theta_W = \frac{M_W}{M_Z}. \tag{2.3}$$

Also known as the *weak mixing angle*, θ_W is a parameter that relates the relative strengths of the weak and electromagnetic couplings. A fit to a variety of experimental measurements yields a value of

$$\sin^2\theta_W = 0.2230 \pm 0.0004. \tag{2.4}$$

As a mediator of the weak force, the charged W bosons couple to fermion pairs that differ in charge by ± 1. Unlike all of the other gauge bosons, the W^+ or W^- possesses the unique ability to change the flavor of fermions with which it interacts. The flavor-changing property of the W^+ boson is illustrated in the diagram below.

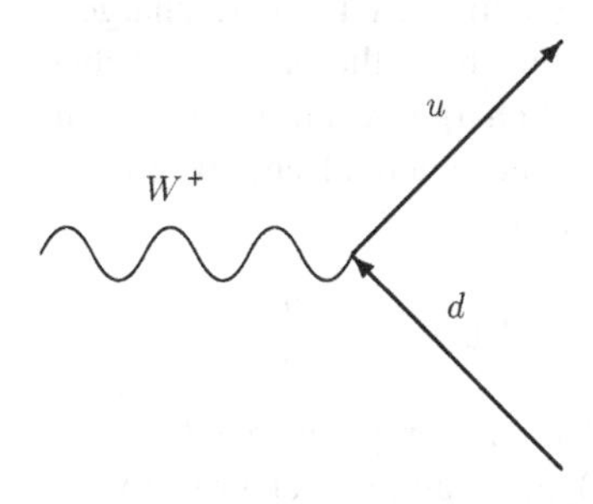

[11]The W^+ and W^- bosons are often written as W bosons, where the superscripts representing electric charge are dropped. Likewise, the Z^0 boson is written as the Z boson.

Since the interaction requires a transfer of electric charge at the vertex, the W boson coupling is said to be associated with a weak *flavor-changing charged current*. This property is of great importance for the production of W bosons in $p\bar{p}$ collisions, where they are an integral part of many ongoing analyses at experiments like CDF and DØ at Fermilab, and have recently been "rediscovered" at CERN in proton-proton collisions produced by the Large Hadron Collider.

2.5.3 Quantum Chromodynamics (QCD)

In 1965, Moo-Young Han, Yoichiro Nambu, and Oscar Greenberg laid the foundation for *quantum chromodynamics* (QCD), the quantum gauge theory that describes the strongest of the four fundamental interactions. The strong interaction is mediated by massless gauge bosons called *gluons* (g). Gluons are the field quanta that carry a unique kind of charge, called *color*, for which the theory is named. Just as electric charge is conserved within the framework of QED, the color charge of QCD is conserved in all interactions between quarks and gluons. There are three distinct values of the color charge – commonly denoted red, green, and blue – together with their corresponding anticolors.

Although both QED and QCD are gauge-invariant field theories, QCD is *non-Abelian*. Physically, this implies a qualitative difference from QED: whereas photons couple only to electrically charged particles, gluons themselves carry the color charge and interact among themselves. In fact, there are actually eight distinct varieties of gluons composed of various permutations of color and anticolor charge. This has important ramifications. Unlike the charge screening of QED, in which virtual electron-positron pairs pop out of the vacuum and align themselves to shield a bare charge, a bare QCD color charge (e.g., a quark) is quickly surrounded by a "sea" of virtual quarks and gluons with the same color. As one probes the bare color charge at shorter and shorter distances, corresponding to higher and higher energies, the observed charge lessens until only the bare charge is seen. This is referred to as *asymptotic freedom*. Farther from the bare color charge, the intervening sea of color increases the observed charge, resulting in a strong attractive force between two distant color charges. The potential energy grows roughly linearly with the separation of charges, according to

$$V(r) = -\frac{4}{3}\frac{\alpha_s}{r} + kr. \tag{2.5}$$

At large distances, the potential energy between two quarks is sufficient to create a real (as opposed to virtual) quark-antiquark pair from the vacuum, thereby breaking the long-distance force and reducing the overall potential energy. This process is known as *fragmentation*. Since fragmentation will always occur as two quarks separate, solitary quarks cannot exist. Instead, quarks must eventually join with other quarks or antiquarks to form colorless bound states. This property of QCD, called *color confinement*, offers an explanation of why no free quarks or gluons have ever been observed in nature.

In Equation 2.5, the quantity α_s is the QCD coupling strength, which describes how the effective charge between two quarks depends on the distance between them. The lowest-order expression for α_s, also known as the *running coupling constant*, is given by

$$\alpha_s(Q) = \frac{6\pi}{(33 - 2n_f)\ln(Q/\Lambda_{\mathrm{QCD}})}. \tag{2.6}$$

Here, Q denotes the square root of the momentum transfer (i.e. the energy of the probe), n_f is the allowed number of quark flavors at that energy, and Λ_{QCD} corresponds roughly to the energy boundary between asymptotically free quarks and hadrons. Measurements of Λ_{QCD} yield a value between 100 and 500 MeV, a scale that coincides well with the masses of the lightest hadrons.

Unlike the QED coupling α, which increases with energy, α_s falls off gradually and approaches an asymptotic value. For $Q \approx \Lambda_{\mathrm{QCD}}$, quarks and gluons interact strongly and arrange themselves into hadrons. As Q becomes much larger than Λ_{QCD}, the effective coupling becomes small and quarks and gluons interact with each other only weakly. The value of α_s is about 0.1 for interactions with a momentum transfer in the 100 GeV to 1 TeV range.

Besides addressing the question of why quarks always appear in bound systems, the notion of color also solved a nagging dilemma in the quark model of hadrons. Baryons were thought to contain either three quarks or three antiquarks, and this recipe successfully described the huge spectrum of newly discovered hadrons in the late 1950s and early 1960s. The Δ^{++} baryon was a peculiar exception. With an electric charge of $+2|e|$, the Δ^{++} could only exist as a combination of three up quarks (uuu) in the lowest orbital momentum state ($l = 0$) with fully aligned spins ($J = S = 3/2$). This configuration violates the Pauli exclusion principle. If, however, each quark carried a different value of the color charge, the fermions would no longer be identical and the exclusion principle would not be violated.

The existence of three unique quark colors is experimentally validated by the measurement of the cross-section ratio:

$$R = \frac{\sigma(e^+e^- \to q\bar{q} \to \text{hadrons})}{\sigma(e^+e^- \to \mu^+\mu^-)} \tag{2.7}$$

For a period of time before 1964, a serious discrepancy between the predicted and measured values of the ratio puzzled theorists and experimentalists alike. The experimental value was three times larger than the predicted value. However, when the numerator was summed over all of the quark colors, the theoretical cross-section ratio reduced to the simple expression

$$R = N_c \sum_i q_i^2, \tag{2.8}$$

where N_c is the number of colors and q_i is the charge of each quark flavor. The sum includes the quark flavors that are kinematically accessible ($2m_i < \sqrt{s}$). A value of $N_c = 3$ brought theory and experiment into excellent agreement.

2.6 The Higgs Boson

As discussed in Sect. 2.5.2, the work of Yang and Mills produced a generalized principle of gauge invariance that led to a new form of quantum field theory. Glashow then developed a technique to link QED with the weak interaction by formulating a gauge field theory with three massless vector bosons in addition to a photon. The problem was that no massless charged field-mediating particles had ever been observed in nature. By 1964, three independent groups (Guralnik, Hagen, and Kibble; Higgs; Brout and Englert) solved this conundrum by combining the principle of gauge invariance with spontaneous symmetry breaking.

Spontaneous symmetry breaking occurs when the true symmetry of a system is hidden by the choice of a particular ground state. As an example, consider a thin, flexible ruler held vertically by its ends, with its edge facing you. Pushing on the ends of the ruler causes it to bend, and the middle shifts either to the left or to the right, breaking the left-right symmetry of the ruler system. This example illustrates a way to generate a discrete set of ground states because there are only two ways for it to bend. However, you could generate a continuous set of ground states if you were to push on the ends of an object like a thin, cylindrical plastic rod rather than a ruler. The rod can bend in the middle, but in any direction. In a field theory, ground states are determined by the minima of a potential (the vacuum states) and the fields are treated as fluctuations about a chosen state. The choice of a single ground state seemingly breaks the symmetry of the system and is considered "spontaneous" because there is no external means by which this occurs. For example, 3-dimensional symmetry on the surface of the earth is broken because we cannot describe up and down in the same way that we can left and right. This is, however, due to the presence of gravity, an external agency.

By 1968, Weinburg and Salam applied it to the electroweak theory. In the simplest application of the general gauge theory, an extra Higgs field is added. Through spontaneous symmetry breaking – choice of a ground state – this field is caused to interact with the weak field to produce masses for the W and Z gauge bosons. The Higgs mechanism can also be applied to the other field theories to demonstrate how quarks and leptons gain mass.

A more intuitive way to think of the Higgs field is to imagine water that fills a pool. If you were to walk through a pool, you would feel heavier because of the water pushing against you. Your inertia is larger because the water makes it harder for you to move, as though you have gained mass. The Higgs field permeates space in the same way. As some particles travel through the universe, they interact with this field and acquire mass as a result. Some particles also interact more strongly with the Higgs field than others and so have a larger mass. Why this is true is still unknown.

Of the scientists that theorized the Higgs mechanism, Peter Higgs mentioned the possibility of a new elementary scalar (spin-0) boson being produced, now called the Higgs boson. This particle is the only standard model particle that has not been observed in nature. Evidence of this particle would verify the existence of the Higgs mechanism and confirm the manner in which particles gain mass.

The formalism of the Higgs mechanism does not provide the actual mass of the Higgs boson, so we experimentalists must search the entire mass range. The Large Electron-Positron Collider at CERN excluded[12] the mass below 114 GeV/c^2 and indirect electroweak measurements constrain the Higgs mass to be below 185 GeV/c^2. Efforts to reduce the range between these masses is ongoing for the CDF and DØ collaborations at Fermilab who have together, as of July 2010, additionally excluded the region between 158 and 175 GeV/c^2. One of the main goals of the Large Hadron Collider (LHC), located at CERN in Switzerland, is to provide evidence for the Higgs. The LHC produced proton-proton collisions at 3.5 TeV for the first time in November 2009, making it the highest energy particle collider in the world. As the CDF and DØ experiments wind down in 2012, it is expected that the LHC will fill in the final gaps of the mass range and, therefore, provide confirmation for or against the existence of the Higgs boson.

2.7 References and Further Reading

The primary sources for the material in this section are [20, 33–35, 37, 40, 56, 89, 93, 97, 102]. For further reading, we also recommend [25].

[12] All mass exclusions quoted here are with a 95% confidence level.

	Leptons		Hadrons			Higgs
	(1, 2, − 1/2)	(1, 1, 1)	(3, 2, 1/6)	(3, 1, − 2/3)	(3, 1, 1/3)	(1, 2, − 1/2)
Generation 1	(Electron neutrino, Electron)	Electron	(Up, Down)	Up	Down	1 Generation only
Generation 2	(Muon neutrino, Muon)	Muon	(Charm, Strange)	Charm	Strange	
Generation 3	(Tau neutrino, Tau)	Tau	(Top, Bottom)	Top	Bottom	

Chapter 3
Algebraic Foundations

3.1 Introduction to Group Theory

In Chap. 1 we spent several pages talking about classical physics, and in doing so we came across three different examples of symmetries – things that don't change while other things do change. These were Noether symmetries in an action, transformations on a vector that preserve the dot product, and gauge transformations in electrodynamics. As we have hinted, this idea of symmetry will play an extremely fundamental role in everything we'll be doing for the rest of this book. In fact it can easily be said that symmetry plays *the* fundamental role in particle physics, so we'll now devote a chapter to understanding it more fully.

So, the primary idea of this chapter is to formulate the mathematics of symmetry. We want a detailed toolbox to use when seeking to understand the role symmetry plays in physical theories. And the name of this mathematical machinery is **Group Theory**.

Group theory is, in short, the mathematics of symmetry. We are going to begin talking about what will seem to be extremely abstract ideas, but eventually we will explain how those ideas relate to physics. As a preface of what is to come, the most foundational idea here is, as we said before, "what doesn't change when something else changes". A group is a precise and well-defined way of specifying the thing or things that change along with what is preserved under that change.

3.1.1 What is a Group?

There are two general descriptions of what a group is intended to do. The first is the notion of symmetry. While this description is a very important part of what groups do, it is not the most obvious. We will discuss how groups describe symmetry later,[1]

[1]Especially when we get to the section on Lie groups in Sect. 3.2.

M. Robinson, *Symmetry and the Standard Model: Mathematics and Particle Physics*,
DOI 10.1007/978-1-4419-8267-4_3,

but for now (as an introductory idea) we will focus on the more intuitive concept behind groups, which is the notion of structure.

To illustrate what we mean by structure, let's look at some examples. The set of integers, denoted $\mathbb{Z}$, possesses a certain structure – specifically, with an appropriate definition of "addition", a given integer "added" to another integer gives a particular integer as a result. The same is true for an appropriate definition of "multiplication" – but addition and multiplication are obviously two different structures. If we take the real number $3 \in \mathbb{Z}$ and add it to another number $4 \in \mathbb{Z}$, we get $7 \in \mathbb{Z}$. Furthermore, we can multiple 3 and 4 and get $12 \in \mathbb{Z}$. This is certainly not difficult – we are extremely familiar with the structure of integer addition and multiplication.

But let's consider an alternative definition of integer addition. Consider letting the "normal" integers be represented by their exponentiated value:

$$a \longrightarrow e^a. \tag{3.1}$$

We can now define our alternative notion of addition and multiplication as follows:

$$\begin{aligned} a + b &\longrightarrow (e^a)(e^b) = e^{a+b}, \\ ab &\longrightarrow (e^a)^b = (e^b)^a = e^{ab}. \end{aligned} \tag{3.2}$$

These definitions are not natural in any way. We are defining them this way. But as a result we have completely preserved the structure of $\mathbb{Z}$ under addition and multiplication.[2] Our previous example is easy to see:

$$\begin{aligned} 3 &\longrightarrow e^3 = 20.0855369, \\ 4 &\longrightarrow e^4 = 54.59815, \\ 3 + 4 &\longrightarrow e^3 e^4 = (20.0855369)(54.59815) = 1016.29101 = e^7 \longrightarrow 7, \\ 3 \cdot 4 &\longrightarrow (e^3)^4 = (20.0855369)^4 = 162754.791 = e^{12} \longrightarrow 12. \end{aligned} \tag{3.3}$$

A few moments of thought should convince you that all we have done is find a new way of *representing* the integers, integer addition, and integer multiplication.

So one might ask: what is the difference between talking about integer addition and multiplication using the normal integers and using our exponentiated version, given that they maintain identical structure in both cases under the operations in question? This question is the heart of abstract algebra, of which group theory is a major component. Abstract algebra, and consequently group theory, are the study of the structure of a set of things along with one or more operations. And because it is often the structure that's important, it doesn't matter how we choose to represent the objects. There is nothing which says it is necessarily better to use one representation or another.

[2] You can no doubt figure out how to incorporate subtraction and division as well.

As a primitive example of an application of this, consider a shepherd thousands of years ago. He knows nothing at all about mathematics. Even counting is beyond him, and addition is completely unknown. But everyday he lets his sheep outside the fence to graze, and every evening he herds them back in. Of course he doesn't want to loose any sheep, but not knowing how to count or add poses a problem. However, the shepherd is very smart. One day, while the sheep were leaving their enclosure, he picked up a handful of pebbles from the ground. Every time a sheep went out, he put a single pebble in a bag. That evening, when they came back in, he took a single pebble out of the bag. He knew that as long as there were pebbles left in the bag, there were still sheep outside the enclosure. So without knowing how to count or add, the shepherd had found a way to make sure he never lost any sheep.

The shepherd was doing abstract algebra. In fact, as we will see, he was doing group theory. He knew nothing of the integers, and he knew nothing about arithmetic. However he was able to realize that the action "one sheep leaves enclosure" had the identical structure as "one pebble goes into the bag" and that "one sheep enters enclosure" has the identical structure as "one pebble is taken out of the bag". The practicality of algebra in this situation is that the shepherd doesn't know anything about counting and arithmetic, but he does know how to put a pebble in a bag. All that really matters is the structure, and one representation of that structure was more attainable than the other.

This example illustrates the underlying point of abstract algebra and group theory. Group theory provides a nice way of understanding structure in a more general sense. The generality is important because, like with the shepherd, one representation may be more useful than another representation for a given application. If we understand the general, abstract details of the structure it becomes much easier to find a representation of that structure for a given task.

3.1.2 Definition of a Group

To begin with, we define a **Group**. This definition may seem cryptic and overly abstract, but the following paragraphs and examples should make it more clear.

A group, denoted $(G, \star)$, is a set of objects, denoted G, and some operation on those objects, denoted $\star$, subject to the following:

(1) For any two elements g_1 and g_2 in G, the element $g_1 \star g_2$ is also in G. This property is called ***closure****.*

(2) For any three elements g_1, g_2 and g_3 in G, the relation $(g_1 \star g_2) \star g_3 = g_1 \star (g_2 \star g_3)$ must hold. This property is called ***associativity****.*

(3) There exists an element of G which we will denote e, that satisfies $e \star g = g \star e = g$ for every element of G. This property is called ***identity****.*

(4) For every element g of G, there is another element of G which we will denote g^{-1} that satisfies $g^{-1} \star g = g \star g^{-1} = e$. This property is called ***inverse****.*

An important note is that the definition of a group doesn't demand that $g_i \star g_j = g_j \star g_i$. This is a very important point, but we will postpone discussing it until later. We mention it now just to bring it to your attention.

Admittedly the above definition is a bit much to take in, especially if you've never seen it before. So we explain each part. By "objects" we literally mean anything. We could be talking about $\mathbb{Z}$, $\mathbb{R}$, or a set of Easter eggs painted different colors. The meaning of "some operation", which we are calling $\star$, can literally be anything you can do to those objects where you take two of them as input and get one as output. We could provide a more formal definition of $\star$ but it will be easier to understand with examples.

To put this in a more intuitive context, we'll look at the above definition another way. Consider a set of locations in some space labelled g_1, g_2, etc. Furthermore, each location g_i is also a set of instructions on how to move relative to a given location g_j. In other words, g_i specifies a location, but it also specifies a particular way of moving from g_j. The element g_i moves you from the location g_j to $g_k = g_j \star g_i$. The four group rules listed above are then intuitively described as follows:

(1) None of the "legal" motions allow you to leave the space consisting of all the locations g_i – you're trapped!
(2) The order of the operations should not matter. (still a bit abstract but this rule won't be terribly important in this text).
(3) Staying where you are is an option.
(4) No matter where you go, you have to be able to get back to where you were.

Now let's consider a few examples. Consider the set of integers with the operation addition: $(G, \star) = (\mathbb{Z}, +)$. We'll check each of the properties one at a time. First we check closure. If you take any two elements of $\mathbb{Z}$ and add them together, is the result in $\mathbb{Z}$? In other words is the sum of two integers an integer? The answer is yes, so closure is met. Next we check associativity. If a, b, and c are in $\mathbb{Z}$, it is trivially true that $(a+b)+c = a+(b+c)$, so associativity is met. Next is identity. Is there an element e in $\mathbb{Z}$ such that when you add e to any other integer you get that same integer? The answer is yes: $e = 0$. So identity is met. And finally we ask if for any integer a in $\mathbb{Z}$ there is another integer that when added to a gives 0? The integer $-a$ satisfies $a + (-a) = 0$ for any integer a, and therefore the inverse requirement is met. Therefore $(\mathbb{Z}, +)$ is a group.

Next consider $(G, \star) = (\mathbb{R}, +)$. Obviously any two real numbers added together is also a real number, so closure is met. Associativity again holds trivially. The identity is again 0. And finally, once again, $-a$ is the inverse of any real number a. So $(G, \star) = (\mathbb{R}, +)$ is a group.

As a third example consider $(G, \star) = (\mathbb{R}, \cdot)$, the real numbers with multiplication. Two real numbers multiplied together is a real number, so closure is met. Associativity holds as usual. The identity is 1. However, we run into a problem with the inverse property. The question is, for any real number is there another real number that you can multiply by it to get 1. The instinctive choice is $a^{-1} = \frac{1}{a}$. But this doesn't work because of $a = 0$. This is the *only* exception, but because there is an exception, $(\mathbb{R}, \cdot)$ is not a group. However, if we take the set to be $\mathbb{R} - \{0\}$, then $(\mathbb{R} - \{0\}, \cdot)$ is a group (you should take a moment to convince yourself of this if it isn't obvious).

As a fourth example consider $(G, \star) = (\mathbb{R}, -)$, the real numbers with subtraction. Obviously the difference between two real numbers is a real number, the identity is 0, and the inverse of a is a (because $a - a = 0$). However, notice that

$$
\begin{aligned}
&7 - (5 - 2) = 7 - 3 = 4, \\
&(7 - 5) - 2 = 2 - 2 = 0, \\
\implies &7 - (5 - 2) \neq (7 - 5) - 2.
\end{aligned}
\tag{3.4}
$$

So $(\mathbb{R}, -)$ is not a group. This illustrates the reason for the associativity rule: it prevents ambiguities in the order of operations.

Another example is $(G, \star) = (\{1\}, \cdot)$. This is the set with only one element, 1, and the operation is normal multiplication. This is a group, but it is extremely uninteresting. It is called the **Trivial Group**.

As a final example consider $(G, \star) = (\mathbb{Z}_3, +)$, the set of integers mod 3. If you're not familiar with this set, it contains only the numbers 0, 1, and 2 (3 mod 3 is 0, 4 mod 3 is 1, 5 mod 3 is 2, and so on). You can check yourself that this is a group. This group is actually very important, and we denote it simply $\mathbb{Z}_3$. More generally, the group of integers mod n is denoted $\mathbb{Z}_n$.

3.1.3 Finite Discrete Groups and Their Organization

We want to make several observations from the examples in the previous section. First of all, there are no rules restricting the number of elements in a group. The trivial group has only one element, the integers with addition have a countably infinite number of elements, and the real numbers with addition have an uncountably infinite number of elements. The number of elements in a group is the **Order** of the group. A group with a finite number of elements is said to have a finite order, and a group with an infinite number of elements has an infinite order. We can also categorize a group based on whether its elements are **Continuous** or **Discrete**. Groups like $(\mathbb{Z}, +)$, $(\{1\}, \cdot)$, and $(\mathbb{Z}_3, +)$ are discrete, whereas $(\mathbb{R}, +)$ and $(\mathbb{R} - \{0\}, \cdot)$ are continuous.

With those categories in mind, we're going to focus on finite discrete groups for now. Specifically, we're going to talk about how to organize them. We use what is called a **Multiplication Table**. A multiplication table is a way of organizing the elements of a group as follows:

$(G,\star)$	e	g_1	g_2	$\cdots$
e	$e\star e$	$e\star g_1$	$e\star g_2$	$\cdots$
g_1	$g_1\star e$	$g_1\star g_1$	$g_1\star g_2$	$\cdots$
g_2	$g_2\star e$	$g_2\star g_1$	$g_2\star g_2$	$\cdots$
$\vdots$	$\vdots$	$\vdots$	$\vdots$	$\ddots$

A multiplication table must contain every element of the group exactly one time in every row and every column. A few minutes thought should convince you that this is necessary to ensure that the definition of a group is satisfied.[3]

As an example, we will draw a multiplication table for the group of order 2. We won't look at specific numbers, but rather call the elements e and g_1. We begin as follows:

$(G,\star)$	e	g_1
e	?	?
g_1	?	?

Three of these are easy to fill in from the identity:

$(G,\star)$	e	g_1
e	e	g_1
g_1	g_1	?

And because we know that every element must appear exactly once, the final question mark must be e. So, there is only one possible group of order 2.

$(G,\star)$	e	g_1
e	e	g_1
g_1	g_1	e

We will consider a few more examples, but we stress at this point that the temptation to plug in numbers should be avoided. For example, you may notice that the integers with addition mod 2 (binary) are an example of the group of order 2. On the other hand, so are the integers 1 and -1 with multiplication. There are several other examples (see if you can think of a few), and it is not good at this point to try to think in terms of specific examples. Recall that the idea behind a group is the *structure*, not the specific numbers and operations that obey this structure. Groups are abstract things with definite structure. We'll talk shortly about how a given group can be represented, but as far as understanding group theory, try to avoid the temptation to plug in numbers.

[3]If this isn't clear, write out several examples where this does not hold and you will quickly see that it isn't possible to satisfy the group axioms.

Moving on, we can proceed with the multiplication table for the group of order 3. You will find that, once again, there is only one option. (Doing this is instructive, and it would be helpful to work this out yourself.)

$(G, \star)$	e	g_1	g_2
e	e	g_1	g_2
g_1	g_1	g_2	e
g_2	g_2	e	g_1

You are encouraged to work out the possibilities for groups of order 4. *(Hint: there are four possibilities.)*

3.1.4 Group Actions

We now (hopefully) have some intuition for how the elements of a group relate to each other. The point, as we have said before, is that a particular group represents a particular structure – there are a set of things, and they relate to each other in a particular way. Now, however, we want to consider how these structures may relate to real life.

Consider three Easter eggs, all painted different colors (say red, orange, and yellow). We'll denote them R, O, and Y. Now, assume that they have been put into a row in the order (ROY). If we want to keep them lined up, not take any eggs away, and not add any eggs, what we can do to them? A few moments thought reveals that there are six "operations" we can perform:

1. Let e be doing nothing to the set, so $e(ROY) = (ROY)$.
2. Let g_1 be a cyclic permutation of the three, $g_1(ROY) = (OYR)$.
3. Let g_2 be a cyclic permutation in the other direction, $g_2(ROY) = (YRO)$.
4. Let g_3 be swapping the first and second, $g_3(ROY) = (ORY)$.
5. Let g_4 be swapping the first and third, $g_4(ROY) = (YOR)$.
6. Let g_5 be swapping the second and third, $g_5(ROY) = (RYO)$.

You can work out the details and find that these six elements are closed, associative, there is an identity, and each has an inverse. We can easily draw a multiplication table for these elements (we strongly encourage you to confirm this table):

$(G, \star)$	e	g_1	g_2	g_3	g_4	g_5
e	e	g_1	g_2	g_3	g_4	g_5
g_1	g_1	g_2	e	g_5	g_3	g_4
g_2	g_2	e	g_1	g_4	g_5	g_3
g_3	g_3	g_4	g_5	e	g_1	g_2
g_4	g_4	g_5	g_3	g_2	e	g_1
g_5	g_5	g_3	g_4	g_1	g_2	e

We should be very careful to draw a distinction between the *elements* of the group and the *objects* the group acts on. The objects in this example are the eggs, and the permutations are the results of the group action. Neither the eggs nor the permutations of the eggs are the elements of the group. The elements of the group are the abstract objects in the multiplication table above which we have assigned to some operation on the eggs, resulting in a new permutation.

Notice something interesting about this group: from the multiplication table we have $g_3 \star g_1 = g_4$, whereas $g_1 \star g_3 = g_5$. So we have the surprising result that in this group it is not necessarily true that $g_i \star g_j = g_j \star g_i$. This leads to a new way of classifying groups. We say a group is **Abelian** if $g_i \star g_j = g_j \star g_i$ for every g_i and g_j in G. If a group is not Abelian, it is **Non-Abelian**. Another term commonly used is **Commute**. If $g_i \star g_j = g_j \star g_i$, then we say that g_i and g_j commute. So an Abelian group is **Commutative**, whereas a non-Abelian group is **Non-Commutative**.

The Easter egg group of order 6 above is an example of a very important type of group. It is denoted S_3 and is a **Symmetric Group**. It is the group corresponding to every permutation of three objects. The more general group of this type is S_n, the symmetric group corresponding to every permutation of n objects. With a few moments thought you can convince yourself that the group S_n will always have order $n!$ (n factorial).

So, the point to take away from the 3 eggs example is that S_3 is the actual *group*, while the eggs are the objects that the group *acts on*. The particular way an element of S_3 changes the eggs around is called the **Group Action** of that element. And each element of S_3 will move the eggs around while leaving them lined up. This ties into our overarching concept of "what doesn't change when something else changes". The fact that there are three eggs with three particular colors lined up doesn't change – these are the preserved properties. The order they appear in, however, does change.

3.1.5 Representations

We suggested above that you think of groups as purely abstract things rather than trying to plug in actual numbers. Now, however, we want to talk about how to see groups, or the elements of groups, in terms of specific numbers. But, we will do this in a very systematic way. The name for a specific set of numbers or objects that form a group is a **Representation**. The remainder of this section (and the next) will primarily be about group representations.

We already discussed a few simple representations when we discussed $(\mathbb{Z}, +)$, $(\mathbb{R} - \{0\}, \cdot)$, and $(\mathbb{Z}_3, +)$. Let's focus on $(\mathbb{Z}_3, +)$ for a moment (the integers mod 3, where $e = 0$, $g_1 = 1$, $g_2 = 2$, with addition). Notice that we could alternatively define $e = 1$, $g_1 = e^{\frac{2\pi i}{3}}$, and $g_2 = e^{\frac{4\pi i}{3}}$, and let $\star$ be multiplication. So, in the "representation" with $(0, 1, 2)$ and addition, we had for example

$$g_1 \star g_2 = (1 + 2) \text{ (mod 3)} = 3 \text{ (mod 3)} = 0 = e$$

whereas now with the multiplicative representation we have[4]

$$g_1 \star g_2 = e^{\frac{2\pi i}{3}} \cdot e^{\frac{4\pi i}{3}} = e^{2\pi i} = e^0 = 1 = e.$$

So the structure of the group is preserved in both representations.

We now seek a more comprehensive way of coming up with representations of a particular group. We begin by introducing some notation. For a group $(G, \star)$ with elements $g_1, g_2, \ldots$, we call the representation of that group $D(G)$, so that the elements of G are $D(e)$, $D(g_1)$, $D(g_2)$ (where each $D(g_i)$ is a matrix of some dimension). We then choose $\star$ to be matrix multiplication. So,

$$D(g_i) \cdot D(g_j) = D(g_i \star g_j). \tag{3.5}$$

It may not seem that we have done anything profound at this point, but we most definitely have. Remember above that we encouraged seeing groups as abstract things, rather than in terms of specific numbers. This is because a group is fundamentally an *abstract* object. A group is not a specific set of numbers, but rather a set of abstract objects with a well-defined structure telling you how those elements relate to each other.

And the beauty of a representation D is that, via normal matrix multiplication, we have a sort of "lens", made of familiar things (like numbers, matrices, or moving Easter eggs), through which we can see into this abstract world. And because $D(g_i) \cdot D(g_j) = D(g_i \star g_j)$, we aren't losing any of the structure of the abstract group by using a representation.

So now that we have some notation, we can develop a formalism to figure out exactly what D is for an arbitrary group.

We will use Dirac vector notation, where the column vector $\mathbf{v}$ is expressed as

$$\mathbf{v} = \begin{pmatrix} v^1 \\ v^2 \\ v^3 \\ \vdots \end{pmatrix} = |v\rangle, \tag{3.6}$$

and the row vector $\mathbf{v}^T$ is expressed as

$$\mathbf{v}^T = \begin{pmatrix} v_1 \ v_2 \ v_3 \ \cdots \end{pmatrix} = \langle v|. \tag{3.7}$$

So, the dot product between two vectors is

$$\mathbf{v}^T \cdot \mathbf{u} = \begin{pmatrix} v_1 \ v_2 \ v_3 \cdots \end{pmatrix} \begin{pmatrix} u^1 \\ u^2 \\ u^3 \\ \vdots \end{pmatrix} = v_1 u^1 + v_2 u^2 + v_3 u^3 + \cdots = \langle v|u\rangle. \tag{3.8}$$

[4]We apologize for the double use of e; one is the identity element and the other is Euler's number. Context should make it easy to tell which we mean.

In the language of Sect. 1.4.1, the metric in this space we're defining is simply the Kronecker delta δ_{ij}. For the remainder of this section, therefore, we will ignore upper vs. lower indices.

Now, we proceed by relating each element of a finite discrete group to one of the standard orthonormal unit vectors:

$$e \to |e\rangle = |\hat{e}_1\rangle \qquad g_1 \to |g_1\rangle = |\hat{e}_2\rangle \qquad g_2 \to |g_2\rangle = |\hat{e}_3\rangle. \tag{3.9}$$

Next, we define the way an element in a representation $D(G)$ acts on these vectors to be

$$D(g_i)|g_j\rangle = |g_i \star g_j\rangle. \tag{3.10}$$

Now we can build our representation. We will (from now on unless otherwise stated) represent the elements of a group G using matrices of various sizes, and the group operation $\star$ will be standard matrix multiplication. The specific matrices that represent a given element g_k of our group will be given by

$$[D(g_k)]_{ij} = \langle g_i|D(g_k)|g_j\rangle. \tag{3.11}$$

As an example, consider again the group of order 2 (we wrote out the multiplication table above on page 56). First, we find the matrix representation of the identity, $[D(e)]_{ij}$,

$$\begin{aligned}
[D(e)]_{11} &= \langle e|D(e)|e\rangle = \langle e|e \star e\rangle = \langle e|e\rangle = 1,\\
[D(e)]_{12} &= \langle e|D(e)|g_1\rangle = \langle e|e \star g_1\rangle = \langle e|g_1\rangle = 0,\\
[D(e)]_{21} &= \langle g_1|D(e)|e\rangle = \langle g_1|e \star e\rangle = \langle g_1|e\rangle = 0,\\
[D(e)]_{22} &= \langle g_1|D(e)|g_1\rangle = \langle g_1|e \star g_1\rangle = \langle g_1|g_1\rangle = 1.
\end{aligned} \tag{3.12}$$

So, the matrix representation of the identity is $D(e) = \begin{pmatrix} 1 & 0 \\ 0 & 1 \end{pmatrix}$. It shouldn't be surprising that the identity element is represented by the identity matrix.

Next we find the representation of $D(g_1)$:

$$\begin{aligned}
[D(g_1)]_{11} &= \langle e|D(g_1)|e\rangle = \langle e|g_1 \star e\rangle = \langle e|g_1\rangle = 0,\\
[D(g_1)]_{12} &= \langle e|D(g_1)|g_1\rangle = \langle e|g_1 \star g_1\rangle = \langle e|e\rangle = 1,\\
[D(g_1)]_{21} &= \langle g_1|D(g_1)|e\rangle = \langle g_1|g_1 \star e\rangle = \langle g_1|g_1\rangle = 1,\\
[D(g_1)]_{22} &= \langle g_1|D(g_1)|g_1\rangle = \langle g_1|g_1 \star g_1\rangle = \langle g_1|e\rangle = 0.
\end{aligned} \tag{3.13}$$

So, the matrix representation of g_1 is $D(g_1) = \begin{pmatrix} 0 & 1 \\ 1 & 0 \end{pmatrix}$. It is straightforward to check that this is a true representation,

$$\begin{aligned}
e \star e &= \begin{pmatrix} 1 & 0 \\ 0 & 1 \end{pmatrix}\begin{pmatrix} 1 & 0 \\ 0 & 1 \end{pmatrix} = \begin{pmatrix} 1 & 0 \\ 0 & 1 \end{pmatrix} = e \quad \checkmark \\
e \star g_1 &= \begin{pmatrix} 1 & 0 \\ 0 & 1 \end{pmatrix}\begin{pmatrix} 0 & 1 \\ 1 & 0 \end{pmatrix} = \begin{pmatrix} 0 & 1 \\ 1 & 0 \end{pmatrix} = g_1 \quad \checkmark \\
g_1 \star e &= \begin{pmatrix} 0 & 1 \\ 1 & 0 \end{pmatrix}\begin{pmatrix} 1 & 0 \\ 0 & 1 \end{pmatrix} = \begin{pmatrix} 0 & 1 \\ 1 & 0 \end{pmatrix} = g_1 \quad \checkmark \\
g_1 \star g_1 &= \begin{pmatrix} 0 & 1 \\ 1 & 0 \end{pmatrix}\begin{pmatrix} 0 & 1 \\ 1 & 0 \end{pmatrix} = \begin{pmatrix} 1 & 0 \\ 0 & 1 \end{pmatrix} = e \quad \checkmark
\end{aligned} \tag{3.14}$$

Instead of considering the next obvious example, the group of order 3, consider the group S_3 from above (the multiplication table is on page 57). The identity representation $D(e)$ is easy; it is just the 6×6 identity matrix. We encourage you to work out the representation of $D(g_1)$ on your own, and check to see that it is

$$D(g_1) = \begin{pmatrix} 0 & 0 & 1 & 0 & 0 & 0 \\ 1 & 0 & 0 & 0 & 0 & 0 \\ 0 & 1 & 0 & 0 & 0 & 0 \\ 0 & 0 & 0 & 0 & 1 & 0 \\ 0 & 0 & 0 & 0 & 0 & 1 \\ 0 & 0 & 0 & 1 & 0 & 0 \end{pmatrix}. \tag{3.15}$$

All 6 matrices can be found this way, and multiplying them out will confirm that they do indeed satisfy the group structure of S_3.

3.1.6 Reducibility and Irreducibility: A Preview

You have probably noticed that equation (3.11) will always produce a set of $n \times n$ matrices, where n is the order of the group. There is actually a name for this particular representation. The $n \times n$ matrix representation of a group of order n is called the **Regular Representation**. More generally, the $m \times m$ matrix representation of a group (of any order) is called the **m-Dimensional Representation**.

But as we have seen, there is more than one representation for a given group (in fact, there are an infinite number of representations).

One thing we can immediately see is that any group that is non-Abelian cannot have a 1×1 matrix representation.[5] This is because scalars (1×1 matrices) always commute, whereas matrices in general do not.

We saw above in equation (3.15) that we can represent the group S_n by $n! \times n!$ matrices. Or more generally, we can represent any group using $m \times m$ matrices, were m equals order(G). This is the regular representation. But it turns out that it is usually possible to find representations that are "smaller" than the regular representation.

To pursue how this might be done, note that we are working with matrix representations of groups. In other words, we are representing groups in *linear spaces*. We will therefore be using a great deal of linear algebra to find smaller representations. This process, of finding a smaller representation, is called **Reducing** a representation. Given an arbitrary representation of some group, the first question that must be asked is "is there a smaller representation?" If the answer is yes, then the representation is said to be **Reducible**. If the answer is no, then it is **Irreducible**.

Before we dive into the more rigorous approach to reducibility and irreducibility, let's consider a more intuitive example, using S_3. In fact, we'll stick with our three painted Easter eggs, R, O, and Y:

1. $e(ROY) = (ROY)$.
2. $g_1(ROY) = (OYR)$.
3. $g_2(ROY) = (YRO)$.
4. $g_3(ROY) = (ORY)$.
5. $g_4(ROY) = (YOR)$.
6. $g_5(ROY) = (RYO)$.

We'll represent the set of eggs by a column vector

$$|E\rangle = \begin{pmatrix} R \\ O \\ Y \end{pmatrix}. \tag{3.16}$$

Now, by inspection, what matrix would do to $|E\rangle$ what g_1 does to (ROY)? In other words, how can we fill in the ?'s in

$$\begin{pmatrix} ? & ? & ? \\ ? & ? & ? \\ ? & ? & ? \end{pmatrix} \begin{pmatrix} R \\ O \\ Y \end{pmatrix} = \begin{pmatrix} O \\ Y \\ R \end{pmatrix} \tag{3.17}$$

[5] For the more mathematically oriented, we of course mean that a non-Abelian group cannot have a *faithful* representation. The trivial representation, where every element is the identity, preserves the group structure but does not do so in a "faithful" way. We aren't interested in such representations and we therefore ignore them.

to make the equality hold? A few moments thought will show that the appropriate matrix is

$$\begin{pmatrix} 0 & 1 & 0 \\ 0 & 0 & 1 \\ 1 & 0 & 0 \end{pmatrix} \begin{pmatrix} R \\ O \\ Y \end{pmatrix} = \begin{pmatrix} O \\ Y \\ R \end{pmatrix}. \tag{3.18}$$

Continuing this reasoning, we can see that the complete set of the matrices is

$$D(e) = \begin{pmatrix} 1 & 0 & 0 \\ 0 & 1 & 0 \\ 0 & 0 & 1 \end{pmatrix}, \quad D(g_1) = \begin{pmatrix} 0 & 1 & 0 \\ 0 & 0 & 1 \\ 1 & 0 & 0 \end{pmatrix}, \quad D(g_2) = \begin{pmatrix} 0 & 0 & 1 \\ 1 & 0 & 0 \\ 0 & 1 & 0 \end{pmatrix},$$

$$D(g_3) = \begin{pmatrix} 0 & 1 & 0 \\ 1 & 0 & 0 \\ 0 & 0 & 1 \end{pmatrix}, \quad D(g_4) = \begin{pmatrix} 0 & 0 & 1 \\ 0 & 1 & 0 \\ 1 & 0 & 0 \end{pmatrix}, \quad D(g_5) = \begin{pmatrix} 1 & 0 & 0 \\ 0 & 0 & 1 \\ 0 & 1 & 0 \end{pmatrix}. \tag{3.19}$$

You can do the matrix multiplication to convince yourself that this is in fact a representation of S_3.

So, in equation (3.15), we had a 6×6 matrix representation. Here, we have a new representation consisting of 3×3 matrices. We have therefore "reduced" the representation. In the next few sections, we will look at more mathematically rigorous ways of thinking about reducible and irreducible representations.

3.1.7 Algebraic Definitions

Hopefully at this point you have some intuition for what a group is, how a group can act on things, and what it means for a group representation to be reducible. Because the concept of reducibility and irreducibility will play a fairly significant role in what is to come (and because the mathematical formalisms we develop here will be useful later), we'll now spend some time going into a bit more detail and mathematical precision with a few ideas. Again, this will seem very abstract, but these ideas will prove to be absolutely essential to understanding particle physics as we proceed through this book and this series.

We'll begin with a few definitions. First, if H is a subset of G, denoted $H \subset G$, such that the elements of H form a group, then we say that H forms a **Subgroup** of G. Though this is a very simple idea, we'll look at a few examples to reinforce it.

Consider (once again) the group S_3, with the elements labeled as before:

1. $g_0(ROY) = (ROY)$.
2. $g_1(ROY) = (OYR)$.
3. $g_2(ROY) = (YRO)$.

4. $g_3(ROY) = (ORY)$.
5. $g_4(ROY) = (YOR)$.
6. $g_5(ROY) = (RYO)$.

where we are labeling $g_0 = e$ for convenience. The multiplication table for this group is given on page 57. Notice that the elements g_0, g_1 and g_2 (which are of course a subset of S_3) are closed, associative, contain the identity, and each element has an inverse. Therefore the set $\{g_0, g_1, g_2\}$ is a subgroup of S_3.[6] Also, notice that you can see that $\{g_0, g_1, g_2\}$ forms a subgroup of S_3 by looking at the multiplication table for S_3: the 9 upper left boxes in the multiplication table (the g_0, g_1, and g_2 rows and columns) all have only g_0's, g_1's, and g_2's.

As another example consider the subset of S_3 consisting of only g_0 and g_3. Both g_0 and g_3 are their own inverses, so then once again there is closure, associativity, identity, and inverse. Looking back at our previous multiplication tables you can see that this subgroup has exactly the same structure as the (unique) group of order 2 we discussed earlier.

Also notice that $\{g_0, g_3\}$, $\{g_0, g_4\}$, and $\{g_0, g_5\}$ are all subgroups of S_3. So, S_3 contains three copies of the group of order 2 *in addition to* the single copy of the group of order 3.

As a final example of subgroups, any group has two subgroups. One is the subgroup consisting of only the identity, $\{g_0\} \subset G$ and the other is the entire group $G \subset G$. We call these two subgroups **Trivial Subgroups**.

Now we introduce a few mathematical ideas that often give students a great deal of trouble the first time they come across them. If you're not already familiar with these ideas, read this section again and again until you are comfortable with them.

If G is a group and H is a subgroup of G, then:

- The set $gH = \{g \star h | h \in H\}$ is called the **Left Coset** of H in G.
- The set $Hg = \{h \star g | h \in H\}$ is called the **Right Coset** of H in G.

There is a right (or left) coset for each element g in G, though they are not necessarily all unique. This definition should be understood as follows: a coset is a *set* consisting of the elements of H all multiplied on the right (or left) by some element of G. Hopefully a few examples will help this make more sense.

Above we discussed the subgroups $\{g_0, g_1, g_2\}$, $\{g_0, g_3\}$, $\{g_0, g_4\}$, and $\{g_0, g_5\}$ of S_3. We can work out the details to find the cosets of each subgroup. For the first subgroup, which we'll denote $G_{012} = \{g_0, g_1, g_2\}$, the left cosets are

$$g_0 \star G_{012} = \{g_0 \star g_0, g_0 \star g_1, g_0 \star g_2\} = \{g_0, g_1, g_2\},$$

$$g_1 \star G_{012} = \{g_1 \star g_0, g_1 \star g_1, g_1 \star g_2\} = \{g_1, g_2, g_0\},$$

$$g_2 \star G_{012} = \{g_2 \star g_0, g_2 \star g_1, g_2 \star g_2\} = \{g_2, g_0, g_1\},$$

[6] We want to reiterate that groups are abstract things. There is only one group S_3 (though there are many representations of it) and there is only one group of order 3 (cf. multiplication for group of order 3 on page 57), and one is a subgroup of the other.

$$\begin{aligned}
g_3 \star G_{012} &= \{g_3 \star g_0, g_3 \star g_1, g_3 \star g_2\} = \{g_3, g_4, g_5\}, \\
g_4 \star G_{012} &= \{g_4 \star g_0, g_4 \star g_1, g_4 \star g_2\} = \{g_4, g_5, g_3\}, \\
g_5 \star G_{012} &= \{g_5 \star g_0, g_5 \star g_1, g_5 \star g_2\} = \{g_5, g_3, g_4\},
\end{aligned} \tag{3.20}$$

and the right cosets are

$$\begin{aligned}
G_{012} \star g_0 &= \{g_0 \star g_0, g_1 \star g_0, g_2 \star g_0\} = \{g_0, g_1, g_2\}, \\
G_{012} \star g_1 &= \{g_0 \star g_1, g_1 \star g_1, g_2 \star g_1\} = \{g_1, g_2, g_0\}, \\
G_{012} \star g_2 &= \{g_0 \star g_2, g_1 \star g_2, g_2 \star g_2\} = \{g_2, g_0, g_1\}, \\
G_{012} \star g_3 &= \{g_0 \star g_3, g_1 \star g_3, g_2 \star g_3\} = \{g_3, g_5, g_4\}, \\
G_{012} \star g_4 &= \{g_0 \star g_4, g_1 \star g_4, g_2 \star g_4\} = \{g_4, g_3, g_5\}, \\
G_{012} \star g_5 &= \{g_0 \star g_5, g_1 \star g_5, g_2 \star g_5\} = \{g_5, g_4, g_3\}.
\end{aligned} \tag{3.21}$$

Notice that for every element g_i of the group, the left coset $g_i \star G_{123}$ is the same as the right coset $G_{123} \star g_i$. For example, the coset $g_3 \star G_{012}$ consists of g_3, g_4, and g_5, while the coset $G_{012} \star g_3$ consists of g_3, g_5, and g_4 – the same three elements (the order they appear doesn't matter – all that matters is that they are the same set). If this is the case for a given subgroup (if all of its left and right cosets contain the same set of elements), we say that $H \subset G$ is a **Normal Subgroup**.

As another example, consider the group of integers under addition, $(\mathbb{Z}, +)$ and the subgroup $(\mathbb{Z}_{even}, +)$. You can take some element outside of $(\mathbb{Z}_{even}, +)$ and act on the left:

$$n_{odd} + \mathbb{Z}_{even} = \{n_{odd} + 0, n_{odd} \pm 2, n_{odd} \pm 4, \ldots\} \tag{3.22}$$

and then on the right:

$$\mathbb{Z}_{even} + n_{odd} = \{0 + n_{odd}, \pm 2 + n_{odd}, \pm 4 + n_{odd}, \ldots\}. \tag{3.23}$$

Clearly

$$n_{even} + \mathbb{Z}_{even} = \mathbb{Z}_{even} + n_{even}. \tag{3.24}$$

The final sets are the same (because addition is commutative). So, $\mathbb{Z}_{even} \subset \mathbb{Z}$ is a normal subgroup. With a little thought, you can convince yourself that *all* subgroups of Abelian groups are normal.

Building on the idea of a normal subgroup, we can now introduce another extremely important algebraic idea, the **Factor Group**. If G is a group and H is a normal subgroup of G, then the factor group of H in G, denoted G/H (read "G mod H"), is the group with elements in the set $G/H = \{gH | g \in G\}$. An example will once again be helpful.

The group structure of this "factor group" comes from the sets themselves. For example, notice that in the cosets of G_{012} there are only two elements – the set $\{g_0, g_1, g_2\}$ and the set $\{g_3, g_4, g_5\}$. If we denote the first of these sets as E and the second as simply G, we can talk about multiplying them together as follows. The multiplication of E and G means that we are taking an element of E and multiplying it by the set G, i.e.

$$\begin{aligned} g_0 \star G &= g_0 \star \{g_3, g_4, g_5\} = \{g_3, g_4, g_5\} = G, \\ g_1 \star G &= g_1 \star \{g_3, g_4, g_5\} = \{g_5, g_3, g_4\} = G, \\ g_2 \star G &= g_2 \star \{g_3, g_4, g_5\} = \{g_4, g_5, g_3\} = G. \end{aligned} \tag{3.25}$$

So no matter what element of E you multiply by G, you get G, and we therefore say that the product of the "elements" E and G is G. You can convince yourself that the same is true had we multiplied G on the right instead. In a similar way, multiplication of G and E means multiplying an element of G by the set E (which as you can show always produces the set G), multiplication of E and E means multiplying an element of E by the set E (which always produces the set E), and multiplication of G and G means multiplication of an element of G by the set G (which always produces the set E). We can therefore draw a multiplication table for this normal subgroup,

S_3/G_{012}	E	G
E	E	G
G	G	E

This is obviously the group of order 2 (cf. the multiplication table on page 56). In other words, we have

$$S_3/G_{012} = \mathbb{Z}_2, \tag{3.26}$$

the integers mod 2.

Note that the reason we were able to make sense of this group was because G_{012} is a *normal* subgroup. To emphasize this, let's consider a non-normal subgroup. Namely, let's consider $G_{03} = \{g_0, g_3\}$. This is obviously a subgroup, but writing out the cosets gives

$$\begin{aligned} g_0 \star \{g_0, g_3\} &= \{g_0, g_3\}, & \{g_0, g_3\} \star g_0 &= \{g_0, g_3\}, \\ g_1 \star \{g_0, g_3\} &= \{g_1, g_5\}, & \{g_0, g_3\} \star g_1 &= \{g_1, g_4\}, \\ g_2 \star \{g_0, g_3\} &= \{g_2, g_4\}, & \{g_0, g_3\} \star g_2 &= \{g_2, g_5\}, \\ g_3 \star \{g_0, g_3\} &= \{g_3, g_0\}, & \{g_0, g_3\} \star g_3 &= \{g_3, g_0\}, \\ g_4 \star \{g_0, g_3\} &= \{g_4, g_2\}, & \{g_0, g_3\} \star g_4 &= \{g_4, g_1\}, \\ g_5 \star \{g_0, g_3\} &= \{g_5, g_1\}, & \{g_0, g_3\} \star g_5 &= \{g_5, g_2\}. \end{aligned} \tag{3.27}$$

You can see that this is not a normal subgroup (i.e. $g_0 \star G_{03} = G_{03} \star g_0$, but $g_1 \star G_{03} \neq G_{03} \star g_1$, etc.). But let's say we want to try to form a group anyway. We can notice that there are 3 elements here, which we'll denote

$$\begin{aligned} E &= \{g_0, g_3\}, \\ G_{15} &= \{g_1, g_5\}, \\ G_{24} &= \{g_2, g_4\}. \end{aligned} \tag{3.28}$$

Now, if we want to define multiplication between, for example, G_{15} and G_{24}, we can think of it as *either* an element of G_{15} times the set G_{24}, or we can think of it as an element of G_{24} times the set G_{15}. Furthermore, for either, we can do left or right multiplication. So let's consider an example of each of these. For an element of G_{15} times the set G_{24} with left multiplication, we have

$$\begin{aligned} g_1 \star \{g_2, g_4\} &= \{g_0, g_3\}, \\ g_5 \star \{g_2, g_4\} &= \{g_4, g_2\}. \end{aligned} \tag{3.29}$$

With right multiplication this is

$$\begin{aligned} \{g_2, g_4\} \star g_1 &= \{g_0, g_5\}, \\ \{g_2, g_4\} \star g_5 &= \{g_3, g_1\}. \end{aligned} \tag{3.30}$$

For the choice of an element of G_{24} times the set G_{15} with left multiplication,

$$\begin{aligned} g_2 \star \{g_1, g_5\} &= \{g_0, g_3\}, \\ g_4 \star \{g_1, g_5\} &= \{g_5, g_1\}. \end{aligned} \tag{3.31}$$

With right multiplication this is

$$\begin{aligned} \{g_1, g_5\} \star g_2 &= \{g_0, g_4\}, \\ \{g_1, g_5\} \star g_4 &= \{g_3, g_2\}. \end{aligned} \tag{3.32}$$

So, with all 4 of these choices, we get inconsistent results. For example, with the first choice (an element of G_{15} times the set G_{24} with left multiplication), we can get either $E = \{g_0, g_3\}$ *or* $G_{24} = \{g_2, g_4\}$ depending on which element of G_{15} we choose to multiply. This means that we don't have a consistent way of multiplying the *set* G_{15} and the *set* G_{24}. You'll notice that with the second choice, one of the resulting sets isn't even a coset!

The intuition behind a normal subgroup is that it is a way of sort of "squinting" at the group and not seeing some of the structure. Specifically, forming the normal subgroup G/H essentially removes H from the group (like dividing it out, hence the notation). You can see this in the normal subgroup example above, S_3/G_{012}.

Any element of G_{012} acts as an identity on the cosets associated with G_{012}. The only non-trivial things that are left are the elements that aren't in G_{012}. Requiring that the subgroup we are mod-ing out be a normal subgroup is simply the necessary condition for multiplication to make sense.

In general, for a normal subgroup H of G, the factor group G/H has elements $g_i H$ (or Hg_i) and the group operation $\star$ is understood to be

$$(g_i H) \star (g_j H) = (g_i \star g_j)H, \tag{3.33}$$

or

$$(Hg_i) \star (Hg_j) = H(g_i \star g_j). \tag{3.34}$$

As another example of this, consider $\mathbb{Z}_{even}$. We can denote $\mathbb{Z}_{even}$ as $2\mathbb{Z}$ because

$$2\mathbb{Z} = 2\{0, \pm 1, \pm 2, \ldots\} = \{0, \pm 2, \pm 4, \ldots\} = \mathbb{Z}_{even}. \tag{3.35}$$

We know that $2\mathbb{Z}$ is a normal subgroup of $\mathbb{Z}$ (because it is Abelian), so we can build the factor group $\mathbb{Z}/2\mathbb{Z}$ as

$$\mathbb{Z}/2\mathbb{Z} = \{0 + 2\mathbb{Z}, \pm 1 + 2\mathbb{Z}, \pm 2 + 2\mathbb{Z}, \ldots\}. \tag{3.36}$$

But, notice that

$$n_{even} + 2\mathbb{Z} = \mathbb{Z}_{even}, \tag{3.37}$$

$$n_{odd} + 2\mathbb{Z} = \mathbb{Z}_{odd}. \tag{3.38}$$

So, the group $\mathbb{Z}/2\mathbb{Z}$ only has two elements; the set of all even integers, and the set of all odd integers. And we know from before that there is only one group of order 2, which we denote $\mathbb{Z}_2$. So, we have found that $\mathbb{Z}/2\mathbb{Z} = \mathbb{Z}_2$. You can convince yourself of the more general result

$$\mathbb{Z}/n\mathbb{Z} = \mathbb{Z}_n. \tag{3.39}$$

As two final examples, consider first the factor groups G/G and G/e. The set G/G is the entire group $G = \{g_0, g_1, g_2, \ldots\}$ with some element of G multiplied by it. And because the group axioms require that the set $g_i G = g_i \star \{g_0, g_1, g_2, \ldots\}$ contain every element of the group for any g_i, this factor group consists of only one element. And therefore we have

$$G/G = e. \tag{3.40}$$

Next, G/e will be the set $\{e\}$ multiplied by some element g_i will contain only that element:

$$g_i\{e\} = \{g_i\} \tag{3.41}$$

and therefore there is a unique coset for every element of G. So,

$$G/e = G. \tag{3.42}$$

Something that might help you understand factor groups better is this: the factor group G/H is the group that is "left over" when everything in H is "collapsed" to the identity element. Thinking about the above examples in terms of this should be helpful.

Finally, we introduce one more definition before moving on. This one is relatively simple so we won't spend much time on it. If G and H are both groups (not necessarily related in any way), then we can form the **Product Group**, denoted $K = G \times H$, where an arbitrary element of K is (g_i, h_j). If the group operation of G is $\star_G$ and the group operation of H is $\star_H$, then two elements of K are multiplied according to the rule

$$(g_i, h_j) \star_K (g_k, h_l) = (g_i \star_G g_k, h_j \star_H h_l). \tag{3.43}$$

3.1.8 Reducibility Revisited

Now that we understand subgroups, cosets, normal subgroups, and factor groups, we can begin a more formal discussion of reducing representations. Recall that in deriving equation (3.11), we made the designation

$$g_0 \to |\hat{e}_1\rangle, \qquad g_1 \to |\hat{e}_2\rangle, \qquad g_2 \to |\hat{e}_3\rangle, \qquad \text{etc.} \tag{3.44}$$

This was used to create an order(G)-dimensional Euclidian space which, while not having any physical meaning, and while obviously not possessing any structure similar to the group, was and will continue to be of great use to us.

So, changing notation slightly, we have an n-dimensional space spanned by the orthonormal vectors $|g_0\rangle, |g_1\rangle, \ldots, |g_{n-1}\rangle$, where g_0 is understood to always refer to the identity element. This brings us to the first definition of this section. For a group $G = \{g_0, g_1, g_2, \ldots\}$, we call the **Algebra** of G the set

$$\mathbb{R}[\mathbf{G}] = \left\{ \sum_{i=0}^{n-1} a_i |g_i\rangle \middle| a_i \in \mathbb{R}\ \forall i \right\}. \tag{3.45}$$

In other words, $\mathbb{R}[\mathbf{G}]$ is the set of all possible linear combinations of the vectors $|g_i\rangle$ with real coefficients. More often algebras are defined over $\mathbb{C}$, but that isn't necessary for our purposes and would only obfuscate the picture we're trying to provide.

Addition of two elements of $\mathbb{R}[\mathbf{G}]$ is merely normal addition of linear combinations,

$$\sum_{i=0}^{n-1} a_i |g_i\rangle + \sum_{i=0}^{n-1} b_i |g_i\rangle = \sum_{i=0}^{n-1} (a_i + b_i)|g_i\rangle. \tag{3.46}$$

This definition amounts to saying that in the n-dimensional Euclidian space we have created with $n = \text{order}(G)$, you can choose any point in the space and this will correspond to a particular linear combination of elements of G.

Now that we have defined an algebra we can talk about group actions. Recall that the g_i's don't act on the $|g_j\rangle$'s, but rather the representation $D(g_i)$ does. We define the action $D(g_i)$ on an element of $\mathbb{R}[\mathbf{G}]$ as follows:

$$\begin{aligned} D(g_i) \cdot \sum_{j=0}^{n-1} a_j |g_j\rangle &= D(g_i) \cdot (a_0|g_0\rangle + a_1|g_1\rangle + \cdots + a_{n-1}|g_{n-1}\rangle) \\ &= a_0|g_i \star g_0\rangle + a_1|g_i \star g_1\rangle + \cdots + a_{n-1}|g_i \star g_{n-1}\rangle \\ &= \sum_{j=0}^{n-1} a_j |g_i \star g_j\rangle. \end{aligned} \tag{3.47}$$

Previously, we discussed how elements of a group act on each other, and we also talked about how elements of a group act on some other object or set of objects (like three painted eggs). We now generalize this notion to a set of q abstract objects a group can act on, denoted $M = \{m_0, m_1, m_2, \ldots, m_{q-1}\}$. Just as before, we build a vector space, similar to the one above used in building an algebra. The orthonormal vectors here will be

$$m_0 \rightarrow |m_0\rangle, \qquad m_1 \rightarrow |m_1\rangle, \qquad \ldots \qquad m_{q-1} \rightarrow |m_{q-1}\rangle. \tag{3.48}$$

This allows us to understand the following definition. The set

$$\mathbb{R}\mathbf{M} = \left\{ \sum_{i=0}^{q-1} a_i |m_i\rangle \middle| a_i \in \mathbb{R}\ \forall i \right\} \tag{3.49}$$

is called the **Module** of M (we don't use the square brackets here to distinguish modules from algebras).

As an example of this, consider once again the group S_3. This time, instead of acting on three painted Easter eggs, the group is going to act on three "objects", m_0,

m_1, and m_2. This means that $\mathbb{R}\mathbf{M}$ is the set of all points in the 3-dimensional space of the form

$$a_0|m_0\rangle + a_1|m_1\rangle + a_2|m_2\rangle \qquad a_i \in \mathbb{R} \text{ for all } i. \tag{3.50}$$

So operating on a given point with, say g_3, gives

$$g_3(a_0|m_0\rangle + a_1|m_1\rangle + a_2|m_2\rangle) = (a_0|g_3m_0\rangle + a_1|g_3m_1\rangle + a_2|g_3m_2\rangle), \tag{3.51}$$

and from the multiplication table on page 57 we know

$$g_3m_0 = m_1, \qquad g_3m_1 = m_0, \qquad g_3m_2 = m_2. \tag{3.52}$$

And so,

$$\begin{aligned}(a_0|g_3m_0\rangle + a_1|g_3m_1\rangle + a_2|g_3m_2\rangle) &= (a_0|m_1\rangle + a_1|m_0\rangle + a_2|m_2\rangle) \\ &= a_1|m_0\rangle + a_0|m_1\rangle + a_2|m_2\rangle. \end{aligned} \tag{3.53}$$

In other words, the effect of g_3 was to swap a_1 and a_0. This can be visualized geometrically as a reflection in the $a_0 = a_1$ plane in the 3-dimensional module space. We can visualize every element of G in this way. They each move points around the module space in a well-defined way.

And this fact brings us to the following, very important, definition. If $\mathbb{R}\mathbf{V}$ is a module and $\mathbb{R}\mathbf{W}$ is a subspace of $\mathbb{R}\mathbf{V}$ that is closed under the action of G, then $\mathbb{R}\mathbf{W}$ is an **Invariant Subspace** of $\mathbb{R}\mathbf{V}$.

Let's consider an example of this. Working (as usual) with S_3, we know from the matrices in (3.19) that S_3 acts on a 3-dimensional space spanned by

$$|m_0\rangle = (1,0,0)^T, \qquad |m_1\rangle = (0,1,0)^T, \qquad \textit{and} \qquad |m_2\rangle = (0,0,1)^T. \tag{3.54}$$

Now consider the subspace spanned by

$$a(|m_0\rangle + |m_1\rangle + |m_2\rangle) \tag{3.55}$$

where a is an element of $\mathbb{R}$. Then the above set of points is the set of all points on the line through the origin defined by $\lambda(\hat{i} + \hat{j} + \hat{k})$, where λ is in $\mathbb{R}$. You can write out the action of any element on S_3 on any point in this subspace, and you will see that they are unaffected. This means that the space spanned by (3.55) is an invariant subspace.

As a note, all modules $\mathbb{R}\mathbf{V}$ have two trivial subspaces – $\mathbb{R}\mathbf{V}$ is a trivial invariant subspace of itself, as is $\mathbb{R}e$.

Finally, we give a more formal definition of reducibility. If a representation D of a group G acts on the space of a module $\mathbb{R}\mathbf{M}$, then the representation D is said to be **Reducible** if $\mathbb{R}\mathbf{M}$ contains a non-trivial invariant subspace. If a representation is not reducible, it is **Irreducible**.

We encouraged you to write out the entire regular representation of S_3 above. If you have done so, you may have noticed that every 6×6 matrix appeared with non-zero elements only in the upper left 3×3 elements, and the lower right 3×3 elements. The upper right and lower left are all 0. This means that, for every element of S_3, there will never be any mixing of the first 3 dimensions with the last 3. So, there are two 3-dimensional invariant subspaces in the module for this particular representation of S_3 (the regular representation).

We can now begin to take advantage of the fact that representations live in linear spaces with the following definition. If V is any n-dimensional space spanned by n linearly independent basis vectors, and U and W are both subspaces of V, then we say that V is the **Direct Sum** of U and W if every vector $\mathbf{v}$ in V can be written as the sum $\mathbf{v} = \mathbf{u} + \mathbf{w}$, where $\mathbf{u}$ is in U and $\mathbf{w}$ is in W, and every operator X acting on elements of V can be separated into parts acting individually on U and W. The notation for this is

$$V = U \oplus W. \tag{3.56}$$

In order to (hopefully) make this clearer, if X_n is an $n \times n$ matrix, it is the direct sum of $m \times m$ matrix A_m and $k \times k$ matrix B_k, denoted $X_n = A_m \oplus B_k$, if and only if X is in **Block Diagonal** form,

$$X_n = \begin{pmatrix} A_m & 0 \\ 0 & B_k \end{pmatrix} \tag{3.57}$$

where $n = m+k$, and A_m, B_k, and the 0's are understood as matrices of appropriate dimension.

We can generalize the previous definition as follows,

$$X_n = A_{n_1} \oplus B_{n_2} \oplus \cdots \oplus C_{n_k} = \begin{pmatrix} A_{n_1} & 0 & \cdots & 0 \\ 0 & B_{n_2} & \cdots & 0 \\ \vdots & \vdots & \ddots & \cdots \\ 0 & 0 & \vdots & C_{n_k} \end{pmatrix} \tag{3.58}$$

where $n = n_1 + n_2 + \cdots + n_k$.

As a simple example of this, let

$$A = \begin{pmatrix} 1 & 1 & -2 \\ -1 & 5 & \pi \\ -17 & 4 & 11 \end{pmatrix}, \tag{3.59}$$

and

$$B = \begin{pmatrix} 1 & 2 \\ 3 & 4 \end{pmatrix}, \tag{3.60}$$

then

$$B \oplus A = \begin{pmatrix} 1 & 2 & 0 & 0 & 0 \\ 3 & 4 & 0 & 0 & 0 \\ 0 & 0 & 1 & 1 & -2 \\ 0 & 0 & -1 & 5 & \pi \\ 0 & 0 & -17 & 4 & 11 \end{pmatrix}. \tag{3.61}$$

To take stock of what we have done so far, we have talked about algebras, which are the vector spaces spanned by the elements of a group, and about modules, which are the vector spaces that representations of groups act on. We have also defined invariant subspaces as follows: given some space and some group that acts on that space, moving the points around in a well-defined way, an invariant subspace is a subspace which always contains the same points. The group doesn't remove any points from that subspace, and it doesn't add any points to it. It merely moves the points around *inside* that subspace. Then, we defined a representation as reducible if there are any non-trivial invariant subspaces in the space that the group acts on.

And what this amounts to is the following: a representation of any group is reducible if it can be written in block diagonal form.

But this leaves the question of what we mean when we say "can be written". How can you "rewrite" a representation? This leads us to the following definition. Given a matrix D and a non-singular (invertible) matrix S, the linear transformation

$$D \longrightarrow D' = S^{-1}DS \tag{3.62}$$

is called a **Similarity Transformation**. Then, we say that two matrices D and D' related by a similarity transformation are **Equivalent**.

And because similarity transformations are linear transformations, if $D(G)$ is a representation of G, then so is $S^{-1}DS$ for literally *any* non-singular matrix S. To see this, if $g_i \star g_j = g_k$, then $D(g_i)D(g_j) = D(g_k)$, and therefore

$$S^{-1}D(g_i)S \cdot S^{-1}D(g_j)S = S^{-1}D(g_i)D(g_j)S = S^{-1}D(g_k)S. \tag{3.63}$$

So, if we have a representation that isn't in block diagonal form, how can we figure out if it is reducible? We must look for a matrix S that will transform it into block diagonal form. If an S exists such that the similarity transformation of D with S puts every element of the representation into block diagonal form, then we know that the representation is reducible. If no such S exists, then the representation is irreducible.

You likely realize immediately that this is not a particularly easy thing to do by inspection. However, it turns out that there is a very straightforward and systematic way of taking a given representation and determining whether or not it is reducible, and if so, what the irreducible representations are.

However, the details of how this can be done, while very interesting, are not necessary for the agenda of this book. Therefore, for the sake of brevity, we will not pursue them. What is important is that you understand not only the details of general group theory and representation theory (which we outlined above), but also the concept of what it means for a group to be reducible or irreducible. You can read more about how to find irreducible representations from reducible representations in the texts listed in the further reading section at the end of this chapter.

We also warn you that you will find the remainder of this book lacking any explicit references to many of the ideas we discussed in these last few sections (invariant subspaces, modules, etc.). However, we introduced them for two reasons. First, they will come up in later books and this was a logical place to include them. Second, the notions of irreducibility, equivalence, etc. that are illustrated by these ideas will be extremely important later in this book, and we hope that our discussion will help with intuition later.

3.2 Introduction to Lie Groups

In Sect. 3.1, we considered groups which are of finite order and discrete, and this allowed us to write out a multiplication table. Here, however, we examine a different type of group. Consider the unit circle, where each point on the circle is specified by an angle θ, measured from the positive x-axis.

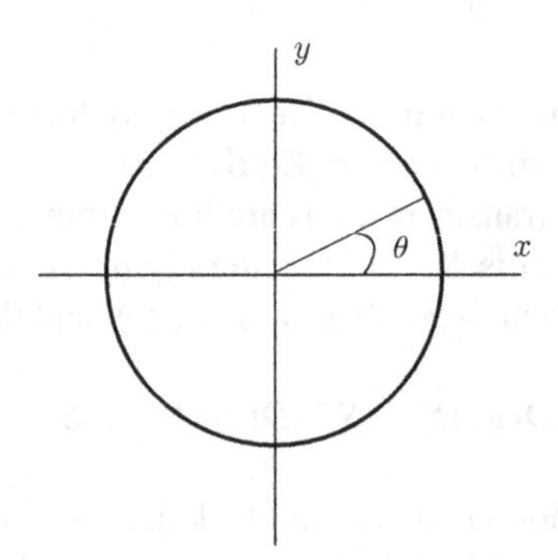

We will refer to the point at $\theta = 0$ as the "starting point" (like *ROY* was for the Easter eggs). Now, just as we considered all possible orientations of (*ROY*) that left the eggs lined up, we consider all possible rotations the wheel can undergo. With the eggs there were only 6 possibilities. Now however, for the wheel there are an infinite number of possibilities for θ (any real number $\in [0, 2\pi)$).

And note that if we denote the set of all angles as G, then all the rotations obey closure ($\theta_1 + \theta_2 = \theta_3 \in G,\ \forall \theta_1, \theta_2 \in G$), associativity (as usual), identity ($0 + \theta = \theta + 0 = \theta$), and inverse (the inverse of θ is $-\theta$).

So, we have a group that is *parameterized* by a continuous variable θ. This means that we are no longer talking about g_i's, but about $g(\theta)$.

Also, notice that this particular group (the circle) is Abelian, which is why we can (temporarily) use addition to represent it. We obviously cannot make a multiplication table because the order of this group is infinite.

One simple representation is the one we used above: taking θ and using addition (mod 2π). A more familiar (and useful) representation is the Euler matrix

$$g(\theta) = \begin{pmatrix} \cos\theta & \sin\theta \\ -\sin\theta & \cos\theta \end{pmatrix} \tag{3.64}$$

with the usual matrix multiplication:

$$\begin{pmatrix} \cos\theta_1 & \sin\theta_1 \\ -\sin\theta_1 & \cos\theta_1 \end{pmatrix} \begin{pmatrix} \cos\theta_2 & \sin\theta_2 \\ -\sin\theta_2 & \cos\theta_2 \end{pmatrix} \tag{3.65}$$

$$= \begin{pmatrix} \cos\theta_1\cos\theta_2 - \sin\theta_1\sin\theta_2 & \cos\theta_1\sin\theta_2 + \sin\theta_1\cos\theta_2 \\ -\sin\theta_1\cos\theta_2 - \cos\theta_1\sin\theta_2 & -\sin\theta_1\sin\theta_2 + \cos\theta_1\cos\theta_2 \end{pmatrix} \tag{3.66}$$

$$= \begin{pmatrix} \cos(\theta_1+\theta_2) & \sin(\theta_1+\theta_2) \\ -\sin(\theta_1+\theta_2) & \cos(\theta_1+\theta_2) \end{pmatrix} \tag{3.67}$$

This will prove to be a much more useful representation than θ with addition.

Groups that are parameterized by one or more continuous variables like this are called **Lie Groups**. The true definition of a Lie group is much more complicated, and that definition should eventually be understood. However, the definition we have given will suffice for the purposes of this book.[7]

3.2.1 Classification of Lie Groups

The usefulness of group theory is that groups represent a mathematical way to make changes to a system while leaving *something* about the system unchanged. For example, we moved (*ROY*) around, but the structure "three eggs with different colors lined up" was preserved. With the circle, we rotated it, but it still maintained its basic structure as a circle. It is in this sense that group theory is a study of **Symmetry**. No matter which of "these" transformations you do to the system, "this" stays the same – this is symmetry.

To see the usefulness of this in physics, recall Noether's Theorem (Sect. 1.2). When you do a *symmetry* transformation to a Lagrangian, you get a conserved quantity. Think back to the Lagrangian for the projectile in equation (1.30). The transformation $x \to x + \epsilon$ was a symmetry because ϵ could take any value, and the Lagrangian was unchanged (note that ϵ forms the Abelian group $(\mathbb{R}, +)$).

[7]We'll talk more about the "real" definition in the next book in this series.

So given a Lagrangian, which represents the structure of a physical system, a symmetry represents a way of changing the Lagrangian while preserving that structure. The particular preserved part of the system is the conserved quantity j we discussed in Sects. 1.2 and 1.5. And as you have no doubt noticed, nearly all physical processes are governed by **Conservation Laws**: conservation of momentum, energy, charge, spin, etc.

So, group theory, and in particular Lie group theory, gives us an extremely powerful way of understanding and classifying symmetries, and therefore conserved charges.[8] And because it allows us to understand conserved charges, group theory (apparently) can be used to understand almost the entirety of the physics in our universe.

We now begin to classify the major types of Lie groups we will be working with in this book. To start, we consider the most general possible Lie group in an arbitrary number of dimensions, n. This will be the group that, for any point p in the n-dimensional space, can continuously take it *anywhere* else in the space. All that is preserved is that the points in the space stay in the space. This means that we can have literally any $n \times n$ matrix, or *linear* transformation, so long as the matrix is invertible (non-singular). Thus, in n dimensions the largest and most general Lie group is the group of all $n \times n$ non-singular matrices. We call this group $GL(n)$, or the **General Linear** group. The most general field of numbers we'll take the elements of $GL(n)$ from is $\mathbb{C}$, so we begin with $GL(n, \mathbb{C})$. This is the group of all $n \times n$ non-singular matrices with complex elements. The preserved quantity is that all points in $\mathbb{C}^n$ stay in $\mathbb{C}^n$.

The most obvious subgroup of $GL(n, \mathbb{C})$ is $GL(n, \mathbb{R})$, or the set of all $n \times n$ invertible matrices with real elements. This leaves all points in $\mathbb{R}^n$ in $\mathbb{R}^n$. So in both $GL(n, \mathbb{C})$ and $GL(n, \mathbb{R})$, the transformations are the matrices and the preserved quantities are the entire spaces (in other words, very little is actually conserved).

To find a further subgroup, recall from linear algebra and vector calculus that in n dimensions, you can take n vectors at the origin such that for a parallelepiped defined by those vectors, the volume can be found from the determinant of the matrix formed by those vectors.

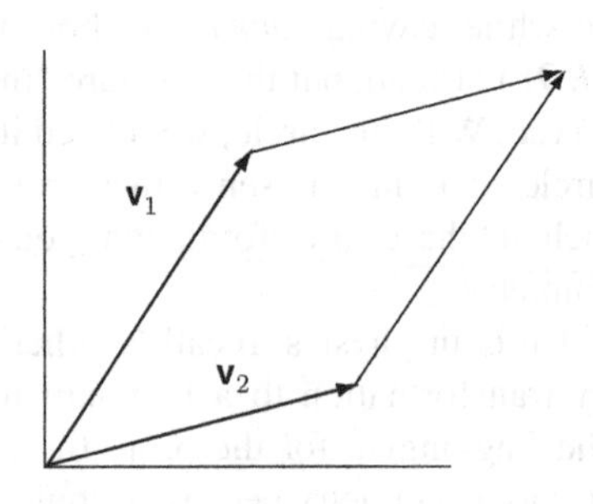

[8]Here we mean "charges" in the general sense of anything conserved under Noether's theorem, not merely electric charge.

So consider the set of all general linear transformations that transform all vectors from the origin (or in other words, points in the space) in such a way that the volume of the corresponding parallelepiped is preserved. This implies that we only consider general linear matrices with determinant 1. Also, the set of all general linear matrices with unit determinant will form a group because of the general rule

$$\det|A \cdot B| = \det|A| \cdot \det|B|. \tag{3.68}$$

So, if $\det|A| = 1$ and $\det|B| = 1$, then $\det|A \cdot B| = 1$. We call this subgroup of $GL(n,\mathbb{C})$ the **Special Linear** group, or $SL(n,\mathbb{C})$. The natural subgroup of this is $SL(n,\mathbb{R})$. This group preserves not only the points in the space (as GL did), but also volumes, as described above.

Now, consider the familiar transformations on vectors in n-dimensional space of generalized Euler angles (generalized in the sense of n dimensions, not just 3). These are transformations that rotate all points around the origin. These rotation transformations leave the radius squared (r^2) invariant. And, because

$$\mathbf{r}^2 = \mathbf{r}^T\mathbf{r}, \tag{3.69}$$

if we transform with a rotation matrix R, then

$$\mathbf{r} \to \mathbf{r}' = R\mathbf{r}, \tag{3.70}$$

and

$$\mathbf{r}^T \to \mathbf{r}'^T = \mathbf{r}^T R^T, \tag{3.71}$$

so

$$\mathbf{r}'^T\mathbf{r}' = \mathbf{r}^T R^T R\mathbf{r}. \tag{3.72}$$

But, as we said, we are demanding that the radius squared be invariant under the action of R, and so we demand

$$\mathbf{r}^T R^T R\mathbf{r} = \mathbf{r}^T\mathbf{r}. \tag{3.73}$$

So, the constraint we are imposing is $R^T R = \mathbb{I}$, which implies $R^T = R^{-1}$. This tells us that the rows and columns of R are orthogonal. Therefore, we call the group of generalized rotations, or generalized Euler angles in n dimensions, $O(n)$, or the **Orthogonal** group. We don't specify $\mathbb{C}$ or $\mathbb{R}$ here because most of the time it will be understood that we are talking about $\mathbb{R}$.[9]

---

[9]If you compare the discussion in this paragraph to the discussion in Sect. 1.4 you'll notice it is very similar. We've really just rederived equations (1.64) in a more general context.

Also, note that because

$$\det|R^T \cdot R| = \det|\mathbb{I}| \tag{3.74}$$

we have

$$(\det|R|)^2 = 1 \tag{3.75}$$

which means

$$\det|R| = \pm 1. \tag{3.76}$$

We again denote the subgroup with $\det|R| = +1$ the **Special Orthogonal** group, or $SO(n)$. To understand what this means, consider an orthogonal matrix with determinant -1, such as

$$M = \begin{pmatrix} 1 & 0 & 0 \\ 0 & 1 & 0 \\ 0 & 0 & -1 \end{pmatrix} \tag{3.77}$$

This matrix is orthogonal, and therefore is an element of the group $O(3)$, but the determinant is -1. This matrix will take the point $(x, y, z)^T$ to the point $(x, y, -z)^T$. This changes the handedness of the system (the right-hand rule will no longer work). So, if we limit ourselves to $SO(n)$, we are preserving the space, the radius, the volume, and the handedness of the space.

For vectors in $\mathbb{C}$ space, we generally do not define orthogonal matrices (although we could). Instead, we discuss the complex version of the radius, where instead of $\mathbf{r}^2 = \mathbf{r}^T\mathbf{r}$, we have

$$\mathbf{r}^2 = \mathbf{r}^\dagger\mathbf{r}, \tag{3.78}$$

where the dagger denotes the Hermitian conjugate,

$$\mathbf{r}^\dagger = (\mathbf{r}^\star)^T, \tag{3.79}$$

where $\star$ denotes complex conjugate.

So, with the elements in R being in $\mathbb{C}$, we have $\mathbf{r} \to R\mathbf{r}$, and $\mathbf{r}^\dagger \to \mathbf{r}^\dagger R^\dagger$. So,

$$\mathbf{r}^\dagger\mathbf{r} \to \mathbf{r}^\dagger R^\dagger R\mathbf{r}, \tag{3.80}$$

and by the same argument as above with the orthogonal matrices, this demands that $R^\dagger R = \mathbb{I}$, or $R^\dagger = R^{-1}$. We denote such matrices **Unitary**, and the set of all such $n \times n$ invertible matrices forms the group $U(n)$. Again, we understand the unitary groups to have elements in $\mathbb{C}$, so we don't specify that. And, we will still have a subset of unitary matrices R with $\det|R| = 1$ called $SU(n)$, the **Special Unitary** groups.

We can summarize the hierarchy we have just described in the following diagram:

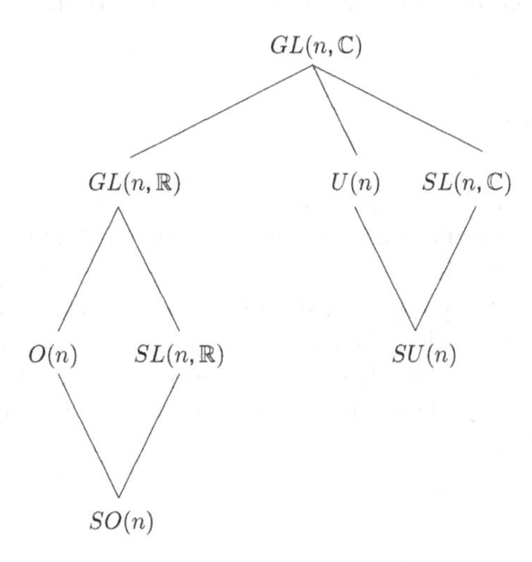

There are several other Lie groups we could talk about, and we will later in this series. For now, however, we only describe one more category of Lie groups before moving on. We saw above that the group $SO(n)$ preserves the radius squared in real space. In orthonormal coordinates, this means that $\mathbf{r}^2 = x_1^2 + x_2^2 + \cdots + x_n^2$, or more generally the dot product $\mathbf{x} \cdot \mathbf{y} = x_1 y_1 + x_2 y_2 + \cdots + x_n y_n$ is preserved.

However we can generalize this to form a group action that preserves not the radius squared, but the value (switching to indicial notation for the dot product)

$$x_i y^i = -x_1 y_1 - x_2 y_2 - \cdots - x_m y_m + x_{m+1} y_{m+1} + \cdots + x_{m+n} y_{m+n} \tag{3.81}$$

where we have used a metric with -1 along the first m entries of the diagonal and $+1$ along the final n entries. We call the group that preserves this quantity $SO(m, n)$. The space we are working in is still $\mathbb{R}^{m+n}$, but we are making transformations that preserve something different than the radius. Note that $SO(m, n)$ will have an $SO(m)$ subgroup and an $SO(n)$ subgroup, consisting of rotations in the first m and last n components separately.

And from our work in Sect. 1.4, we know that the transformations on only the m-dimensional subspace or only the n dimensional subspace will involve sin's and cos's as in (1.73), while transformations that mix the two subspaces will involve the hyperbolic trig functions sinh and cosh as in (1.75).

Finally, notice that the specific group of this type, $SO(1, 3)$, is the group that preserves the value $s^2 = -x_1 y_1 + x_2 y_2 + x_3 y_3 + x_4 y_4$, or written more suggestively,

$$s^2 = -c^2 t^2 + x^2 + y^2 + z^2. \tag{3.82}$$

Therefore, the group $SO(1, 3)$ is the **Lorentz Group**. Any action that is invariant under $SO(1, 3)$ is said to be a Lorentz invariant theory (as all physical theories of our universe should be). We will find that thinking of special relativity in these terms (as a Lie group), rather than in the terms of Sect. 1.4 will be much more useful.[10]

3.2.2 Generators

Now that we have a good "birds eye view" of Lie groups, we can begin to pick apart the details of how they work. We'll find that these details provide a very powerful toolbox for physical systems.

As we said before, a Lie group is a group that is parameterized by a set of continuous parameters, which we call α_i for $i = 1, \ldots, n$, where n is the number of parameters the group depends on. The elements of the group will then be denoted $g(\alpha_i)$.

Because all groups include an identity element, we will choose to parameterize them in such a way that

$$g(\alpha_i)\big|_{\alpha_i=0} = e, \tag{3.83}$$

the identity element. So, if we are going to talk about representations,

$$D_n(g(\alpha_i))\big|_{\alpha_i=0} = \mathbb{I}, \tag{3.84}$$

where $\mathbb{I}$ is the $n \times n$ identity matrix for whatever dimension (n) representation we want.

Now, take $\delta\alpha_i$ to be very small ($\delta\alpha_i << 1$). So, $D_n(g(0 + \delta\alpha_i))$ can be Taylor expanded:

$$D_n(g(\delta\alpha_i)) = \mathbb{I} + \delta\alpha_i \frac{\partial D_n(g(\alpha_i))}{\partial \alpha_i}\Big|_{\alpha_i=0} + \cdots. \tag{3.85}$$

The terms $\frac{\partial D_n}{\partial \alpha_i}\big|_{\alpha_i=0}$ are extremely important, and we give them their own label:

$$X_i = -i \frac{\partial D_n}{\partial \alpha_i}\Big|_{\alpha_i=0} \tag{3.86}$$

(we have included the $-i$ in order to make X_i Hermitian, which will be convenient later). So, the representation for infinitesimal $\delta\alpha_i$ is then

$$D_n(\delta\alpha_i) = \mathbb{I} + i\delta\alpha_i X_i + \cdots \tag{3.87}$$

(where we have switched our notation from $D_n(g(\alpha))$ to $D_n(\alpha)$ for brevity), where X_i's are constant matrices which we will determine later.

---

[10] In fact, the Lie group picture of special relativity is what's actually going on physically. We will eventually see (though it will be a while) that it is this way of thinking about *special* relativity that makes it being a special case of *general* relativity the most obvious.

Now, let's say that we want to see what the representation will look like for a finite value of α_i rather than an infinitesimal value. A finite transformation will be the result of an infinite number of infinitesimal transformations. Or in other words, $\alpha_i = N\delta\alpha_i$ as $N \to \infty$. So, $\delta\alpha_i = \frac{\alpha_i}{N}$, and an infinite number of infinitesimal transformations is

$$\lim_{N\to\infty}(1 + i\delta\alpha_i X_i)^N = \lim_{N\to\infty}\left(1 + i\frac{\alpha_i}{N}X_i\right)^N. \tag{3.88}$$

If you expand this out for several values of N, you will see that it is exactly

$$\lim_{N\to\infty}\left(1 + i\frac{\alpha_i}{N}X_i\right)^N = e^{i\alpha_i X_i}. \tag{3.89}$$

We call the X_i's the **Generators** of the group, and there is one for each parameter required to specify a particular element of the group. For example, consider $SO(3)$, the group of rotations in three dimensions. We know from vector calculus that an element of $SO(3)$ requires three angles, usually denoted θ, ϕ, and ψ. Therefore, $SO(3)$ will require three generators, which will be denoted X_θ, X_ϕ, and X_ψ. We will discuss how the generators can be found soon.

In general, there will be several (in fact, infinite) different sets of X_i's that define a given group (just as there are an infinite number of representations of any finite group). What we will find is that up to a similarity transformation, a particular set of generators defines a particular representation of a group.

So, any element of a group in a particular representation can be written as

$$D_n(\alpha_i) = e^{i\alpha_i X_i} \tag{3.90}$$

for any group (the i index in the exponent is understood to be summed over all parameters and generators). The entire representation is defined completely by the generators X_i.

The best way to think of the parameter space for the group is as a vector space, where the generators describe the behavior near the identity, but form a basis for the entire vector space. By analogy, think of the unit vectors $\hat{i}$, $\hat{j}$, and $\hat{k}$ in $\mathbb{R}^3$. They are defined at the origin, but they can be combined with real numbers/parameters to specify any arbitrary point in $\mathbb{R}^3$. In the same way, the generators are the "unit vectors" of the parameter space (which in general is a much more complicated space than Euclidian space), and the parameters (like θ, ϕ, and ψ) specify where in the parameter space you are in terms of the generators. That point in the parameter space will then correspond to a particular element of the group. We'll talk about this perspective much more in later volumes.

We call the number of generators of a group (or equivalently the number of parameters necessary to specify an element), the **Dimension** of the group. For example, the dimension of $SO(3)$ is 3. The dimension of $SO(2)$ (rotations in the plane) however is only 1 (only θ is needed), so there will be only one generator.

3.2.3 Lie Algebras

In Sect. 3.1 we discussed algebras. An algebra is a space spanned by elements of the group with $\mathbb{C}$ coefficients parameterizing the Euclidian space we defined. Obviously we can't define an algebra in the same way for Lie groups, because the elements are continuous. But, as discussed in the last section, a particular element of a Lie group is defined by the values of the parameters in the parameter space spanned by the generators. We will see that the generators will form the algebras for Lie groups.

Consider two elements of the same group with generators X_i, one with parameter values α_i and the other with parameter values β_i. The product of the two elements will then be $e^{i\alpha_i X_i}e^{i\beta_j X_j}$. Because we are assuming this is a group, we know that the product must be an element of the group (due to the closure property of groups), and therefore the product must be specified by some set of parameters δ_k, so

$$e^{i\alpha_i X_i}e^{i\beta_j X_j} = e^{i\delta_k X_k}. \tag{3.91}$$

Note that the product won't necessarily simply be

$$e^{i\alpha_i X_i}e^{i\beta_j X_j} = e^{i(\alpha_i X_i + \beta_j X_j)} \tag{3.92}$$

because the generators are matrices and therefore don't in general commute.

So, we want to figure out what δ_i will be in terms of α_i and β_i. We do this as follows.

$$i\delta_i X_k = \ln(e^{i\delta_k X_k}) = \ln(e^{i\alpha_i X_i}e^{\beta_j X_j}) = \ln(1 + e^{i\alpha_i X_i}e^{\beta_j X_j} - 1) = \ln(1+x)$$

where we have defined

$$x = e^{i\alpha_i X_i}e^{\beta_j X_j} - 1. \tag{3.93}$$

We will proceed by expanding only to second order in α_i and β_j, though the result we will obtain will hold at arbitrary order.[11] By Taylor expanding the exponential terms,

$$\begin{aligned}
e^{i\alpha_i X_i}e^{i\beta_j X_j} - 1 &= (1 + i\alpha_i X_i + \frac{1}{2}(i\alpha_i X_i)^2 + \cdots) \\
&\quad \times(1 + i\beta_j X_j + \frac{1}{2}(i\beta_j X_j)^2 + \cdots) - 1 \\
&= 1 + i\beta_j X_j - \frac{1}{2}(\beta_j X_j)^2 + i\alpha_i X_i - \alpha_i X_i \beta_j X_j - \frac{1}{2}(\alpha_i X_i)^2 - 1 \\
&= i(\alpha_i X_i + \beta_j X_j) - \alpha_i X_i \beta_j X_j - \frac{1}{2}\big((\alpha_i X_i)^2 + (\beta_j X_j)^2\big).
\end{aligned} \tag{3.94}$$

[11] We encourage you to confirm this.

Then, using the general Taylor expansion $\ln(1+x) = x - \frac{x^2}{2} + \frac{x^3}{3} - \frac{x^4}{4} + \cdots$, and again keeping terms only to second order in α and β, we have

$$
\begin{aligned}
x - \frac{x^2}{2} &= \left[i(\alpha_i X_i + \beta_j X_j) - \alpha_i X_i \beta_j X_j - \frac{1}{2}[(\alpha_i X_i)^2 + (\beta_j X_j)^2] \right] \\
&\quad -\frac{1}{2}\left[i(\alpha_i X_i + \beta_j X_j) - \alpha_i X_i \beta_j X_j - \frac{1}{2}[(\alpha_i X_i)^2 + (\beta_j X_j)^2] \right]^2 \\
&= i(\alpha_i X_i + \beta_j X_j) - \alpha_i X_i \beta_j X_j - \frac{1}{2}\left[(\alpha_i X_i)^2 + (\beta_j X_j)^2 \right] \\
&\quad -\frac{1}{2}\left[-(\alpha_i X_i + \beta_j X_j)(\alpha_i X_i + \beta_j X_j) \right] \\
&= i(\alpha_i X_i + \beta_j X_j) - \alpha_i X_i \beta_j X_j - \frac{1}{2}\left[(\alpha_i X_i)^2 + (\beta_j X_j)^2 \right] \\
&\quad +\frac{1}{2}\left[(\alpha_i X_i)^2 + (\beta_j X_j)^2 + \alpha_i \beta_j (X_i X_j + X_j X_i) \right] \\
&= i(\alpha_i X_i + \beta_j X_j) + \frac{1}{2}\alpha_i \beta_j (X_j X_i - X_i X_j) \\
&= i(\alpha_i X_i + \beta_j X_j) - \frac{1}{2}\alpha_i \beta_j [X_i, X_j] \\
&= i(\alpha_i X_i + \beta_j X_j) - \frac{1}{2}[\alpha_i X_i, \beta_j X_j],
\end{aligned} \tag{3.95}
$$

where we are using the commutator defined as

$$[A, B] = AB - BA. \tag{3.96}$$

So finally we can see

$$i\delta_k X_k = i(\alpha_i X_i + \beta_j X_j) - \frac{1}{2}[\alpha_i X_i, \beta_j, X_j], \tag{3.97}$$

or

$$e^{i\alpha_i X_i} e^{i\beta_j X_j} = e^{i(\alpha_i X_i + \beta_j X_j) - \frac{1}{2}[\alpha_i X_i, \beta_j X_j]}. \tag{3.98}$$

Equation (3.98) is called the **Baker-Campbell-Hausdorff** formula, and it is one of the most important relations in group theory and in physics. Notice that if the generators commute, this reduces to the normal equation for multiplying exponentials. You can think of equation (3.98) as the generalization of the normal exponential multiplication rule.

Now, it is clear that the commutator $[X_i, X_j]$ must be proportional to some linear combination of the generators of the group (because of closure). So, it must be the case that

$$[X_i, X_j] = if_{ijk}X_k \tag{3.99}$$

for some set of constants f_{ijk}. These constants are called the **Structure Constants** of the group, and if they are completely known, the commutation relations between all the generators are known, and so the entire group can be determined in any representation you want.[12] The generators, under the specific commutation relations defined by the structure constants, form the **Lie Algebra** of the group, and it is this commutation structure which forms the structure of the Lie group.

3.2.4 The Adjoint Representation

We will talk about several representations for each group we discuss, but we will mention a very important one now. We mentioned before that the structure constants f_{ijk} completely determine the entire structure of the group. We begin by using the Jacobi identity,

$$\big[X_i, [X_j, X_k]\big] + \big[X_j, [X_k, X_i]\big] + \big[X_k, [X_i, X_j]\big] = 0 \tag{3.100}$$

(if you aren't familiar with this identity, try multiplying it out. You will find that it is identically true; all the terms cancel exactly). But, from equation (3.99), we can write

$$\big[X_i, [X_j, X_k]\big] = if_{jka}[X_i, X_a] = if_{jka}f_{iab}X_b.$$

Plugging this into (3.100) we get

$$\begin{aligned}
&if_{jka}f_{iab}X_b + if_{kia}f_{jab}X_b + if_{ija}f_{kab}X_b = 0\\
\Rightarrow\ &(f_{jka}f_{iab} + f_{kia}f_{jab} + f_{ija}f_{kab})iX_b = 0\\
\Rightarrow\ &f_{jka}f_{iab} + f_{kia}f_{jab} + f_{ija}f_{kab} = 0.
\end{aligned} \tag{3.101}$$

So, if we define the matrices

$$[T^a]_{bc} = -i\ f_{abc} \tag{3.102}$$

[12]Note that we are using raised and lowered indices here, which from Sect. 1.4.1 means that we're assuming a metric. For now, we can get by simply using the Kronecker metric, but we will see (much) later that this isn't always the case. For a preview see [94], Sect. 15.2.

then it is easy to show that (3.101) leads to

$$[T^a, T^b] = i\, f_{abc} T^c. \tag{3.103}$$

So, the structure constants themselves form a representation of the group (as defined by (3.102)). We call this representation the **Adjoint Representation**, and it will prove to be extremely important.

Notice that the indices labeling the rows and columns in (3.102) each run over the same values as the indices labeling the T matrices. This tells us that the adjoint representation is made of $n \times n$ matrices, where n is the dimension of the group, or the number of parameters in the group. For example, $SO(3)$ requires three parameters to specify an element (θ, ϕ, ψ), so the adjoint representation of $SO(3)$ will consist of 3×3 matrices. $SO(2)$ on the other hand is Abelian, and therefore all of the structure constants vanish. Therefore there is no adjoint representation of $SO(2)$.

We now go on to consider several specific groups in detail.

3.2.5 *SO*(2)

We start by looking at an extremely simple group, $SO(2)$. This is the group of rotations in the plane that leaves

$$\mathbf{r}^2 = x^2 + y^2 = \begin{pmatrix} x & y \end{pmatrix} \cdot \begin{pmatrix} x \\ y \end{pmatrix} = \mathbf{v}^T \mathbf{v} \tag{3.104}$$

invariant. So for some generator X (which we will now find) of $SO(2)$,

$$\mathbf{v} \to R(\theta)\mathbf{v} = e^{i\theta X}\mathbf{v}, \tag{3.105}$$

and

$$\mathbf{v}^T \to \mathbf{v}^T e^{i\theta X^T}. \tag{3.106}$$

So, expanding to first order only (in θ),

$$\mathbf{v}^T e^{i\theta X^T} e^{i\theta X}\mathbf{v} = \mathbf{v}^T(1 + i\theta X^T + i\theta X)\mathbf{v} = \mathbf{v}^T\mathbf{v} + \mathbf{v}^T i\theta(X + X^T)\mathbf{v}. \tag{3.107}$$

And because we demand that r^2 be invariant, we demand that $X + X^T = 0 \Rightarrow X = -X^T$. So, X must be antisymmetric. Therefore we take

$$X = \frac{1}{i}\begin{pmatrix} 0 & 1 \\ -1 & 0 \end{pmatrix} \tag{3.108}$$

(the $\frac{1}{i}$ is included to balance the i we inserted in equation (3.86) to ensure that X is Hermitian). So, an arbitrary element of $SO(2)$ will be

$$\begin{aligned}
e^{i\theta X} &= e^{i\theta\frac{1}{i}\begin{pmatrix}0 & 1\\ -1 & 0\end{pmatrix}} = e^{\theta\begin{pmatrix}0 & 1\\ -1 & 0\end{pmatrix}} \\
&= \begin{pmatrix}0 & 1\\ -1 & 0\end{pmatrix}^0 + \theta\begin{pmatrix}0 & 1\\ -1 & 0\end{pmatrix}^1 + \frac{1}{2}\theta^2\begin{pmatrix}0 & 1\\ -1 & 0\end{pmatrix}^2 + \cdots \\
&= \begin{pmatrix}1 & 0\\ 0 & 1\end{pmatrix} + \theta\begin{pmatrix}0 & 1\\ -1 & 0\end{pmatrix} - \frac{1}{2}\theta^2\begin{pmatrix}1 & 0\\ 0 & 1\end{pmatrix} - \frac{1}{3!}\theta^3\begin{pmatrix}0 & 1\\ -1 & 0\end{pmatrix} + \cdots \\
&= \begin{pmatrix}1 - \frac{1}{2}\theta^2 + \cdots & \theta - \frac{1}{3!}\theta^3 + \cdots\\ -(\theta - \frac{1}{3!}\theta^3 + \cdots) & 1 - \frac{1}{2}\theta^2 + \cdots\end{pmatrix} \\
&= \begin{pmatrix}\cos\theta & \sin\theta\\ -\sin\theta & \cos\theta\end{pmatrix},
\end{aligned} \tag{3.109}$$

which is exactly what we would expect for a matrix describing rotations in the plane. Also, notice that because $SO(2)$ is Abelian, the commutation relations trivially vanish ($[X, X] = 0$), and so all of the structure constants are zero.

Now that we have found an explicit example of a generator and seen an example of how generators relate to group elements, we move on to slightly more complicated examples.

3.2.6 *SO*(3)

In order to better illustrate how generators work, we will approach $SO(3)$ differently by working backwards. Above, we found the generators and used them to calculate the group elements. Here, we begin with the known group elements of $SO(3)$, which are just the standard Euler matrices for rotations in 3-dimensional space:

$$R_x(\phi) = \begin{pmatrix}1 & 0 & 0\\ 0 & \cos\phi & \sin\phi\\ 0 & -\sin\phi & \cos\phi\end{pmatrix}, \tag{3.110}$$

$$R_y(\psi) = \begin{pmatrix}\cos\psi & 0 & -\sin\psi\\ 0 & 1 & 0\\ \sin\psi & 0 & \cos\psi\end{pmatrix}, \tag{3.111}$$

$$R_z(\theta) = \begin{pmatrix}\cos\theta & \sin\theta & 0\\ -\sin\theta & \cos\theta & 0\\ 0 & 0 & 1\end{pmatrix}. \tag{3.112}$$

Now recall the definition of the generators, equation (3.86). We can use it to find the generators of $SO(3)$, which we will denote J_x, J_y, and J_z, giving

$$J_x = \frac{1}{i}\frac{dR_x(\phi)}{d\phi}\bigg|_{\phi=0} = \frac{1}{i}\begin{pmatrix} 0 & 0 & 0 \\ 0 & -\sin\phi & \cos\phi \\ 0 & -\cos\phi & \sin\phi \end{pmatrix}\bigg|_{\phi=0} = \frac{1}{i}\begin{pmatrix} 0 & 0 & 0 \\ 0 & 0 & 1 \\ 0 & -1 & 0 \end{pmatrix}. \quad (3.113)$$

And similarly

$$J_y = \frac{1}{i}\begin{pmatrix} 0 & 0 & -1 \\ 0 & 0 & 0 \\ 1 & 0 & 0 \end{pmatrix}, \qquad J_z = \frac{1}{i}\begin{pmatrix} 0 & 1 & 0 \\ -1 & 0 & 0 \\ 0 & 0 & 0 \end{pmatrix}. \quad (3.114)$$

You can plug these into the exponentials with the appropriate parameters (ϕ, ψ, or θ) and find that $e^{i\phi J_x}$, $e^{i\psi J_y}$, and $e^{i\theta J_z}$ reproduce (3.110), (3.111), and (3.112), respectively.

Furthermore, you can multiply out the commutators to find

$$[J_x, J_y] = iJ_z, \qquad [J_y, J_z] = iJ_x, \qquad [J_z, J_x] = iJ_y, \quad (3.115)$$

or

$$[J_i, J_j] = i\epsilon_{ijk}J_k. \quad (3.116)$$

This tells us that the structure constants for $SO(3)$ are

$$f_{ijk} = \epsilon_{ijk}, \quad (3.117)$$

where ϵ_{ijk} is the totally antisymmetric tensor. The structure constants being non-zero is consistent with $SO(3)$ being a non-Abelian group.

3.2.7 $SU(2)$

We approached $SO(2)$ by looking at what needed to be preserved. From that we derived the form of the generator and consequently the form of the group elements. We approached $SO(3)$ by starting with the group elements, and from that we derived the form of the generators. We'll approach $SU(2)$ yet another way: by starting with the structure constants. It turns out they are the same as the structure constants for $SO(3)$:

$$f_{ijk} = \epsilon_{ijk}. \quad (3.118)$$

To see why, recall from Sect. 3.2.1 that $SU(2)$ are rotations in two complex dimensions. The most general form of such a matrix $U \in SU(2)$ is $U = \begin{pmatrix} a & b \\ c & d \end{pmatrix}$. The "Special" part of $SU(2)$ demands that the determinant be equal to 1, or

$$ad - bc = 1, \tag{3.119}$$

and the "Unitary" part demands that $U^{-1} = U^{\dagger}$. So,

$$U^{-1} = \begin{pmatrix} d & -b \\ -c & a \end{pmatrix} = U^{\dagger} = \begin{pmatrix} a^{\star} & c^{\star} \\ b^{\star} & d^{\star} \end{pmatrix}. \tag{3.120}$$

Or in other words,

$$U = \begin{pmatrix} a & b \\ -b^{\star} & a^{\star} \end{pmatrix}, \tag{3.121}$$

where we demand $|a|^2 + |b|^2 = 1$.

Both a and b are in $\mathbb{C}$ and therefore have two real components each, so U has four real parameters. The constraint $|a|^2 + |b|^2 = 1$ fixes one of them, leaving three real parameters, just like in $SO(3)$. This is a loose illustration of why $SU(2)$ and $SO(3)$ have the same structure constants. They are both rotational groups with 3 real parameters. But the point is that $SU(2)$ will have 3 generators.

3.2.8 *SU(2) and Physical States*

The elements of any Lie group (in a d-dimensional representation consisting of $d \times d$ matrices) will act on vectors, just like the 3×3 matrices representing S_3 acted on $\begin{pmatrix} R & O & Y \end{pmatrix}^T$ in Sect. 3.1.6. In order to understand the space a Lie group acts on, we need a basis for the space. We would like a basis that fits nicely with the Lie group elements (or operators) that are acting on the space. Rather than choose an arbitrary basis for the space, we'll take advantage of our linear algebra toolbox and use the eigenvectors of the representation of the Lie group. The eigenvectors will always form a basis of the eigenspace (or physical space) the group is acting on.

Using similarity transformations, one or more of the generators of a Lie group can be diagonalized. For now, trust us that with $SU(2)$, it is only possible to diagonalize one of the three generators at a time (you may convince yourself of this by studying the commutation relations). We will call the generators of $SU(2)$ J^1, J^2, and J^3, and by convention we take J^3 to be the diagonal one. So, consequently, the eigenvectors of J^3 will be the basis vectors of the physical vector space upon which $SU(2)$ acts.[13]

[13] We'll look at specific examples of this, so if it isn't entirely clear now, keep reading.

Now, we know that J^3 (whatever it is ... we don't know at this point) will in general have more than one eigenvalue. Let's call the largest eigenvalue of J^3 (whatever it is) j, and the eigenvectors of J^3 will be denoted $|j;m\rangle$ (the first j is merely a label - the second value describes the vector), where m is the eigenvalue of the eigenvector. The eigenvector corresponding to the greatest eigenvalue j will obviously then be $|j;j\rangle$. So, $J^3|j;j\rangle = j|j;j\rangle$, or more generally $J^3|j;m\rangle = m|j;m\rangle$.

Now let's assume that we know $|j;j\rangle$. There is a trick we can employ to find the rest of the states. Define the following linear combinations of the generators[14]:

$$J^{\pm} = \frac{1}{\sqrt{2}}(J^1 \pm iJ^2). \tag{3.122}$$

Now, using the fact that the $SU(2)$ generators obey the commutation relations in equation (3.117), it is easy to show the following relations,

$$[J^3, J^{\pm}] = \pm J^{\pm} \qquad \text{and} \qquad [J^+, J^-] = J^3. \tag{3.123}$$

Notice that because by definition J^i are all Hermitian, we have

$$(J^{\pm})^{\dagger} = J^{\mp}. \tag{3.124}$$

We now use these tools to find another eigenvector of J^3.

Consider some arbitrary eigenvector $|j;m\rangle$. We know the eigenvalue of this will be m, so $J^3|j;m\rangle = m|j;m\rangle$. But now let's create some new state by acting on $|j;m\rangle$ with either of the operators (3.122). The new state will be $J^{\pm}|j;m\rangle$, but what will the J^3 eigenvalue be? Using the commutation relations in (3.123),

$$J^3 J^{\pm}|j;m\rangle = (\pm J^{\pm} + J^{\pm}J^3)|j;m\rangle = (m \pm 1)J^{\pm}|j;m\rangle. \tag{3.125}$$

So, the vector $J^+|j;m\rangle$ is the eigenvector with eigenvalue $m+1$, and the vector $J^-|j;m\rangle$ is the eigenvector with eigenvalue $m-1$.

So at this point, we know that if we have some arbitrary eigenvector $|j;m\rangle$, we can use $J^{\pm}$ to move up or down to the eigenvector with the next highest or lowest eigenvalue. For this reason, $J^{\pm}$ are called the **Raising** and **Lowering** operators. They raise and lower the eigenvalue of the state by one.

The eigenvector with the largest eigenvalue j, $|j;j\rangle$, cannot be raised any higher, so we define $J^+|j;j\rangle = 0$. We will see that there is also a lowest eigenvalue j', so we similarly define $J^-|j;j'\rangle = 0$.

Now, considering once again $|j;j\rangle$, we know that if we operate on this state with J^-, we will get the eigenvector with the eigenvalue $j-1$. But, we don't know exactly what this state will be – we're missing its coefficient. But, we know it will

[14]This likely seems entirely arbitrary. We'll derive it later. For now just accept that we're defining these new expressions and making use of them.

be proportional to $|j;j-1\rangle$. So, we set $J^-|j;j\rangle = N_j|j;j-1\rangle$, where N_j is the proportionality constant. To find N_j, we take the inner product (and using (3.124)):

$$\begin{aligned}\left|J^-|j;j\rangle\right|^2 &= (J^-|j;j\rangle)^\dagger(J^-|j;j\rangle) = \langle j;j|J^+J^-|j;j\rangle \\ &= |N_j|^2\langle j;j-1|j;j-1\rangle.\end{aligned} \tag{3.126}$$

But we can also write

$$\begin{aligned}\langle j;j|J^+J^-|j;j\rangle &= \langle j;j|(J^+J^- - J^-J^+)|j;j\rangle = \langle j;j|[J^+,J^-]|j;j\rangle \\ &= \langle j;j|J^3|j;j\rangle = j\langle j;j|j;j\rangle = j,\end{aligned} \tag{3.127}$$

where we used the fact that $J^+|j;j\rangle = 0$ to get the first equality, and (3.123) to get the third equality. We also assumed that $|j;j\rangle$ is normalized. In other words, that

$$\langle j;j|j;j\rangle = 1. \tag{3.128}$$

So, (3.127) tells us

$$\langle j;j-1|j;j-1\rangle = 1 \iff N_j = \sqrt{j}. \tag{3.129}$$

And our normalized state is therefore

$$\frac{J^-}{N_j}|j;j\rangle = \frac{J^-}{\sqrt{j}}|j;j\rangle = |j;j-1\rangle. \tag{3.130}$$

or

$$J^-|j;j\rangle = \sqrt{j}|j;j-1\rangle. \tag{3.131}$$

Repeating this to find N_{j-1}, we have

$$\begin{aligned}|N_{j-1}|^2\langle j;j-2|j;j-2\rangle &= \langle j;j-1|J^+J^-|j;j-1\rangle \\ &= \left\langle j;j|\frac{J^+}{\sqrt{j}}J^+J^-\frac{J^-}{\sqrt{j}}|j;j\right\rangle \\ &= \frac{1}{j}\langle j;j|J^+J^+J^-J^-|j;j\rangle \\ &= \frac{1}{j}\langle j;j|J^+(J^3 + J^-J^+)J^-|j;j\rangle \\ &= \frac{1}{j}\langle j;j|(J^+J^3J^- + J^+J^-J^+J^-)|j;j\rangle\end{aligned}$$

$$
\begin{aligned}
&= \frac{1}{j}\langle j;j|(J^+(-J^- + J^-J^3) \\
&\qquad + J^+J^-(J^3 + J^-J^+))|j;j\rangle \\
&= \frac{1}{j}[\langle j;j|(-J^+J^- + jJ^+J^- + jJ^+J^-)|j;j\rangle] \\
&= \frac{1}{j}\langle j;j|(-[J^+,J^-] + 2j[J^+,J^-])|j;j\rangle \\
&= \frac{1}{j}\langle j;j|(-J^3 + 2jJ^3)|j;j\rangle \\
&= \frac{1}{j}(2j^2 - j) = 2j - 1. \qquad (3.132)
\end{aligned}
$$

So, $|N_{j-1}|^2 = 2j - 1$, or $N_{j-1} = \sqrt{2j-1}$.

We can continue this process, and we will find that the general result is

$$N_{j-k} = \frac{1}{\sqrt{2}}\sqrt{(2j-k)(k+1)}. \qquad (3.133)$$

The general states are then defined by

$$|j;j-k\rangle = \frac{1}{N_{j-k}}(J^-)^k|j;j\rangle. \qquad (3.134)$$

Notice that these expressions recover (3.129) and (3.132) for $k = 0$ and $k = 1$, respectively. Furthermore, notice that when $k = 2j$,

$$N_{j-2j} = \frac{1}{\sqrt{2}}\sqrt{(2j-2j)(2j+1)} = 0. \qquad (3.135)$$

So, the state

$$|j;j-k\rangle\big|_{k=2j} = |j;-j\rangle \qquad (3.136)$$

is the state with the lowest eigenvalue, and by definition

$$J^-|j;-j\rangle = 0. \qquad (3.137)$$

So, in a general representation of $SU(2)$, we have $2j+1$ states with eigenvalues

$$\{j, j-1, j-2, \dots, -j+2, -j+1, -j\}. \qquad (3.138)$$

This therefore demands that $j = \frac{n}{2}$ for some integer n. In other words, the highest eigenvalue of an $SU(2)$ eigenvector can be $0, \frac{1}{2}, 1, \frac{3}{2}, 2$, etc.

Furthermore, using these states, it is easy to show

$$\begin{aligned}\langle j;m'|J^3|j;m\rangle &= m\,\delta_{m',m},\\ \langle j;m'|J^+|j;m\rangle &= \frac{1}{\sqrt{2}}\sqrt{(j+m+1)(j-m)}\,\delta_{m',m+1},\\ \langle j;m'|J^-|j;m\rangle &= \frac{1}{\sqrt{2}}\sqrt{(j+m)(j-m+1)}\,\delta_{m',m-1}.\end{aligned} \tag{3.139}$$

3.2.9 $SU(2)$ for $j = \frac{1}{2}$

In the last section we talked about representations of $SU(2)$ for an arbitrary "maximum" value of j, which we discovered must be an integer or half integer. Now we want to look at a few examples of $SU(2)$ for specific values of j to see how the general ideas of the previous section work. The $j = 0$ case is trivial, so we won't bother with it for now (though we will discuss it later when we talk about physical theories). We then begin with the next lowest value, $j = \frac{1}{2}$.

For $j = \frac{1}{2}$, the two eigenvalues of J^3 will be $\frac{1}{2}$ and $\frac{1}{2} - 1 = -\frac{1}{2}$. So, denoting the J^3 generator of $SU(2)$ when $j = \frac{1}{2}$ as $J^3_{1/2}$, we have

$$J^3_{1/2} = \begin{pmatrix} 1/2 & 0 \\ 0 & -1/2 \end{pmatrix}. \tag{3.140}$$

Because we have assumed this to be the diagonal generator we were able to just put the known eigenvalues in the diagonal spots.

Now, inverting (3.122) to get

$$J^1 = \frac{1}{\sqrt{2}}(J^- + J^+) \qquad \text{and} \qquad J^2 = \frac{i}{\sqrt{2}}(J^- - J^+), \tag{3.141}$$

and using the standard matrix equation $[J^a_j]_{m',m} = \langle j,m'|J^a|j,m\rangle$, and the explicit products in (3.139), we can find (for example)

$$\left\langle \frac{1}{2};-\frac{1}{2}\right|J^1\left|\frac{1}{2};-\frac{1}{2}\right\rangle = \left\langle \frac{1}{2};-\frac{1}{2}\right|\frac{1}{\sqrt{2}}(J^- + J^+)\left|\frac{1}{2};-\frac{1}{2}\right\rangle = \cdots = 0. \tag{3.142}$$

So $[J^1]_{11} = 0$. Then,

$$\left\langle \frac{1}{2};-\frac{1}{2}\right|J^1\left|\frac{1}{2};\frac{1}{2}\right\rangle = \left\langle \frac{1}{2};-\frac{1}{2}\right|\frac{1}{\sqrt{2}}(J^- + J^+)\left|\frac{1}{2};\frac{1}{2}\right\rangle = \cdots = \frac{1}{2}. \tag{3.143}$$

So $[J^1]_{12} = \frac{1}{2}$.

We can continue this to find all the elements for each generator for $j = 1/2$. The final result will be

$$J^1_{1/2} = \frac{1}{2}\begin{pmatrix} 0 & 1 \\ 1 & 0 \end{pmatrix} = \frac{\sigma^1}{2}, \quad J^2_{1/2} = \frac{1}{2}\begin{pmatrix} 0 & -i \\ i & 0 \end{pmatrix} = \frac{\sigma^2}{2}, \quad J^3_{1/2} = \frac{1}{2}\begin{pmatrix} 1 & 0 \\ 0 & -1 \end{pmatrix} = \frac{\sigma^3}{2} \tag{3.144}$$

where the σ^i matrices are the **Pauli Spin Matrices** which you should recognize from any introductory quantum course. This is no accident! We will discuss these matrices in much, much more detail later (you'll likely be sick of them by the end of the book). For now recall that $SU(2)$ is the group of transformations in 2-dimensional complex space. The $SU(2)$ matrix has one of the real parameters fixed, leaving three real parameters. We are going to see that $SU(2)$ is the group which represents quantum mechanical spin, where j is the value of the spin of the particle. In other words, particles with spin $1/2$ are described by the $j = 1/2$ representation (the 2×2 representation in (3.144)), and particles with spin 1 are described by the $j = 1$ representation, and so on. In other words, $SU(2)$ describes quantum mechanical spin in three dimensions in the same way that $SO(3)$ describes normal "spin" in three dimensions. We will talk about the physical implications, reasons, and meaning of this later.

However, as a warning, be careful at this point not to think too much in terms of physics. You have likely covered $SU(2)$ in great detail in a quantum mechanics course (though you may not have known it was called "$SU(2)$"), but the approach we are taking here has a different goal than what you have likely seen before. The properties of $SU(2)$ we are seeing here are actually very, very specific and simplified illustrations of much deeper concepts in Lie groups. In order to understand particle physics, we must understand Lie groups in this specialized way. So for now, try to fight the temptation to merely understand everything we are doing in terms of the physics you have seen before, and learn this as we are presenting it – pure mathematics. We will focus on how it applies to physics later, in a more full and more fundamental way than introductory quantum mechanics makes apparent.

3.2.10 SU(2) for j = 1

You can follow the same procedure we used above to find

$$J^1_1 = \frac{1}{\sqrt{2}}\begin{pmatrix} 0 & 1 & 0 \\ 1 & 0 & 1 \\ 0 & 1 & 0 \end{pmatrix}, \quad J^2_1 = \frac{1}{\sqrt{2}}\begin{pmatrix} 0 & -i & 0 \\ i & 0 & -i \\ 0 & i & 0 \end{pmatrix}, \quad J^3_1 = \begin{pmatrix} 1 & 0 & 0 \\ 0 & 0 & 0 \\ 0 & 0 & -1 \end{pmatrix}. \tag{3.145}$$

Notice that only J^3_1 is diagonal (as before), and that the eigenvalues are $\{1, 0, -1\}$, or $\{j, j-1, j-2 = -j\}$ as we'd expect.

3.2.11 $SU(2)$ for Arbitrary j

As we've seen, any representation of $SU(2)$ will have three generators, J_j^1, J_j^2, and J_j^3, and this will always be a $d = 2j+1$-dimensional representation (it will consist of $d \times d$ matrices). These representations, for each value of j, are the irreducible representations of $SU(2)$. This means that the physical space this representation acts on will be d-dimensional, and there will therefore be d J_j^3 eigenvectors that span the physical space. These basis eigenvectors will be

$$|j;j\rangle = \begin{pmatrix} 1 \\ 0 \\ 0 \\ \vdots \\ 0 \end{pmatrix}, \quad |j;j-1\rangle = \begin{pmatrix} 0 \\ 1 \\ 0 \\ \vdots \\ 0 \end{pmatrix}, \quad |j;j-2\rangle = \begin{pmatrix} 0 \\ 0 \\ 1 \\ \vdots \\ 0 \end{pmatrix}, \quad \cdots \quad |j;-j\rangle = \begin{pmatrix} 0 \\ 0 \\ 0 \\ \vdots \\ 1 \end{pmatrix},$$

with eigenvalues $\{j, j-1, j-2, \cdots, -j\}$, respectively.

Then, for any j, equation (3.122) tells us that we can form the linear combinations $J_j^{\pm} = \frac{1}{\sqrt{2}}(J_j^1 \pm iJ_j^2)$ to be our raising and lowering operators. We've shown that $J^{\pm}$ are raising and lowering operators algebraically (cf. equation (3.125)), but it is instructive to see this explicitly. For example, for $j = 1/2$, equation (3.122) gives

$$J_{1/2}^{+} = \frac{1}{\sqrt{2}}\left[\frac{1}{2}\begin{pmatrix} 0 & 1 \\ 1 & 0 \end{pmatrix} + \frac{i}{2}\begin{pmatrix} 0 & -i \\ i & 0 \end{pmatrix}\right] = \frac{1}{2\sqrt{2}}\begin{pmatrix} 0 & 2 \\ 0 & 0 \end{pmatrix} = \frac{1}{\sqrt{2}}\begin{pmatrix} 0 & 1 \\ 0 & 0 \end{pmatrix}, \tag{3.146}$$

and similarly

$$J_{1/2}^{-} = \cdots = \frac{1}{\sqrt{2}}\begin{pmatrix} 0 & 0 \\ 1 & 0 \end{pmatrix}. \tag{3.147}$$

So, the two $j = 1/2$ eigenvectors will be $|\frac{1}{2};\frac{1}{2}\rangle = \begin{pmatrix} 1 \\ 0 \end{pmatrix}$ and $|\frac{1}{2};-\frac{1}{2}\rangle = \begin{pmatrix} 0 \\ 1 \end{pmatrix}$, and acting on these as follows gives

$$\begin{aligned} J_{1/2}^{+}\left|\frac{1}{2};\frac{1}{2}\right\rangle &= \frac{1}{\sqrt{2}}\begin{pmatrix} 0 & 1 \\ 0 & 0 \end{pmatrix}\begin{pmatrix} 1 \\ 0 \end{pmatrix} = 0, \\ J_{1/2}^{-}\left|\frac{1}{2};-\frac{1}{2}\right\rangle &= \frac{1}{\sqrt{2}}\begin{pmatrix} 0 & 0 \\ 1 & 0 \end{pmatrix}\begin{pmatrix} 0 \\ 1 \end{pmatrix} = 0, \end{aligned} \tag{3.148}$$

and similarly

$$J_{1/2}^{-}\left|\frac{1}{2};\frac{1}{2}\right\rangle = \left|\frac{1}{2};-\frac{1}{2}\right\rangle,$$
$$J_{1/2}^{+}\left|\frac{1}{2};-\frac{1}{2}\right\rangle = \left|\frac{1}{2};\frac{1}{2}\right\rangle, \tag{3.149}$$

which is exactly what we would expect.

The same calculation can be done for the $j = 1$ case, and we will get the same results. The only difference is that the $m = j = 1$ state (the first eigenvector) can be lowered *twice*. The first time J_1^- acts it takes the $m = 1$ state to the state with eigenvalue 0 (the $m = 0$ state), and the second time it acts it takes the $m = 0$ state to the state with eigenvalue -1 (the $m = -1$ state). Acting a third time will destroy the $m = -1$ state (take it to 0). Analogously, the lowest state, with eigenvalue $m = -1$ can be raised twice. We can do the same analysis for any $j =$ integer or half integer.

As we said before, we interpret j as the quantum mechanical spin of a particle, and the group $SU(2)$ describes that rotation. It is important to recognize that quantum spin is not a rotation through spacetime (it would be described by $SO(3)$ if it was), but rather through the mathematically constructed spinor space. We will talk more about this space later.

So for a given particle with spin, we can talk about both its rotation through physical spacetime using $SO(3)$, as well as its rotation through complex spinor space using $SU(2)$. Both values will be physically measurable and will be conserved quantities. The total angular momentum of the particle will be the combination of both spin and spacetime angular momentum. Again, we will talk much more about the spin of physical particles when we return to a discussion of physics. We only mention this now to give a preview of where this is going. However, spin is not the only thing $SU(2)$ describes. We will also find that it is the group which governs the weak nuclear force (whereas $U(1)$ describes the electromagnetic force, and $SU(3)$ describes the strong force ... much, much more on this later).

3.2.12 Root Space

As a comment before beginning this section, it is likely that you will find this to be among the most difficult parts of this book. However, while the material here is extremely difficult (especially the first time it is encountered), it is also extremely important to the development of particle physics. In fact, this section contains what is in many ways the most central topic of particle physics (mathematically speaking). If the contents are not clear you are encouraged to read this section multiple times until it becomes clear. It may also be helpful to study this section while looking closely at the examples in the following sections.

We saw in the previous section that we can view the physical space that a group is acting on by using the eigenvectors of the diagonal generators as a basis. And for $SU(2)$, these eigenvectors can be arranged in order of decreasing eigenvalue. Then, the non-diagonal generators can be used to form linear combinations that act as raising and lowering operators, which transform one eigenvector to another, changing the eigenvalue by an amount defined by the commutation relations of the generators. We now demonstrate how this generalizes to an arbitrary Lie group.

An arbitrary Lie group is defined in terms of its generators. As we said at the end of Sect. 3.2.2, it is best to think of the generators as being analogous to the basis vectors spanning some space. Of course, the space the generators span is much more complicated than $\mathbb{R}^n$ in general, but the generators span the space the same way. In this sense, the generators form a linear vector space. So, we must define an inner product for them. Recall from linear algebra that an inner product should map two vectors to (in this case) the reals $\mathbb{R}$, and should satisfy the following:

linearity: $\langle \alpha(a+b), c\rangle = \alpha(\langle a,c\rangle + \langle b,c\rangle)$.
commutativity: $\langle a,b\rangle = \langle b,a\rangle$.
non-negativity: $\langle a,a\rangle \geq 0$ with equality only for $a = 0$.

We can satisfy all of these properties if we choose the inner product so that for generators T^i and T^j,

$$\langle T^i, T^j\rangle = \frac{1}{\kappa} Tr(T^i T^j) = \delta^{ij}, \tag{3.150}$$

where κ is some normalization constant. Notice that this is also independent of the choice of basis. For example, choose some arbitrary basis T^i and then do a similarity transformation to choose another basis,

$$T^i \longrightarrow U^{-1} T^i U. \tag{3.151}$$

The inner product is unchanged due to the invariance of the trace under cyclic permutations,

$$\begin{aligned} Tr(T^i T^j) &\longrightarrow Tr(U^{-1} T^i U U^{-1} T^j U) \\ &= Tr(U^{-1} T^i T^j U) \\ &= Tr(U U^{-1} T^i T^j) \\ &= Tr(T^i T^j). \end{aligned} \tag{3.152}$$

So with an inner product defined (we'll worry about κ later), we set up a definition. In the set of generators of a Lie group, there will be a closed subalgebra of generators which all commute with each other, but not with generators outside of this subalgebra. In other words, this is the set of generators which can be simultaneously diagonalized through some similarity transformation (diagonal matrices will always commute with each other, but not necessarily with any non-diagonal matrices). For $SU(2)$, we saw that there was only one generator in this subalgebra which we chose to be J_j^3.

Let's say that a particular Lie group has N generators total, or is an N-dimensional group. Then, let's say that there are $M < N$ generators in this mutually commuting subalgebra. We call those M generators the **Cartan Subalgebra**, and the generators in it are called **Cartan Generators**. We define the number M as the **Rank** of the group. So, $SU(2)$ was an $N = 3$-dimensional group, and was rank $M = 1$ because only 1 generator could be diagonalized at a time.

By convention we will denote the Cartan generators H^i for $i = 1, \ldots, M$ and the non-Cartan generators E^i for $i = 1, \ldots, N - M$. With $SU(2)$ we had $H^1 = J_j^3$, and $E^1 = J_j^1$, $E^2 = J_j^2$.

Before moving on, we point out that this should have a familiar feel to it. If you think back to an introductory class in quantum mechanics, you can recall that we always choose some set of variables that all commute with each other (usually we choose *either* position or momentum because $[x, p] \neq 0$). Then, we expand the physical states in terms of the position *or* momentum eigenvectors. Here, we are doing the exact same thing, only in a much more general context (and keep in mind that we're technically doing math here, not physics).

Moving on, the H^i's are simultaneously diagonalized (because we've defined them that way). Therefore we will write the physical states in terms of their eigenvalues. In an n-dimensional representation D_n, the generators are $n \times n$ matrices, so the eigenvectors are n-dimensional. And because the eigenvectors are n-dimensional, this means that there will be a total of n eigenvectors, and each will have one eigenvalue with each of the M Cartan generators H^i. So, for each of these eigenvectors, which we temporarily denote $|j\rangle$ (these are simply the standard basis unit vectors), for $j = 1, \ldots, n$, we have the M eigenvalues with M Cartan generators, which we call t_j^i (where $j = 1, \ldots, n$ labels the eigenvectors, and $i = 1, \ldots, M$ labels the eigenvalues). We then form what is called a **Weight Vector**

$$\mathbf{t}_j = \begin{pmatrix} t_j^1 \\ t_j^2 \\ t_j^3 \\ \vdots \\ t_j^M \end{pmatrix} \tag{3.153}$$

where $j = 1, \ldots, n$. The individual components of these vectors, the t_j^i's, are called the **Weights**.

For a j representation[15] of $SU(2)$ there are $N = 3$ generators, each of which are $(2j+1)\times(2j+1)$ matrices which act on vectors with $2j+1$ components. So there are $2j + 1$ weight vectors, each of which are $M = 1$ dimensional (making them scalars – though we'll still write them as vectors). For the $j = \frac{1}{2}$ representation these vectors are

$$\mathbf{t}_1 = \begin{pmatrix} \frac{1}{2} \end{pmatrix}, \quad \text{and} \quad \mathbf{t}_2 = \begin{pmatrix} -\frac{1}{2} \end{pmatrix}. \tag{3.154}$$

For the $j = 1$ representation they are

$$\mathbf{t}_1 = \begin{pmatrix} 1 \end{pmatrix}, \quad \mathbf{t}_2 = \begin{pmatrix} 0 \end{pmatrix}, \quad \text{and} \quad \mathbf{t}_3 = \begin{pmatrix} -1 \end{pmatrix}. \tag{3.155}$$

You can continue this for larger representations of $SU(2)$.

Now back to the case of an arbitrary Lie group with an arbitrary representation D_n. We'll denote the state $|D_n; \mathbf{t}_j\rangle$ (instead of $|j\rangle$). So, our eigenvalues will be

$$H^i |D_n; \mathbf{t}_j\rangle = t^i_j |D_n; \mathbf{t}_j\rangle. \tag{3.156}$$

(note that there is no sum on j here).

All we've really done at this point is introduce some new notation. But as we'll see in just a few lines, this new way of thinking can help us quite a bit. We proceed by focusing on the adjoint representation (go reread Sect. 3.2.4 if you don't remember what this is). The generators are defined by equation (3.102), $[T^a]_{bc} = -if_{abc}$. Recall that each index runs from 1 to N, so that the generators in the adjoint representation are $N \times N$ matrices, and the eigenvectors are N-dimensional. The number of weight vectors will depend on the dimension of the representation, and the dimension of each weight vector will depend on the rank of the group.

As a point of nomenclature, weights in the adjoint representation are called **Roots**, and the corresponding vectors (as in (3.153)) are called **Root Vectors**.

So for $SU(2)$, which is a 3-dimensional group (because there are 3 generators), the adjoint will be the 3×3, or $j = 1$ representation. The root vectors are written above in equation (3.155).

As a comment, the fact that we're using the adjoint representation will not limit our results. Recall the point that we stressed when we first introduced groups: a group is a way of designating a particular structure. That structure shouldn't depend on the way we choose to represent the group. Certain aspects of the group may be more or less clear depending on how we represent it, but the structure remains unchanged. Therefore, much of what we can discover about a group from the adjoint representation will be true for any representation.

Moving on (read this paragraph and the ones following it very, very carefully), we can make some observations about adjoint representations. We know that there

---

[15]In this paragraph, j specifies the representation of $SU(2)$ in question. In the proceeding and following paragraphs, j is merely an index.

are N generators (in any representation), each of which is an $N \times N$ matrix (in the adjoint representation). Therefore, these act on N-dimensional vectors. Obviously there will be N eigenvectors (basis vectors) to span this space. In other words, there is a one-to-one correspondence between the eigenvectors and the generators. So, in equation (3.153), $j = 1, \ldots, N$. This is clear in the $j = 1$ (adjoint) representation of $SU(2)$ discussed above. There are three eigenvectors (corresponding to eigenvalues 1, 0, and -1), just as there are three generators. We can therefore take them to be in one-to-one correspondence. Note that we aren't saying that there is an automatic way of assigning a generator to a eigenvector; we're simply pointing out that in the adjoint representation we can easily make such a one-to-one assignment.[16]

We actually make this assignment by explicitly assigning each eigenvector to a generator as follows. First, we take advantage of the fact that we now have the same number of generators, eigenvectors, and root vectors (all equal to N). Before we were labeling the generators T^j for $j = 1, \ldots, N$. Now, we can also label the root vectors $\mathbf{t}_j$ for $j = 1, \ldots, N$. Because j runs over the same values in each case, we now simply denote our generators as $T^{\mathbf{t}_j}$ instead of T^j.

Then, we refer to general eigenstates[17] as $|Adj; T^{\mathbf{t}_j}\rangle$, where $j = 1, \ldots, N$ and $\mathbf{t}_j$ is the M-dimensional root vector corresponding to $T^{\mathbf{t}_j}$. And, we also divide the states $|Adj; T^{\mathbf{t}_j}\rangle$ into two groups: those corresponding to the M Cartan generators $|Adj; H^{\mathbf{h}_j}\rangle$ (where $j = 1, \ldots, M$ and $\mathbf{h}_j$ is the M-dimensional root vector corresponding to $H^{\mathbf{h}_j}$), and those corresponding to the $N - M$ non-Cartan generators $|Adj; E^{\mathbf{e}_j}\rangle$ (where $j = 1, \ldots N - M$ and $\mathbf{e}_j$ is the M-dimensional root vector corresponding to $E^{\mathbf{e}_j}$).

Don't be alarmed by the superscripts being vectors. We are using this notation for later convenience, and $T^{\mathbf{t}_i}$ here means the same thing T^j did before (the jth generator). This notation, which we use only for the adjoint representation, is simply taking advantage of the fact that in the adjoint representation, the following four things are all equal: the total number of generators, the number of eigenvectors of the Cartan generators, the dimension of the representation, and the number of weight/root vectors are all the same.

As two final results before moving on to the main results, notice that with the adjoint representation eigenstates $|Adj; T^{\mathbf{t}_j}\rangle$, we can use equation (3.150) to define the inner product between states as

$$\langle Adj; T^{\mathbf{t}_j}|Adj; T^{\mathbf{t}_k}\rangle = \frac{1}{\kappa} Tr(T^{\mathbf{t}_j} T^{\mathbf{t}_k}) = \delta^{jk}. \tag{3.157}$$

---

[16]There actually is a natural way of making this assignment that we'll see soon. We're just not saying that now.

[17]We are using the term "eigenstate" here in place of the word "eigenvector". They mean the same thing, although "eigenvector" is traditionally the math term while "eigenstate" is the physics term. An eigenstate is simply a particular eigenstate, or eigenvector, of a given operator.

Also, the matrix elements of a given generator will then be given by the familiar equation

$$-if_{abc} = [T^{\mathbf{t}_a}]_{bc} = \langle Adj; T^{\mathbf{t}_b}|T^{\mathbf{t}_a}|Adj; T^{\mathbf{t}_c}\rangle. \tag{3.158}$$

Now to get our main results for this section, we want to know what an arbitrary generator $T^{\mathbf{t}_a}$ will do to an arbitrary state $|Adj; T^{\mathbf{t}_b}\rangle$ in the adjoint representation. To do this we use the identity operator familiar from an introductory quantum mechanics course.

$$\mathbb{I} = \sum_i |Adj; T^{\mathbf{t}_i}\rangle\langle Adj; T^{\mathbf{t}_i}|. \tag{3.159}$$

Plugging this in, we have

$$\begin{aligned} T^{\mathbf{t}_a}|Adj; T^{\mathbf{t}_b}\rangle &= \sum_c |Adj; T^{\mathbf{t}_c}\rangle\langle Adj; T^{\mathbf{t}_c}|T^{\mathbf{t}_a}|Adj; T^{\mathbf{t}_b}\rangle = \sum_c |Adj; T^{\mathbf{t}_c}\rangle[T^{\mathbf{t}_a}]_{cb} \\ &= \sum_c |Adj; T^{\mathbf{t}_c}\rangle(-if_{acb}) = \sum_c if_{abc}|Adj; T^{\mathbf{t}_c}\rangle \\ &= \sum_c |Adj; if_{abc}T^{\mathbf{t}_c}\rangle. \end{aligned} \tag{3.160}$$

Then, from equation (3.99) we know

$$\sum_c if_{abc}T^{\mathbf{t}_c} = [T^{\mathbf{t}_a}, T^{\mathbf{t}_b}]. \tag{3.161}$$

Plugging this in gives

$$T^{\mathbf{t}_a}|Adj; T^{\mathbf{t}_b}\rangle = |Adj; [T^{\mathbf{t}_a}, T^{\mathbf{t}_b}]\rangle, \tag{3.162}$$

where $[T^{\mathbf{t}_a}, T^{\mathbf{t}_b}]$ is simply the commutator.

The derivation of equation (3.162) is extremely important, and it is vital that you understand it. However, it is also one of the more difficult results of this already difficult section. You are therefore encouraged (again) to read through this section, comparing it with examples several times until it becomes clear.

We derived (3.162) in terms of any adjoint generators $T^{\mathbf{t}_i}$. We now consider (3.162) for both the Cartan and non-Cartan generators, $H^{\mathbf{h}_a}$'s and $E^{\mathbf{e}_a}$'s. If we have a Cartan generator acting on a state corresponding to a Cartan generator, we have (from equation (3.156))

$$H^{\mathbf{h}_a}|Adj; H^{\mathbf{h}_b}\rangle = h^a_b|Adj; H^{\mathbf{h}_b}\rangle \tag{3.163}$$

(where there is no sum on b). But from (3.162) we have

$$H^{\mathbf{h}_a}|Adj; H^{\mathbf{h}_b}\rangle = |Adj; [H^{\mathbf{h}_a}, H^{\mathbf{h}_b}]\rangle. \tag{3.164}$$

By definition, the Cartan generators commute, so $[H^{\mathbf{h}_a}, H^{\mathbf{h}_b}] = 0$, and therefore

$$\mathbf{h}_b = 0. \tag{3.165}$$

So we can drop them from our notation, leaving the eigenstates corresponding to non-Cartan generators denoted $|Adj; H^j\rangle$.

On the other hand if we have a Cartan generator acting on an eigenstate corresponding to a non-Cartan generator, equation (3.156) gives

$$H^a|Adj; E^{\mathbf{e}_b}\rangle = e_b^a|Adj; E^{\mathbf{e}_b}\rangle. \tag{3.166}$$

And equation (3.162) gives

$$H^a|Adj; E^{\mathbf{e}_b}\rangle = |Adj; [H^a, E^{\mathbf{e}_b}]\rangle. \tag{3.167}$$

Now, we don't know *a priori* what $[H^a, E^{\mathbf{e}_b}]$ is, but comparing (3.166) and (3.167), we see

$$|Adj; e_b^a E^{\mathbf{e}_b}\rangle = |Adj; [H^a, E^{\mathbf{e}_b}]\rangle. \tag{3.168}$$

And because we know that each of these vectors corresponds directly to the generators, we have the final result

$$[H^a, E^{\mathbf{e}_b}] = e_b^a E^{\mathbf{e}_b}. \tag{3.169}$$

Now we want to know what a non-Cartan generator does to a given eigenstate. Consider an arbitrary state $|Adj; T^{\mathbf{t}_b}\rangle$ with H^c eigenvalue t_b^c. We can act on this with $E^{\mathbf{e}_a}$ to create the new state $E^{\mathbf{e}_a}|Adj; T^{\mathbf{t}_b}\rangle$. So what will the H^c eigenvalue of this new state be? Using (3.169),

$$\begin{aligned} H^c E^{\mathbf{e}_a}|Adj; T^{\mathbf{t}_b}\rangle &= (H^c E^{\mathbf{e}_a} - E^{\mathbf{e}_a} H^c + E^{\mathbf{e}_a} H^c)|Adj; T^{\mathbf{t}_b}\rangle \\ &= ([H^c, E^{\mathbf{e}_a}] + E^{\mathbf{e}_a} H^c)|Adj; T^{\mathbf{t}_b}\rangle \\ &= (e_a^c E^{\mathbf{e}_a} + E^{\mathbf{e}_a} t_b^c)|Adj; T^{\mathbf{t}_b}\rangle = (t_b^c + e_a^c) E^{\mathbf{e}_a}|Adj; T^{\mathbf{t}_b}\rangle \\ &= (\mathbf{t}_b + \mathbf{e}_a)^c E^{\mathbf{e}_a}|Adj; T^{\mathbf{t}_b}\rangle \end{aligned} \tag{3.170}$$

(where there is no sum on b). So, by acting on one of the eigenstates with a non-Cartan generator $E^{\mathbf{e}_a}$, we have shifted the H^c eigenvalue by one of the coordinates of the root vector. What this means is that the non-Cartan generators play a role analogous to the raising and lowering operators we saw in $SU(2)$, except instead of merely shifting the state "up" and "down", it moves the states around through some M-dimensional space. We will see examples of this for $SU(2)$ and $SU(3)$ in the following sections.[18]

From this, we can also see that if there is an operator that can transform from one state to another, there must be a corresponding operator that will make the opposite transformation. Therefore, for every operator $E^{\mathbf{e}_a}$, we expect to have the operator $E^{-\mathbf{e}_a}$, and corresponding eigenstate $|Adj; E^{-\mathbf{e}_a}\rangle$. So, let's consider the state $E^{\mathbf{e}_a}|Adj; E^{-\mathbf{e}_a}\rangle$. We know from (3.162) that

$$E^{\mathbf{e}_a}|Adj; E^{-\mathbf{e}_a}\rangle = |Adj; [E^{\mathbf{e}_a}, E^{-\mathbf{e}_a}]\rangle. \tag{3.171}$$

The eigenvalue of this state can be found using equation (3.170):

$$H^b E^{\mathbf{e}_a}|Adj; E^{-\mathbf{e}_a}\rangle = (-\mathbf{e}_a + \mathbf{e}_a)^b E^{\mathbf{e}_a}|Adj; E^{-\mathbf{e}_a}\rangle = 0. \tag{3.172}$$

But according to equation (3.165), states with 0 eigenvalue are states corresponding to Cartan generators. Therefore we conclude that the state $E^{\mathbf{e}_a}|Adj; E^{-\mathbf{e}_a}\rangle$ is proportional to some linear combination of the Cartan states,

$$E^{\mathbf{e}_a}|Adj; E^{-\mathbf{e}_a}\rangle = \sum_b N_a^b|Adj; H^b\rangle, \tag{3.173}$$

where the N_a^b's are the constants of proportionality. To find the constants N_a^b, we follow an approach similar to the one we used in deriving (3.133). Taking the inner product and using (3.162),

$$\begin{aligned}\langle Adj; H^c|E^{\mathbf{e}_a}|Adj; E^{-\mathbf{e}_a}\rangle &= \sum_b N_a^b\langle Adj; H^c|Adj; H^b\rangle = \sum_b N_a^b\delta^{cb} = N_a^c \\ &\Rightarrow \langle Adj; H^c|Adj; [E^{\mathbf{e}_a}, E^{-\mathbf{e}_a}]\rangle = N_a^c.\end{aligned} \tag{3.174}$$

Then, using (3.157)

$$\begin{aligned}\langle Adj; H^c|Adj; [E^{\mathbf{e}_a}, E^{-\mathbf{e}_a}]\rangle &= \frac{1}{\kappa}Tr(H^c[E^{\mathbf{e}_a}, E^{-\mathbf{e}_a}]) \\ &= \frac{1}{\kappa}Tr(E^{-\mathbf{e}_a}[H^c, E^{\mathbf{e}_a}]) \\ &= \frac{1}{\kappa}e_a^c Tr(E^{-\mathbf{e}_a}E^{\mathbf{e}_a}) \\ &= e_a^c\delta^{aa} \\ &= e_a^c.\end{aligned} \tag{3.175}$$

[18] Actually you've already seen the $SU(2)$ example in Sect. 3.2.8. The raising and lowering operators moved states around the $M = 1$-dimensional space we saw there (i.e. from $-1/2$ to $+1/2$, etc.).

So,

$$N_a^c = e_a^c. \tag{3.176}$$

And therefore equation (3.173) is now

$$E^{\mathbf{e}_a}|Adj; E^{-\mathbf{e}_a}\rangle = |Adj; [E^{\mathbf{e}_a}, E^{-\mathbf{e}_a}]\rangle = \sum_b e_a^b |Adj; H^b\rangle, \tag{3.177}$$

This leads to our final result,

$$[E^{\mathbf{e}_a}, E^{-\mathbf{e}_a}] = \sum_b e_a^b H^b. \tag{3.178}$$

Though we did all of this using the adjoint representation, this structure is the same in any representation, and therefore the key results are valid in any representation D_n. We worked in the adjoint simply because that makes the results easiest to obtain. The extensive use we made of labeling the eigenvectors with the generators can only be done in the adjoint representation because only in the adjoint does the number of eigenvectors equal the number of generators. However, this will not be a problem. The important results from this section are (3.169) and (3.178), which are true in any representation. Part of what we will do later is find these structures in other representations.

The importance of the ideas in this section cannot be stressed enough. However, the material is somewhat abstract. So, we consider a few examples of how all this works. Our first goal will be to simply show how this structure can be found in an arbitrary representation, and then later we will see why it is a useful structure.

3.2.13 Adjoint Representation of $SU(2)$

While going through the previous section we briefly discussed specific $SU(2)$ examples. However we now want to go through the adjoint representation of $SU(2)$ again in more detail in order to emphasize the ideas. As we've said a few times the adjoint representation of $SU(2)$ is the $j = 1$, or 3×3 representation written out in equations (3.145). As a first step, notice that (3.150) and (3.157) hold when we choose $\kappa = 2$.

Next in order to apply the "technology" we discovered in the previous section, we need to make use of the fundamental relationships we found, namely (3.169) and (3.178), or

$$[H^a, E^{\mathbf{e}_b}] = e_b^a E^{\mathbf{e}_b},$$
$$[E^{\mathbf{e}_a}, E^{-\mathbf{e}_a}] = e_a^b H^b. \tag{3.179}$$

Taking stock of what we actually have, there is one Cartan generator H^1 and two non-Cartan generators E^1 and E^2 (cf. (3.145)):

$$H^1 = J_1^3 = \begin{pmatrix} 1 & 0 & 0 \\ 0 & 0 & 0 \\ 0 & 0 & -1 \end{pmatrix},$$

$$E^1 = J_1^1 = \frac{1}{2}\begin{pmatrix} 0 & 1 & 0 \\ 1 & 0 & 1 \\ 0 & 1 & 0 \end{pmatrix}, \qquad E^2 = J_1^2 = \frac{1}{2}\begin{pmatrix} 0 & -i & 0 \\ i & 0 & -i \\ 0 & i & 0 \end{pmatrix}. \tag{3.180}$$

We also have three eigenvectors which for now we'll denote $\mathbf{v}_i$:

$$\mathbf{v}_1 = \begin{pmatrix} 1 \\ 0 \\ 0 \end{pmatrix}, \quad \mathbf{v}_2 = \begin{pmatrix} 0 \\ 1 \\ 0 \end{pmatrix}, \quad \mathbf{v}_3 = \begin{pmatrix} 0 \\ 0 \\ 1 \end{pmatrix}. \tag{3.181}$$

And then, finally, we have the (one-dimensional) root vectors in equations (3.155). There are three of them because this is an $n = 3$-dimensional representation (corresponding to the fact that $SU(2)$ is a three-dimensional group; we are focusing on adjoint representations for now), and each of them is 1-dimensional, corresponding to the fact that $SU(2)$ is a rank $M = 1$ group.[19] Writing out the root vectors again, we have

$$\mathfrak{t}_1 = \begin{pmatrix} 1 \end{pmatrix}, \qquad \mathfrak{t}_2 = \begin{pmatrix} 0 \end{pmatrix}, \qquad \text{and} \qquad \mathfrak{t}_3 = \begin{pmatrix} -1 \end{pmatrix}. \tag{3.182}$$

We can graph these on the real line as shown below,

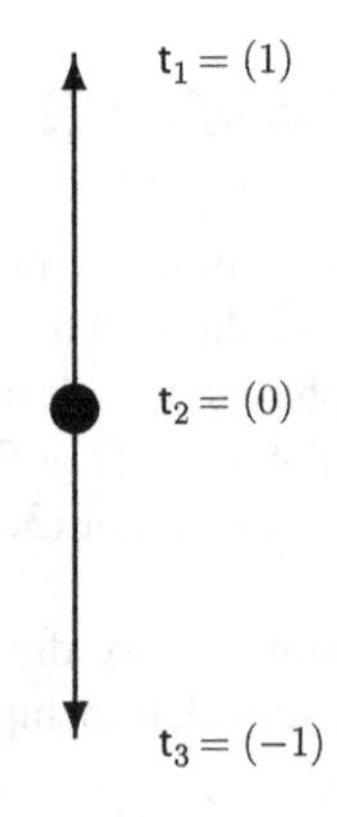

[19]If this isn't ringing a bell, reread Sect. 3.2.12 again more carefully.

So, we have three generators (3.180), three eigenvectors (3.181), and three root vectors (3.182). A major theme in the previous section was that there are the same number of generators, root vectors, and eigenvectors. In order to exploit the technology of the previous section and impose (3.169) and (3.178) we will need to make the correlations between these three sets of things explicit (each generator corresponds to a particular root vector and a particular eigenvector).

So let's begin with (3.169) (or the first of (3.179)). Leaving the tedious (but straightforward) matrix multiplication to you, we have

$$[H^1, E^1] = \cdots = \frac{1}{2}\begin{pmatrix} 0 & 1 & 0 \\ -1 & 0 & 1 \\ 0 & -1 & 0 \end{pmatrix}, \tag{3.183}$$

$$[H^1, E^2] = \cdots = -\frac{i}{2}\begin{pmatrix} 0 & 1 & 0 \\ 1 & 0 & 1 \\ 0 & 1 & 0 \end{pmatrix}. \tag{3.184}$$

But this is a problem, because according to (3.169), $[H^1, E^i]$ should be proportional to E^i, but this is not the case here. However, notice that from the definitions of the generators (3.180), equation (3.183) gives us

$$\frac{1}{2}\begin{pmatrix} 0 & 1 & 0 \\ -1 & 0 & 1 \\ 0 & -1 & 0 \end{pmatrix} = iE^2, \tag{3.185}$$

and equation (3.184) gives us

$$-\frac{i}{2}\begin{pmatrix} 0 & 1 & 0 \\ 1 & 0 & 1 \\ 0 & 1 & 0 \end{pmatrix} = -iE^1 \tag{3.186}$$

Or writing this more suggestively,

$$\begin{aligned} [H^1, E^1] &= iE^2, \\ [H^1, E^2] &= -E^1. \end{aligned} \tag{3.187}$$

Taking linear combinations of (3.187) we have

$$[H^1, \alpha E^1 \pm \beta i E^2] = \beta E^1 \pm \alpha i E^2, \tag{3.188}$$

where α and β are some constants. Note that this has the correct form for equation (3.169) as long as $\alpha = \beta$. So, we are now working with the following generators:

$$H^1 = \begin{pmatrix} 1 & 0 & 0 \\ 0 & 0 & 0 \\ 0 & 0 & -1 \end{pmatrix}, \tag{3.189}$$

$$\alpha(E^1 + iE^2) = \cdots = \alpha \begin{pmatrix} 0 & 1 & 0 \\ 0 & 0 & 1 \\ 0 & 0 & 0 \end{pmatrix}, \tag{3.190}$$

$$\alpha(E^1 - iE^2) = \cdots = \alpha \begin{pmatrix} 0 & 0 & 0 \\ 1 & 0 & 0 \\ 0 & 1 & 0 \end{pmatrix}. \tag{3.191}$$

Recall from Sect. 3.2.2 that we encouraged you to think of the generators as basis vectors of a vector space. In working with (3.189)–(3.191) instead of (3.180), all we're doing is taking a different linear combination of the basis vectors. This doesn't change the structure of the algebra any more than changing your basis does in any vector space.

In order to make more sense of what we have, let's act on our eigenvectors $\mathbf{v}_i$ (in (3.181)) with each of these:

$$H^1\mathbf{v}_1 = \begin{pmatrix} 1 & 0 & 0 \\ 0 & 0 & 0 \\ 0 & 0 & -1 \end{pmatrix}\begin{pmatrix} 1 \\ 0 \\ 0 \end{pmatrix} = \begin{pmatrix} 1 \\ 0 \\ 0 \end{pmatrix} = (1)\mathbf{v}_1, \tag{3.192}$$

$$H^1\mathbf{v}_2 = \begin{pmatrix} 1 & 0 & 0 \\ 0 & 0 & 0 \\ 0 & 0 & -1 \end{pmatrix}\begin{pmatrix} 0 \\ 1 \\ 0 \end{pmatrix} = 0 = (0)\mathbf{v}_2, \tag{3.193}$$

$$H^1\mathbf{v}_3 = \begin{pmatrix} 1 & 0 & 0 \\ 0 & 0 & 0 \\ 0 & 0 & -1 \end{pmatrix}\begin{pmatrix} 0 \\ 0 \\ 1 \end{pmatrix} = \begin{pmatrix} 0 \\ 0 \\ -1 \end{pmatrix} = (-1)\mathbf{v}_3, \tag{3.194}$$

$$\alpha(E^1 + iE^2)\mathbf{v}_1 = \alpha \begin{pmatrix} 0 & 1 & 0 \\ 0 & 0 & 1 \\ 0 & 0 & 0 \end{pmatrix}\begin{pmatrix} 1 \\ 0 \\ 0 \end{pmatrix} = 0, \tag{3.195}$$

$$\alpha(E^1 + iE^2)\mathbf{v}_2 = \alpha \begin{pmatrix} 0 & 1 & 0 \\ 0 & 0 & 1 \\ 0 & 0 & 0 \end{pmatrix}\begin{pmatrix} 0 \\ 1 \\ 0 \end{pmatrix} = \alpha \begin{pmatrix} 1 \\ 0 \\ 0 \end{pmatrix} = \alpha\mathbf{v}_1, \tag{3.196}$$

$$\alpha(E^1 + iE^2)\mathbf{v}_3 = \alpha \begin{pmatrix} 0 & 1 & 0 \\ 0 & 0 & 1 \\ 0 & 0 & 0 \end{pmatrix} \begin{pmatrix} 0 \\ 0 \\ 1 \end{pmatrix} = \alpha \begin{pmatrix} 0 \\ 1 \\ 0 \end{pmatrix} = \alpha \mathbf{v}_2, \tag{3.197}$$

$$\alpha(E^1 - iE^2)\mathbf{v}_1 = \alpha \begin{pmatrix} 0 & 0 & 0 \\ 1 & 0 & 0 \\ 0 & 1 & 0 \end{pmatrix} \begin{pmatrix} 1 \\ 0 \\ 0 \end{pmatrix} = \alpha \begin{pmatrix} 0 \\ 1 \\ 0 \end{pmatrix} = \alpha \mathbf{v}_2, \tag{3.198}$$

$$\alpha(E^1 - iE^2)\mathbf{v}_2 = \alpha \begin{pmatrix} 0 & 0 & 0 \\ 1 & 0 & 0 \\ 0 & 1 & 0 \end{pmatrix} \begin{pmatrix} 0 \\ 1 \\ 0 \end{pmatrix} = \alpha \begin{pmatrix} 0 \\ 0 \\ 1 \end{pmatrix} = \alpha \mathbf{v}_3, \tag{3.199}$$

$$\alpha(E^1 - iE^2)\mathbf{v}_3 = \alpha \begin{pmatrix} 0 & 0 & 0 \\ 1 & 0 & 0 \\ 0 & 1 & 0 \end{pmatrix} \begin{pmatrix} 0 \\ 0 \\ 1 \end{pmatrix} = 0. \tag{3.200}$$

This is beginning to help. Recall from (3.182) that the three root vectors were $+1$, 0, and -1. Then, notice that in every case where H^1 acts on an eigenvector it leaves it unchanged (moves it 0). Then, notice that when $\alpha(E^1 + iE^2)$ acts on an eigenvector it moves it "up" one spot (or if it is the "max" eigenvector it takes it to 0). And, finally, when $\alpha(E^1 - iE^2)$ acts on an eigenvector it moves it "down" one spot (or if it is the "min" eigenvector it takes it to 0). This gives us a very good indication of how we should relate the root vectors (3.182) to the generators: H^1 is associated with 0, $\alpha(E^1+iE^2)$ is associated with $+1$, and $\alpha(E^1-iE^2)$ is associated with -1:

$$\begin{aligned} H^1 &= H^1, \\ \alpha(E^1 + iE^2) &= \alpha E^{\mathbf{t}_1} = \alpha E^{\mathbf{e}_1} = \alpha E^{+1}, \\ \alpha(E^1 - iE^2) &= \alpha E^{\mathbf{t}_3} = \alpha E^{\mathbf{e}_3} = \alpha E^{-1}. \end{aligned} \tag{3.201}$$

This is completely consistent with what we had in Sect. 3.2.12. Namely, from equation (3.165) and the note following it we have the Cartan generator H^1 corresponding to the 0 root vector, and the non-Cartan generators are of the form $E^{\mathbf{e}_a}$ and $E^{-\mathbf{e}_a}$ as (3.178) demands.

Also, from equation (3.164) we know that the state with the 0 eigenvalue with H^1 will be the state corresponding to the Cartan generator, so from (3.193) we can say

$$\mathbf{v}_2 = |Adj; H^1\rangle = \begin{pmatrix} 0 \\ 1 \\ 0 \end{pmatrix} \tag{3.202}$$

Then, from (3.166) and (3.167) we can see that

$$[H^1, E^{+1}] = +E^{+1} \Rightarrow H^1|Adj; E^{+1}\rangle = +|Adj; E^{+1}\rangle,$$
$$[H^1, E^{-1}] = -E^{-1} \Rightarrow H^1|Adj; E^{-1}\rangle = -|Adj; E^{-1}\rangle, \tag{3.203}$$

which from (3.192) and (3.194) tells us that we should set

$$\mathbf{v}_1 = |Adj; E^{+1}\rangle = \begin{pmatrix} 1 \\ 0 \\ 0 \end{pmatrix},$$
$$\mathbf{v}_3 = |Adj; E^{-1}\rangle = \begin{pmatrix} 0 \\ 0 \\ 1 \end{pmatrix} \tag{3.204}$$

Now all that's left to do is find α. We do this by (finally) imposing (3.178) (or the second of equations (3.179)). Writing out the left side,

$$\begin{aligned}
[E^{+1}, E^{-1}] &= \alpha^2[E^1 + iE^2, E^1 - iE^2] \\
&= \alpha^2\big([E^1, E^1] - i[E^1, E^2] + i[E^2, E^1] + [E^2, E^2]\big) \\
&= -2i\alpha^2[E^1, E^2] = \cdots \\
&= -2i\alpha^2 iH^1 = 2\alpha^2 H^1.
\end{aligned} \tag{3.205}$$

And from (3.178) the right side should be $(+1)H^1$, which means

$$\alpha = \frac{1}{\sqrt{2}}.$$

So, finally, writing everything in terms of the root vectors, we have that our generators are

$$\begin{aligned}
H^1 &= H^1, \\
E^{\mathbf{e}_1} &= E^{+1} = \frac{1}{\sqrt{2}}(E^1 + iE^2), \\
E^{\mathbf{e}_3} &= E^{-1} = \frac{1}{\sqrt{2}}(E^1 - iE^2).
\end{aligned} \tag{3.206}$$

Notice that these are exactly what we had above in (3.122). So, we have derived from first principles the "pure math" version of the trick used to understand quantum mechanical spin in an introductory quantum theory course!

Summarizing, the eigenstates are

$$|Adj; E^{+1}\rangle = \mathbf{v}_1 = \begin{pmatrix} 1 \\ 0 \\ 0 \end{pmatrix},$$

$$|Adj; H^1\rangle = \mathbf{v}_2 = \begin{pmatrix} 0 \\ 1 \\ 0 \end{pmatrix},$$

$$|Adj; E^{-1}\rangle = \mathbf{v}_3 = \begin{pmatrix} 0 \\ 0 \\ 1 \end{pmatrix}. \tag{3.207}$$

So, we've managed to find an exact one-to-one correspondence between the three generators, the three root vectors, and the three eigenvectors. And in doing so, we have seen that equations (3.169) and (3.178) both hold. And therefore we have access to all of the technology developed in Sect. 3.2.12.

We'll briefly break our theme of only discussing mathematics to discuss how this will show up in physics. The idea of all of this is that a given eigenvector, or eigenstate, will correspond to a physical particle. The different points of the root space correspond to physical charges. And, as we have seen, a given eigenvector corresponds to a particular point in the root space, and therefore to a particular physical charge. In other words, a physical state has a charge that depends on which point in the root space it corresponds to. In this $SU(2)$ example, the root space consists of only the three points $+1$, 0, and -1 on the real line. Then the state $\mathbf{v}_1 = |Adj; E^{+1}\rangle$ corresponds to the point $+1$, $\mathbf{v}_2 = |Adj; H^1\rangle$ corresponds to the point 0, and $\mathbf{v}_3 = |Adj; E^{-1}\rangle$ corresponds to the point -1. Then, the generators are the "tools" that change one physical state to another. For example, if you have the particle corresponding to $\mathbf{v}_2 = |Adj; H^1\rangle$, acting on it with the generator E^{+1} will move it to the particle corresponding to $\mathbf{v}_1 = |Adj; E^{+1}\rangle$.

Physically, this means that the eigenstates are physical particles with charge, and the generators are the things that describe how those particles interact. All of this should be made much, much more clear as we consider actual physical examples later in the book.

3.2.14 *SU(2) for Arbitrary j ... Again*

Now that we have a deeper understanding of the algebraic structure of $SU(2)$ in the adjoint representation, let's see how this translates to other representations. As we saw in Sect. 3.2.11, we can form the linear combinations in equation (3.206) for any j = integer or half integer. The weight vectors will always look like those in the diagram on page 104 (in other words, raising and lowering operators always raise or lower their eigenvalue by 1).

The space of physical states, on the other hand, changes for each representation. For $j = \frac{1}{2}$, there are only two physical states, for $j = 1$ there are three physical states, for $j = \frac{3}{2}$ there are four physical states, and so on. We can graph the root spaces for these representations below (the dark dots represent the eigenvalues for the physical states in each representation).

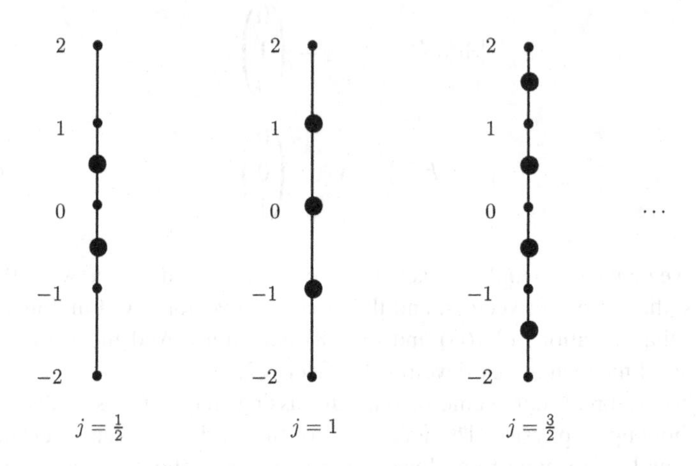

and so on.

Now, compare these diagrams to the one on page 104. Notice we drew that one with vectors and these merely with dots. The reason for this is very important. Each representation has a unique number of eigenvectors, and as we saw above each of these eigenvectors have their own H^1 eigenvalue (recall that for the j representation these eigenvalues range from j to $-j$ by integer steps). That is what is indicated by these diagrams – the $j = \frac{1}{2}$ representation has two eigenstates with values $\frac{1}{2}$ and $-\frac{1}{2}$. The $j = 1$ representation has three eigenstates with values 1, 0, and −1. The $j = \frac{3}{2}$ representation has four eigenstates with values $\frac{3}{2}$, $\frac{1}{2}$, $-\frac{1}{2}$, and $-\frac{3}{2}$. This pattern continues forever.

The point is that these diagrams correspond to the eigenvalues of the eigenvectors, and therefore the number of dots changes for each representation. However, what doesn't change for each representation is the number of generators; there are always three: one Cartan generator H^1 and two non-Cartan generators $E^{\pm 1}$. And as we mentioned in the previous section and saw briefly in Sect. 3.2.11, the particular "raising" and "lowering" combinations always act to move from one eigenvector to another. In other words, for any eigenvector in any representation, the raising and lowering generators move you up one or down one position on the root space. Or put in terms of what we've just seen, the raising and lowering generators move you up or down one position on the root space diagrams we just drew. And, as we've seen, for a given eigenvector there are only three options – stay at that vector, move up, or move down – corresponding to the three generators H^1, E^{+1}, and E^{-1}.

And that is why the root space for the adjoint representation has a unique importance. For any representation, the root diagram for the adjoint representation tells you how you can move from one vector to another in any representation. Here, you can always move up, down, or stay where you are, as indicated by the adjoint ($j = 1$) diagram.

We'll look next at $SU(3)$ and hopefully this will be made a bit clearer.

3.2.15 *SU*(3)

We have two goals in this section. The first is to see an example of the same structure we've been looking at in a non-adjoint representation, and the second is to understand the structure of $SU(3)$, which will play a major role in particle physics.

So, in the interest of meeting both of these goals, we'll study $SU(3)$ in a non-adjoint representation. To do this we'll start with the structure constants (cf. Sect. 3.2.3). The non-zero structure constants are

$$f_{123} = 1, \quad f_{147} = f_{165} = f_{246} = f_{257} = f_{345} = f_{376} = \frac{1}{2}, \quad f_{458} = f_{678} = \frac{\sqrt{3}}{2}.$$

As we said above, the structure constants are true for any representation, so if we wanted we could go through the tedious work of finding solutions for representations of various sizes, including the adjoint. However, as we said, we want to see how all of this works in a non-adjoint representation, so instead of the adjoint, we'll work in what is called the **Fundamental Representation**, which consists of 3×3 matrices. They are $T^a = \frac{1}{2}\lambda^a$ for $a = 1, \ldots, 8$, where

$$\lambda^1 = \begin{pmatrix} 0 & 1 & 0 \\ 1 & 0 & 0 \\ 0 & 0 & 0 \end{pmatrix}, \; \lambda^2 = \begin{pmatrix} 0 & -i & 0 \\ i & 0 & 0 \\ 0 & 0 & 0 \end{pmatrix}, \; \lambda^3 = \begin{pmatrix} 1 & 0 & 0 \\ 0 & -1 & 0 \\ 0 & 0 & 0 \end{pmatrix}, \; \lambda^4 = \begin{pmatrix} 0 & 0 & 1 \\ 0 & 0 & 0 \\ 1 & 0 & 0 \end{pmatrix},$$
$$\lambda^5 = \begin{pmatrix} 0 & 0 & -i \\ 0 & 0 & 0 \\ i & 0 & 0 \end{pmatrix}, \; \lambda^6 = \begin{pmatrix} 0 & 0 & 0 \\ 0 & 0 & 1 \\ 0 & 1 & 0 \end{pmatrix}, \; \lambda^7 = \begin{pmatrix} 0 & 0 & 0 \\ 0 & 0 & -i \\ 0 & i & 0 \end{pmatrix}, \; \lambda^8 = \frac{1}{\sqrt{3}} \begin{pmatrix} 1 & 0 & 0 \\ 0 & 1 & 0 \\ 0 & 0 & -2 \end{pmatrix}. \tag{3.208}$$

These matrices are called the **Gell-Mann Matrices**. Seeing which are Cartan (diagonal) and which are non-Cartan (not diagonal) is easy: λ^3 and λ^8 are Cartan and the rest are non-Cartan. So, because there are two Cartan generators, the rank of $SU(3)$ is 2.

Before moving on, we summarize a few results, though we won't bother to prove them (the proofs can be found in any introductory text on Lie groups). An arbitrary

$SU(n)$ group will always have n^2-1 generators, and will be rank $n-1$. An arbitrary $SO(n)$ group will always have $\frac{n(n-1)}{2}$ generators. (We won't worry about the rank of the orthogonal groups until later in the book, when we examine $SO(10)$ in particular.) So, the adjoint representation of $SU(n)$ will consist of $(n^2-1)\times(n^2-1)$ matrices, and the fundamental representation will consist of $n \times n$ matrices.

Note that if we did want to work in the adjoint we would have to deal with 8×8 matrices, which would be extremely tedious. The 3×3 fundamental representation defined by (3.208) will be much nicer.

We'll approach this in a similar way as we did $SU(2)$. Because this is a three-dimensional representation, we will have three eigenvectors, which we'll label (for now)

$$\mathbf{v}_1 = \begin{pmatrix} 1 \\ 0 \\ 0 \end{pmatrix}, \qquad \mathbf{v}_2 = \begin{pmatrix} 0 \\ 1 \\ 0 \end{pmatrix}, \qquad \mathbf{v}_3 = \begin{pmatrix} 0 \\ 0 \\ 1 \end{pmatrix}. \tag{3.209}$$

Next, recall from our discussion above that part of the point of all this is the correspondence between the eigenvectors $\mathbf{v}_i$ and points in the root space, which correspond to the physical charges. Specifically, a given eigenvector corresponds to a point in the root space depending on its eigenvalues with the Cartan generators. Here we have the same thing, but the complication is that we have *two* Cartan generators, $\frac{1}{2}\lambda^3$ and $\frac{1}{2}\lambda^8$. So, each of the eigenvectors (3.209) will have eigenvalues with *two* Cartan generators. This means we'll have to incorporate the material from Sect. 3.2.12 very carefully. If we label the non-Cartan generators according to

$$E^1 = T^1, \quad E^2 = T^2, \quad E^3 = T^4, \quad E^4 = T^5, \quad E^5 = T^6, \quad E^6 = T^7, \tag{3.210}$$

and the Cartan generators

$$H^1 = \frac{1}{2}\lambda^3 = \frac{1}{2}\begin{pmatrix} 1 & 0 & 0 \\ 0 & -1 & 0 \\ 0 & 0 & 0 \end{pmatrix}, \qquad H^2 = \frac{1}{2}\lambda^8 = \frac{1}{2\sqrt{3}}\begin{pmatrix} 1 & 0 & 0 \\ 0 & 1 & 0 \\ 0 & 0 & -2 \end{pmatrix}, \tag{3.211}$$

we can find the eigenvalues of the eigenvectors as follows:

$$\begin{aligned} &H^1\mathbf{v}_1 = \left(\frac{1}{2}\right)\mathbf{v}_1, \qquad H^1\mathbf{v}_2 = \left(-\frac{1}{2}\right)\mathbf{v}_2, \qquad H^1\mathbf{v}_3 = (0)\mathbf{v}_3, \\ &H^2\mathbf{v}_1 = \left(\frac{1}{2\sqrt{3}}\right)\mathbf{v}_1, \qquad H^2\mathbf{v}_2 = \left(\frac{1}{2\sqrt{3}}\right)\mathbf{v}_2, \qquad H^2\mathbf{v}_3 = \left(-\frac{1}{\sqrt{3}}\right)\mathbf{v}_3. \end{aligned} \tag{3.212}$$

So the eigenvector $\mathbf{v}_1$ has eigenvalue $\frac{1}{2}$ with H^1 and eigenvalue $\frac{1}{2\sqrt{3}}$ with H^2. This is consistent with the general fact that the dimension of the root space is always equal to the rank, which in this case is $n - 1 = 3 - 1 = 2$ (for $SU(3)$). So, in the two-dimensional root space, $\mathbf{v}_1$ corresponds to the point $\left(\frac{1}{2}, \frac{1}{2\sqrt{3}}\right)$. The points that $\mathbf{v}_2$ and $\mathbf{v}_3$ correspond to are easy to read off, and we can therefore easily write down the weight vectors (they aren't called "root" vectors because we're not in the adjoint representation):

$$\mathbf{t}_1 = \begin{pmatrix} \frac{1}{2} \\ \frac{1}{2\sqrt{3}} \end{pmatrix}, \quad \mathbf{t}_2 = \begin{pmatrix} -\frac{1}{2} \\ \frac{1}{2\sqrt{3}} \end{pmatrix}, \quad \mathbf{t}_3 = \begin{pmatrix} 0 \\ -\frac{1}{\sqrt{3}} \end{pmatrix}. \tag{3.213}$$

And we can graph them in the two-dimensional (H^1, H^2) plane as below (this picture is exactly analogous to the $SU(2)$ picture on page 104):

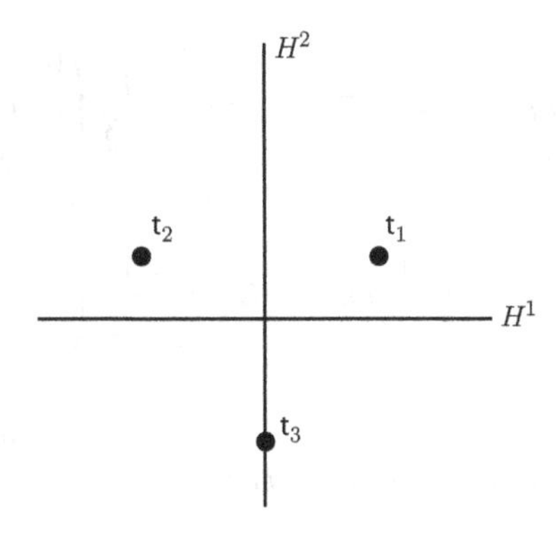

So these are the "charge" values corresponding to the three physical eigenstates $\mathbf{v}_i$. (Notice we could have seen these points by simply looking at the elements of the matrices H^1 and H^2). So, now we have a good understanding of the "charge lattice" in this representation of $SU(3)$, which depended only on the Cartan generators H^1 and H^2. Now we need to understand what the non-Cartan generators are doing. We could do this by going through the tedium of imposing (3.169) and (3.178) again, but we can take a shortcut by learning a lesson from what we've seen already.

The idea is that we're looking for "raising" and "lowering" operators – things that will move one physical state (one of the vectors $\mathbf{v}_i$) to another. A quick glance at the generators (3.208) make it clear that they aren't currently in the form of raising and lowering operators, so once again we'll have to find particular linear combinations (like $E^{\pm 1}$) for this to work. But, again, a quick glance at (3.208) makes finding these fairly easy. For example, looking at λ^1, notice it only has non-zero components in two elements. And if you look at the rest of the generators, you can notice that *only* λ^2 has non-zero elements in the same locations. This indicates that a good guess would be (we're including the normalization out front with foresight; normally you can check it yourself by checking the commutation relations):

$$\frac{1}{\sqrt{2}}(T^1 + iT^2) = \frac{1}{\sqrt{2}}(E^1 + iE^2) = \frac{1}{\sqrt{2}}\begin{pmatrix} 0 & 1 & 0 \\ 0 & 0 & 0 \\ 0 & 0 & 0 \end{pmatrix},$$

$$\frac{1}{\sqrt{2}}(T^1 - iT^2) = \frac{1}{\sqrt{2}}(E^1 - iE^2) = \frac{1}{\sqrt{2}}\begin{pmatrix} 0 & 0 & 0 \\ 1 & 0 & 0 \\ 0 & 0 & 0 \end{pmatrix}. \tag{3.214}$$

It should be clear that the first of these will raise $\mathbf{v}_2$ to $\mathbf{v}_1$, and the second will lower $\mathbf{v}_1$ to $\mathbf{v}_2$. Continuing, we find that the remaining combinations will be

$$\frac{1}{\sqrt{2}}(T^4 + iT^5) = \frac{1}{\sqrt{2}}(E^3 + iE^4) = \frac{1}{\sqrt{2}}\begin{pmatrix} 0 & 0 & 1 \\ 0 & 0 & 0 \\ 0 & 0 & 0 \end{pmatrix},$$

$$\frac{1}{\sqrt{2}}(T^4 - iT^5) = \frac{1}{\sqrt{2}}(E^3 - iE^4) = \frac{1}{\sqrt{2}}\begin{pmatrix} 0 & 0 & 0 \\ 0 & 0 & 0 \\ 1 & 0 & 0 \end{pmatrix},$$

$$\frac{1}{\sqrt{2}}(T^6 + iT^7) = \frac{1}{\sqrt{2}}(E^5 + iE^6) = \frac{1}{\sqrt{2}}\begin{pmatrix} 0 & 0 & 0 \\ 0 & 0 & 1 \\ 0 & 0 & 0 \end{pmatrix},$$

$$\frac{1}{\sqrt{2}}(T^6 - iT^7) = \frac{1}{\sqrt{2}}(E^5 - iE^6) = \frac{1}{\sqrt{2}}\begin{pmatrix} 0 & 0 & 0 \\ 0 & 0 & 0 \\ 0 & 1 & 0 \end{pmatrix}. \tag{3.215}$$

Now, let's think about how these will act on each of the physical state eigenvectors. We'll start with the first generator, the first matrix in (3.214). We can apply this to all three of the eigenvectors represented in the diagram on page 113. The point in that diagram labeled $\mathbf{t}_1$ corresponds to $\mathbf{v}_1$ (cf. (3.212)), so acting on $\mathbf{v}_1$ with $\frac{1}{\sqrt{2}}(E^1 + iE^2)$ gives

$$\frac{1}{\sqrt{2}}(E^1 + iE^2)\mathbf{v}_1 = \frac{1}{\sqrt{2}}\begin{pmatrix} 0 & 1 & 0 \\ 0 & 0 & 0 \\ 0 & 0 & 0 \end{pmatrix}\begin{pmatrix} 1 \\ 0 \\ 0 \end{pmatrix} = 0. \tag{3.216}$$

Continuing with the other two vectors, this is

$$\frac{1}{\sqrt{2}}(E^1 + iE^2)\mathbf{v}_2 = \frac{1}{\sqrt{2}}\begin{pmatrix}0 & 1 & 0\\ 0 & 0 & 0\\ 0 & 0 & 0\end{pmatrix}\begin{pmatrix}0\\1\\0\end{pmatrix} = \frac{1}{\sqrt{2}}\begin{pmatrix}1\\0\\0\end{pmatrix} = \frac{1}{\sqrt{2}}\mathbf{v}_1,$$

$$\frac{1}{\sqrt{2}}(E^1 + iE^2)\mathbf{v}_3 = \frac{1}{\sqrt{2}}\begin{pmatrix}0 & 1 & 0\\ 0 & 0 & 0\\ 0 & 0 & 0\end{pmatrix}\begin{pmatrix}0\\0\\1\end{pmatrix} = 0. \tag{3.217}$$

So looking again at the diagram on page 113 the generator $\frac{1}{\sqrt{2}}(E^1+iE^2)$ annihilates the $\mathbf{t}_1$ and $\mathbf{t}_3$ vectors, and sends $\mathbf{t}_2$ to $\mathbf{t}_1$ (up to a constant factor). You can work out the details to show that the second generator in the list, $\frac{1}{\sqrt{2}}(E^1 - iE^2)$, is the exact opposite; it sends $\mathbf{t}_1$ to $\mathbf{t}_2$ and annihilates $\mathbf{t}_2$ and $\mathbf{t}_3$. And, just as with $SU(2)$ we had that the vectors that move you around the one-dimensional root space were simply ± 1, the vectors that move you back and forth from $\mathbf{t}_1$ and $\mathbf{t}_2$ are $\pm\begin{pmatrix}1\\0\end{pmatrix}$. So, we write

$$\frac{1}{\sqrt{2}}(T^1 \pm iT^2) = \frac{1}{\sqrt{2}}(E^1 \pm iE^2) = E^{\pm\begin{pmatrix}1\\0\end{pmatrix}} = E^{\pm\mathbf{e}_1}. \tag{3.218}$$

You can (and should) repeat this for the other four generators in (3.215) and you will find that they correspond to

$$\frac{1}{\sqrt{2}}(T^4 \pm iT^5) = \frac{1}{\sqrt{2}}(E^3 \pm iE^4) = E^{\pm\begin{pmatrix}1/2\\ \sqrt{3}/2\end{pmatrix}} = E^{\pm\mathbf{e}_2},$$

$$\frac{1}{\sqrt{2}}(T^6 \pm iT^7) = \frac{1}{\sqrt{2}}(E^5 \pm iE^6) = E^{\pm\begin{pmatrix}-1/2\\ \sqrt{3}/2\end{pmatrix}} = E^{\pm\mathbf{e}_3}. \tag{3.219}$$

So, our weight vectors (same as root vectors but not in adjoint representation) can be drawn as below:

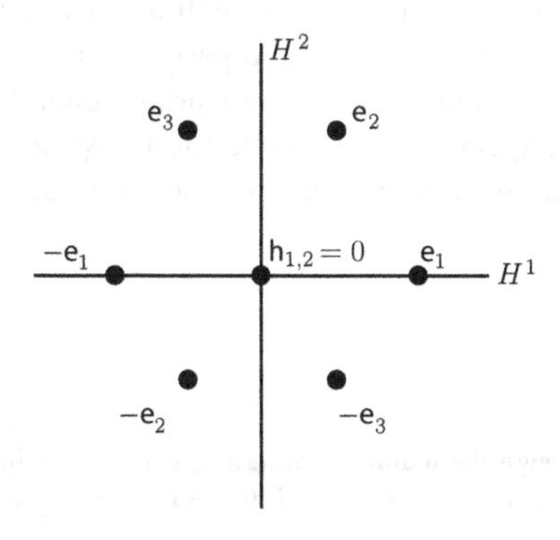

(the dot in the center indicates the two Cartan generators, which leave an eigenvector alone). Comparing this diagram to the one on page 113, it is easy to see that these six vectors are exactly the vectors necessary to move from one state to another.

As a comment, we did this in a very geometrical way. We could have followed what we did in Sect. 3.2.13 and imposed (3.169) and (3.178), and we would have gotten exactly the same results. However, we followed this approach for two reasons: it would have been a bit trickier working in a non-adjoint representation, and because we wanted to obviate the deeper geometrical structure. We strongly encourage you to go through and check (3.169) and (3.178) to confirm that they do work. One of the biggest differences you'll find is that here, because we aren't using the adjoint representation, there won't be a nice one-to-one relationship between the root vectors, eigenvectors, and generators. Here there are actually *two* different weight vector diagrams: one representing the eigenvalues of the state vectors with the Cartan generators (cf. diagram on page 113), and another representing the actual root vectors (cf. diagram on page 136).[20] The reason for multiple diagrams is the same as why all the $SU(2)$ diagrams were different (cf. diagram on page 110), but the adjoint diagram (cf. diagram on page 104) had a unique importance to all of them.

So, the diagram above (on page 136) is the *root* space diagram for the *adjoint* representation of $SU(3)$. And while it does not represent the eigenvalues of the state vectors in all representations, it does represent the way the state vectors in any representation transform from one to another. For example, consider the adjoint representation where the diagram on page 136 *does* represent the eigenvalues of the physical eigenvectors. It is still true that these are the very same vectors that will transform between each other – i.e. $\mathbf{e}_2$ will move $-\mathbf{e}_1$ to $\mathbf{e}_3$, or $-\mathbf{e}_3$ to $\mathbf{e}_1$, just as $-\mathbf{e}_1$ will move from $\mathbf{e}_2$ to $\mathbf{e}_3$ or $-\mathbf{e}_3$ to $-\mathbf{e}_2$, etc.

3.2.16 What is the Point of All of This?

Before finally getting back to physics, we give a spoiler of how Lie theory is used in physics. What we are going to find is that some physical interaction (electromagnetism, weak force, strong force) will ultimately be described by a Lie group in some particular representation. The particles that interact with that force will be described by the eigenvectors of the Cartan generators of the group, and the eigenvalues of those eigenvectors will be the physically measurable charges. The number of charges associated with the interaction is equal to the dimension of

[20] Had we actually gone through the tedium of working with the adjoint representation, we root diagram would have been exactly the diagram on 136, not the one on page 113.

the fundamental representation. So, for example, the strong force (as we will see) is in a 3-dimensional representation of the group that describes it. And, as we will see, there will be three different charges associated with the strong force, called the three "colors" (red, green, and blue).

Recall that we said that while the eigenvectors correspond to the physical particles, the generators describe how those physical particles/eigenvectors change. We can better say this by pointing out that the generators actually correspond to the force carrying particles. Photons, gluons, and W and Z bosons will all be described by the generators of their respective Lie group. The Cartan generators will be force-carrying particles that can interact with any particle charged under that group by transferring energy and momentum, but do not change the charge (photons and Z bosons). This is because, as we've seen, the Cartan generators are *not* the "raising" and "lowering" operators. On the other hand, the non-Cartan generators will be the force carrying particles which interact with any particle charged under that group by not only transferring energy and momentum, but also changing the charge (W bosons and gluons).

We won't be able to come back to discussing how this works until much later, and until examples are worked out, this may not be clear. We merely wanted to give an idea of where we are going with this.

3.3 The Lorentz Group

We discussed the Lorentz transformations in Sect. 1.4, but we took a very "physical" approach. As we mentioned in Sect. 3.2.1, the Lorentz transformations are actually the elements of the Lorentz group $SO(1,3)$. Now that we've gained some understanding of Lie groups in general, we can approach Lorentz transformations (and consequently the entire structure of special relativity) with much greater mathematical depth.

At first we'll limit our discussion to four spacetime dimensions. The case of arbitrary dimensions gets a bit complicated, so we'll hold off on that until later. For now understand that our discussion in the next few sections holds only for $SO(1,3)$, not $SO(1,n)$ for general n.

3.3.1 The Lorentz Algebra

We'll begin by looking at the algebra of the Lorentz group, otherwise known as the Lorentz algebra. There are countless ways of deriving the Lorentz algebra, all with varying degrees of mathematical rigor and sophistication. For our purposes, however, it will suffice to simply use the generators directly from the known transformations given in Sect. 1.4. This is sort of working backwards from a known representation of the group to the generators. Because the commutation relations are the same for any representation, however, we will still be able to extract some general information from this approach.

We'll change our notation a bit from Sect. 1.4 to make it clear exactly what we're talking about. Rather than using Λ we'll use R for rotations, B for boosts, J for the rotation generators, and K for the boost generators. We'll reserve Λ for a generic Lorentz transformation (when we're not trying to specify what kind of transformation). So including the matrix indices on the left-hand side, from 1.4 there are three rotations:

$$[R_x(\theta_x)]^\mu_\nu = \begin{pmatrix} 1 & 0 & 0 & 0 \\ 0 & 1 & 0 & 0 \\ 0 & 0 & \cos\theta_x & \sin\theta_x \\ 0 & 0 & -\sin\theta_x & \cos\theta_x \end{pmatrix},$$

$$[R_y(\theta_y)]^\mu_\nu = \begin{pmatrix} 1 & 0 & 0 & 0 \\ 0 & \cos\theta_y & 0 & -\sin\theta_y \\ 0 & 0 & 1 & 0 \\ 0 & \sin\theta_y & 0 & \cos\theta_y \end{pmatrix},$$

$$[R_z(\theta_z)]^\mu_\nu = \begin{pmatrix} 1 & 0 & 0 & 0 \\ 0 & \cos\theta_z & \sin\theta_z & 0 \\ 0 & -\sin\theta_z & \cos\theta_z & 0 \\ 0 & 0 & 0 & 1 \end{pmatrix}, \tag{3.220}$$

and three boosts

$$[B_x(\phi_x)]^\mu_\nu = \begin{pmatrix} \cosh\phi_x & -\sinh\phi_x & 0 & 0 \\ -\sinh\phi_x & \cosh\phi_x & 0 & 0 \\ 0 & 0 & 1 & 0 \\ 0 & 0 & 0 & 1 \end{pmatrix},$$

$$[B_y(\phi_y)]^\mu_\nu = \begin{pmatrix} \cosh\phi_y & 0 & -\sinh\phi_y & 0 \\ 0 & 1 & 0 & 0 \\ -\sinh\phi_y & 0 & \cosh\phi_y & 0 \\ 0 & 0 & 0 & 1 \end{pmatrix},$$

$$[B_z(\phi_z)]^\mu_\nu = \begin{pmatrix} \cosh\phi_y & 0 & 0 & -\sinh\phi_y \\ 0 & 1 & 0 & 0 \\ 0 & 0 & 1 & 0 \\ -\sinh\phi_y & 0 & 0 & \cosh\phi_y \end{pmatrix}. \tag{3.221}$$

We can use equation (3.86) directly to find the rotation generators J_i:

$$[J_x]^\mu_\nu = \left[-i \frac{dR_x(\theta_x)}{d\theta_x}\bigg|_{\theta_x=0} \right]^\mu_\nu = \begin{pmatrix} 0 & 0 & 0 & 0 \\ 0 & 0 & 0 & 0 \\ 0 & 0 & 0 & -i \\ 0 & 0 & i & 0 \end{pmatrix}, \tag{3.222}$$

and similarly,

$$[J_y]^\mu_\nu = \begin{pmatrix} 0 & 0 & 0 & 0 \\ 0 & 0 & 0 & i \\ 0 & 0 & 0 & 0 \\ 0 & -i & 0 & 0 \end{pmatrix}, \qquad [J_z]^\mu_\nu = \begin{pmatrix} 0 & 0 & 0 & 0 \\ 0 & 0 & -i & 0 \\ 0 & i & 0 & 0 \\ 0 & 0 & 0 & 0 \end{pmatrix}. \tag{3.223}$$

Then the boost generators K_i are

$$[K_x]^\mu_\nu = \begin{pmatrix} 0 & i & 0 & 0 \\ i & 0 & 0 & 0 \\ 0 & 0 & 0 & 0 \\ 0 & 0 & 0 & 0 \end{pmatrix}, \quad [K_y]^\mu_\nu = \begin{pmatrix} 0 & 0 & i & 0 \\ 0 & 0 & 0 & 0 \\ i & 0 & 0 & 0 \\ 0 & 0 & 0 & 0 \end{pmatrix}, \quad [K_z]^\mu_\nu = \begin{pmatrix} 0 & 0 & 0 & i \\ 0 & 0 & 0 & 0 \\ 0 & 0 & 0 & 0 \\ i & 0 & 0 & 0 \end{pmatrix}. \tag{3.224}$$

We can work out the (tedious) matrix algebra to find the commutation relations

$$\begin{aligned} [J_i, J_j] &= i\epsilon_{ijk} J_k, \\ [J_i, K_j] &= i\epsilon_{ijk} K_k, \\ [K_i, K_j] &= -i\epsilon_{ijk} J_k. \end{aligned} \tag{3.225}$$

So, we have two types of transformations: rotations and boosts. An interesting thing to notice is that these two things do not commute. Rotations are closed under commutation, but the commutator of two boosts is a rotation. This makes sense if you recall from Sect. 1.4 that boosts are rotations mixing space and time. Just as rotations do not commute (i.e. $SO(3)$ is non-Abelian), boosts and pure rotations do not commute. Obviously a general Lorentz transformation with boost parameter $\boldsymbol{\phi}$ and rotation parameter $\boldsymbol{\theta}$ will be of the form

$$L = e^{i\mathbf{J}\cdot\boldsymbol{\theta} + i\mathbf{K}\cdot\boldsymbol{\phi}}. \tag{3.226}$$

As a final comment, while the matrices as we have written them (with one index up and the other down) are often more useful because they are what actually acts on vectors, it will be worthwhile to write down the matrices with both indices lowered. Doing this with the generators we get

$$\begin{aligned} [J_x]_{\mu\nu} &= \eta_{\mu\rho}[J_x]^\rho_\nu = \cdots &= [J_x]^\mu_\nu, \\ [J_y]_{\mu\nu} &= \eta_{\mu\rho}[J_y]^\rho_\nu = \cdots &= [J_y]^\mu_\nu, \\ [J_z]_{\mu\nu} &= \eta_{\mu\rho}[J_z]^\rho_\nu = \cdots &= [J_z]^\mu_\nu, \end{aligned} \tag{3.227}$$

$$[K_x]_{\mu\nu} = \eta_{\mu\rho}[K_x]^{\rho}_{\nu} = \begin{pmatrix} 0 & -i & 0 & 0 \\ i & 0 & 0 & 0 \\ 0 & 0 & 0 & 0 \\ 0 & 0 & 0 & 0 \end{pmatrix},$$

$$[K_y]_{\mu\nu} = \eta_{\mu\rho}[K_y]^{\rho}_{\nu} = \begin{pmatrix} 0 & 0 & -i & 0 \\ 0 & 0 & 0 & 0 \\ i & 0 & 0 & 0 \\ 0 & 0 & 0 & 0 \end{pmatrix},$$

$$[K_z]_{\mu\nu} = \eta_{\mu\rho}[K_z]^{\rho}_{\nu} = \begin{pmatrix} 0 & 0 & 0 & -i \\ 0 & 0 & 0 & 0 \\ 0 & 0 & 0 & 0 \\ i & 0 & 0 & 0 \end{pmatrix}. \tag{3.228}$$

Notice that when we write the generators like this they are all antisymmetric and pure imaginary, and therefore they are all Hermitian. However, you can work out the details to show that when they are all Hermitian they no longer satisfy the commutation relations (3.225).

3.3.2 *The Underlying Structure of the Lorentz Group*

In order to get a better feel for the structure of this group, let's consider the following linear combinations of the generators:

$$N_i^{\pm} = \frac{1}{2}(J_i \pm iK_i). \tag{3.229}$$

If you work out the (relatively straightforward) details to find the commutation relations of these new generators you will find that they are[21]

$$\begin{aligned} [N_i^+, N_j^+] &= i\epsilon_{ijk}N_k^+, \\ [N_i^-, N_j^-] &= i\epsilon_{ijk}N_k^-, \\ [N_i^-, N_j^+] &= 0. \end{aligned} \tag{3.230}$$

Comparing this to the commutation relations for $SU(2)$ from (3.118) we see that we have two copies of $SU(2)$. So, this linear combination of generators has revealed that the Lorentz group is actually made up of two copies of $SU(2)$. We want to reiterate that this is only true in $1+3$ spacetime dimensions.

[21]Note that this assumes that we are using the standard form of the generators in (3.222)–(3.224), not the (Hermitian) form with the lowered indices in (3.228).

This is a very useful fact because we discussed $SU(2)$ at great length earlier in this chapter. One of the things we learned is that representations of $SU(2)$ are specified by a value j which is either an integer or a half-integer. Representations of the Lorentz group in $1 + 3$ dimensions will be specified by two values of j – one for each copy of $SU(2)$. In other words, a given representation of $SO(1, 3)$ will be specified by the doublet (j, j'), where j corresponds to the $SU(2)$ generated by the N_i^+'s and j' corresponds to the $SU(2)$ generated by the N_i^-'s. Because the j representation of $SU(2)$ is made up of $(2j + 1) \times (2j + 1)$ matrices, the (j, j') representation of the Lorentz group $SO(1, 3)$ will be made up of $(2j + 1)(2j' + 1)\times$ $(2j + 1)(2j' + 1)$ matrices.

3.3.3 *Representations of the Lorentz Group*

In this section, we will look at various representations of the Lorentz group in $1 + 3$ spacetime dimensions. Before doing so we'll make a somewhat subtle point. We found the algebra of the Lorentz group in $1 + 3$ dimensions above (cf. (3.225)) by looking at the vector representation (the representation that acts on spacetime vectors) that we're familiar with from introductory physics. And it was *this* representation that showed us that $SO(1, 3)$ contains two copies of $SU(2)$. Furthermore, because we already know something about $SU(2)$, we know that an *arbitrary* representation of $SO(1, 3)$ should be given by the two j values (one for each $SU(2)$). However, at this point, we have no idea what a given (j, j') combination means. This includes us not knowing which values for j and j' correspond to the vector representation that indicated the double $SU(2)$ structure to begin with.

So, our strategy for this section will be to try different values for j and j' and see what the representations look like. And as we plug in values for j and j', all we know is the commutation relations for the six generators; we know nothing about what form they take (i.e. we can't assume up front that they look anything like the vector representation generators (3.222)–(3.224)). Hopefully, however, we'll eventually come to a representation that we can identify as the vector representation.

3.3.3.1 The (0, 0) Representation

The smallest representation of the Lorentz group will be given by $(j, j') = (0, 0)$. This will consist of 1×1 matrices and therefore acts on 1 dimensional vectors, or scalars. In other words, this is the trivial representation – it describes things that don't change under Lorentz transformations.

3.3.3.2 The $(\frac{1}{2}, 0)$ Representation

The next representation is the $(\frac{1}{2}, 0)$ representation. In this representation only the N_i^+ generators are used. This is the representation where N_i^- are taken to be zero:

$$N_i^- = \frac{1}{2}(J_i - iK_i) = 0 \Longrightarrow J_i = iK_i. \tag{3.231}$$

Because we know that the N_i^+ generators are the $j = \frac{1}{2}$ representation of $SU(2)$, they must be equal to the Pauli matrices (cf. Sect. 3.2.9),

$$N_i^+ = \frac{1}{2}(J_i + iK_i) = \frac{1}{2}(iK_i + iK_i) = iK_i = \frac{1}{2}\sigma_i, \tag{3.232}$$

where we plugged in (3.231) to get the second equality. Writing out both rotation and boost generators, we have

$$K_i = -i\frac{1}{2}\sigma_i, \qquad J_i = iK_i = \frac{1}{2}\sigma_i. \tag{3.233}$$

We can use (3.233) to work out how an object in the $(\frac{1}{2}, 0)$ representation transforms under rotations and boosts. In general these are

$$\begin{aligned} R(\boldsymbol{\theta}) &= e^{i\boldsymbol{\theta}\cdot\boldsymbol{J}} = e^{i\boldsymbol{\theta}\cdot\frac{\boldsymbol{\sigma}}{2}}, \\ B(\boldsymbol{\phi}) &= e^{i\boldsymbol{\phi}\cdot\boldsymbol{K}} = e^{\boldsymbol{\phi}\cdot\frac{\boldsymbol{\sigma}}{2}}. \end{aligned} \tag{3.234}$$

Breaking these into components to get the explicit forms, for rotations we have

$$\begin{aligned} R_x(\theta_x) &= e^{i\theta_x J_x} = e^{i\theta_x \frac{1}{2}\sigma_x} \\ &= 1 + \frac{i}{2}\theta_x\sigma_x + \frac{1}{2}\left(\frac{i}{2}\theta_x\sigma_x\right)^2 + \cdots \\ &= \begin{pmatrix} 1 & 0 \\ 0 & 1 \end{pmatrix} + \frac{i}{2}\theta_x \begin{pmatrix} 0 & 1 \\ 1 & 0 \end{pmatrix} - \frac{1}{2}\left(\frac{\theta_x}{2}\right)^2 \begin{pmatrix} 1 & 0 \\ 0 & 1 \end{pmatrix} + \cdots \\ &= \begin{pmatrix} 1 - \frac{1}{2}\left(\frac{\theta_x}{2}\right)^2 + \cdots & i\frac{\theta_x}{2} + \cdots \\ i\frac{\theta_x}{2} + \cdots & 1 - \frac{1}{2}\left(\frac{\theta_x}{2}\right)^2 + \cdots \end{pmatrix} \\ &= \begin{pmatrix} \cos\frac{\theta_x}{2} & i\sin\frac{\theta_x}{2} \\ i\sin\frac{\theta_x}{2} & \cos\frac{\theta_x}{2} \end{pmatrix}. \end{aligned} \tag{3.235}$$

Similarly,

$$R_y(\theta_y) = \begin{pmatrix} \cos\frac{\theta_y}{2} & \sin\frac{\theta_y}{2} \\ -\sin\frac{\theta_y}{2} & \cos\frac{\theta_y}{2} \end{pmatrix}, \tag{3.236}$$

and

$$R_z(\theta_z) = \begin{pmatrix} \cos\frac{\theta_z}{2} + i\sin\frac{\theta_z}{2} & 0 \\ 0 & \cos\frac{\theta_z}{2} - i\sin\frac{\theta_z}{2} \end{pmatrix} = \begin{pmatrix} e^{i\frac{\theta_z}{2}} & 0 \\ 0 & e^{-i\frac{\theta_z}{2}} \end{pmatrix}. \tag{3.237}$$

For the boosts we have

$$B_x(\phi_x) = \begin{pmatrix} \cosh\frac{\phi_x}{2} & \sinh\frac{\phi_x}{2} \\ \sinh\frac{\phi_x}{2} & \cosh\frac{\phi_x}{2} \end{pmatrix}, \tag{3.238}$$

$$B_y(\phi_y) = \begin{pmatrix} \cosh\frac{\phi_y}{2} & -i\sinh\frac{\phi_y}{2} \\ i\sinh\frac{\phi_y}{2} & \cosh\frac{\phi_y}{2} \end{pmatrix}, \tag{3.239}$$

and

$$B_z(\phi_z) = \begin{pmatrix} \cosh\frac{\phi_z}{2} + \sinh\frac{\phi_z}{2} & 0 \\ 0 & \cosh\frac{\phi_z}{2} - \sinh\frac{\phi_z}{2} \end{pmatrix} = \begin{pmatrix} e^{\frac{\phi_z}{2}} & 0 \\ 0 & e^{-\frac{\phi_z}{2}} \end{pmatrix}. \tag{3.240}$$

We'll look at what types of objects these transformations act on later. It should be clear, however, that this is certainly not the vector representation.

3.3.3.3 The $(0, \frac{1}{2})$ Representation

Next we work out the details for the $(0, \frac{1}{2})$ representation. Now only the N_i^- part is used, so we have

$$N_i^+ = \frac{1}{2}(J_i + iK_i) = 0 \implies J_i = -iK_i. \tag{3.241}$$

Then

$$N_i^- = \frac{1}{2}(J_i - iK_i) = \frac{1}{2}(-iK_i - iK_i) = -iK_i = \frac{1}{2}\sigma_i, \tag{3.242}$$

so

$$J_i = \frac{1}{2}\sigma_i, \qquad K_i = i\frac{1}{2}\sigma_i. \tag{3.243}$$

Thus, the general transformations in this representation are

$$\begin{aligned} R(\boldsymbol{\theta}) &= e^{i\boldsymbol{\theta}\cdot\boldsymbol{J}} = e^{i\boldsymbol{\theta}\cdot\frac{\boldsymbol{\sigma}}{2}}, \\ B(\boldsymbol{\phi}) &= e^{i\boldsymbol{\phi}\cdot\boldsymbol{K}} = e^{-\boldsymbol{\phi}\cdot\frac{\boldsymbol{\sigma}}{2}}. \end{aligned} \tag{3.244}$$

Written out explicitly in each component, under rotations we have

$$R_x(\theta_x) = \begin{pmatrix} \cos\frac{\theta_x}{2} & i\sin\frac{\theta_x}{2} \\ i\sin\frac{\theta_x}{2} & \cos\frac{\theta_x}{2} \end{pmatrix}, \tag{3.245}$$

$$R_y(\theta_y) = \begin{pmatrix} \cos\frac{\theta_y}{2} & \sin\frac{\theta_y}{2} \\ -\sin\frac{\theta_y}{2} & \cos\frac{\theta_y}{2} \end{pmatrix}, \tag{3.246}$$

and

$$R_z(\theta_z) = \begin{pmatrix} e^{i\frac{\theta_z}{2}} & 0 \\ 0 & e^{-i\frac{\theta_z}{2}} \end{pmatrix}. \tag{3.247}$$

Notice that these are exactly the same as the rotation elements for the $(\frac{1}{2}, 0)$ representation. For the boosts, however, we have

$$B_x(\phi_x) = \begin{pmatrix} \cosh\frac{\phi_x}{2} & -\sinh\frac{\phi_x}{2} \\ -\sinh\frac{\phi_x}{2} & \cosh\frac{\phi_x}{2} \end{pmatrix}, \tag{3.248}$$

$$B_y(\phi_y) = \begin{pmatrix} \cosh\frac{\phi_y}{2} & i\sinh\frac{\phi_y}{2} \\ -i\sinh\frac{\phi_y}{2} & \cosh\frac{\phi_y}{2} \end{pmatrix}, \tag{3.249}$$

$$and B_z(\phi_z) = \begin{pmatrix} e^{-\frac{\phi_z}{2}} & 0 \\ 0 & e^{\frac{\phi_z}{2}} \end{pmatrix}. \tag{3.250}$$

So, the two representations $(\frac{1}{2}, 0)$ and $(0, \frac{1}{2})$ are identical under rotations, but slightly different under boosts.

3.3.3.4 The Relationship Between $(\frac{1}{2}, 0)$ and $(0, \frac{1}{2})$

We typically call the $(\frac{1}{2}, 0)$ representation the **Left-Handed Spinor** representation, and the $(0, \frac{1}{2})$ representation the **Right-Handed Spinor** representation. The reason for these names will become apparent later in the book when we discuss the physical meaning of these representations. For now, we're just doing math. Simply understand that the word "spinor" is the name for the representation, and the name for the thing that the representation acts on. In other words, "spinor" is just the word we use instead of "vector" when we're talking about this representation.

Consider a two component complex spinor ψ_L that transforms under the left-handed representation. All this means is that if you do a Lorentz transformation (a rotation or a boost) to your frame, ψ_L will transform according to the left-handed spinor transformation rules, rather than the vector transformation rules. Under rotations ψ_L transforms as

$$\psi_L \longrightarrow \psi_L' = e^{i\frac{1}{2}\boldsymbol{\theta}\cdot\boldsymbol{\sigma}}\psi_L, \tag{3.251}$$

and under boosts it transforms as

$$\psi_L \longrightarrow \psi_L' = e^{\frac{1}{2}\boldsymbol{\phi}\cdot\boldsymbol{\sigma}}\psi_L. \tag{3.252}$$

Now (for reasons that will be clear shortly) take the complex conjugate of ψ_L and multiply it by $i\sigma^2$, giving the new spinor

$$\bar{\psi}_L = i\sigma^2\psi_L^\star. \tag{3.253}$$

Consider its transformation under rotations:

$$\begin{aligned}
\bar{\psi}_L \longrightarrow \bar{\psi}_L' &= i\sigma^2(\psi_L')^\star \\
&= i\sigma^2\big(e^{i\frac{1}{2}\boldsymbol{\theta}\cdot\boldsymbol{\sigma}}\psi_L\big)^\star \\
&= i\sigma^2\big(e^{-i\frac{1}{2}\boldsymbol{\theta}\cdot\boldsymbol{\sigma}^\star}\mathbb{I}\big)\psi_L^\star \\
&= i\sigma^2\big(e^{-i\frac{1}{2}\boldsymbol{\theta}\cdot\boldsymbol{\sigma}^\star}(-i\sigma^2)(i\sigma^2)\big)\psi_L^\star \\
&= e^{i\frac{1}{2}\boldsymbol{\theta}\cdot\boldsymbol{\sigma}}i\sigma^2\psi_L^\star \\
&= e^{i\frac{1}{2}\boldsymbol{\theta}\cdot\boldsymbol{\sigma}}\bar{\psi}_L,
\end{aligned} \tag{3.254}$$

where we used the identity $(-i\sigma^2)(i\sigma^2) = \mathbb{I}$ to get the fourth line, and the identity $(i\sigma^2)\boldsymbol{\sigma}^\star(-i\sigma^2) = -\boldsymbol{\sigma}$ to get the second to last line.[22] The point is that $\bar{\psi}_L$ transforms the same as ψ_L under rotations. Under boosts, however,

$$\begin{aligned}
\bar{\psi}_L \longrightarrow \bar{\psi}_L' &= i\sigma^2(\psi_L')^\star \\
&= i\sigma^2\big(e^{\frac{1}{2}\boldsymbol{\phi}\cdot\boldsymbol{\sigma}}\psi_L\big)^\star \\
&= i\sigma^2\big(e^{\frac{1}{2}\boldsymbol{\phi}\cdot\boldsymbol{\sigma}^\star}(-i\sigma^2)(i\sigma^2)\big)\psi_L^\star \\
&= e^{-\frac{1}{2}\boldsymbol{\phi}\cdot\boldsymbol{\sigma}}i\sigma^2\psi_L^\star \\
&= e^{-\frac{1}{2}\boldsymbol{\phi}\cdot\boldsymbol{\sigma}}\bar{\psi}_L,
\end{aligned} \tag{3.255}$$

which you can identify as the transformation law for the *right* handed $(0, \frac{1}{2})$ representation.

---

[22]You can work out these identities yourself. For our purposes they're not particularly profound, just useful for what we're doing.

You can work out the details to show that if ψ_R is a spinor in the right-handed representation, then $\bar{\psi}_R = i\sigma^2\psi_R^\star$ will transform as a left-handed spinor ψ_R. If we do this twice (using $-i\sigma^2 = (i\sigma^2)^{-1}$ going back) we get

$$\begin{aligned} \psi_L &= \bar{\psi}_R \\ &= i\sigma^2\psi_R^\star, \\ \Longrightarrow i\sigma^2(\psi_L)^\star &= -i\sigma^2(i\sigma^2\psi_R^\star)^\star \\ &= -i\sigma^2(-i)(-\sigma^2)\psi_R \\ &= \psi_R, \end{aligned} \tag{3.256}$$

which is of course a left-handed spinor.

So, what we have found is that the left- and right-handed representations are, in a sense, conjugates of each other. A spinor in either representation can be written as a spinor in the other representation by merely taking its complex conjugate multiplying it by $\pm i\sigma^2$ ($-$ when going from left to right-handed and $+$ when going from right to left-handed).

3.3.3.5 The $(\frac{1}{2}, \frac{1}{2})$ Representation

With this representation, both N_i^+ and N_i^- form an $SU(2)$. So, a given state in this representation will have two copies of $SU(2)$ acting on it: a right-handed copy and a left-handed copy. So, we'll give such a state two indices, one that transforms under the left-handed $SU(2)$ and another that transforms under the right-handed $SU(2)$. Because of the commutation relations (3.230) we know that these transformations won't interfere with each other ((3.230) indicates that they are non-overlapping copies of $SU(2)$).

So if we denote the $+$ indices (left-handed indices transforming under N_i^+) without a dot and the $-$ indices (right-handed indices transforming under N_i^-) with a dot, a state will be written as $v^{\dot{a}b}$. Notice that because this object has two indices that each take on two values, it has a total of four components. So we can interpret it in different ways. The most natural (in a sense) would be as a 2×2 complex matrix.

Another way would be to "force" this into a vector form by choosing an arbitrary basis. For example, because there are two $SU(2)$ indices, and each can take on two values ($\frac{1}{2}$ or $-\frac{1}{2}$), the matrix representation could be written as ($\uparrow$ for $+\frac{1}{2}$ and $\downarrow$ for $-\frac{1}{2}$)

$$v^{\dot{a}b} = \begin{pmatrix} \uparrow\uparrow & \uparrow\downarrow \\ \downarrow\uparrow & \downarrow\downarrow \end{pmatrix}, \tag{3.257}$$

and we can assign a basis as follows:

$$| \uparrow\uparrow\rangle = \begin{pmatrix} 1 \\ 0 \\ 0 \\ 0 \end{pmatrix}, \quad | \uparrow\downarrow\rangle = \begin{pmatrix} 0 \\ 1 \\ 0 \\ 0 \end{pmatrix}, \quad | \downarrow\uparrow\rangle = \begin{pmatrix} 0 \\ 0 \\ 1 \\ 0 \end{pmatrix}, \quad | \downarrow\downarrow\rangle = \begin{pmatrix} 0 \\ 0 \\ 0 \\ 1 \end{pmatrix}. \tag{3.258}$$

Obviously this choice is arbitrary. However, we will see shortly that there is a more "natural" choice.

But for now we'll stick with the 2×2 complex matrix. Recall that just as any real matrix can be written as the sum of a symmetric matrix and an antisymmetric matrix, any complex matrix can be written as the sum of a Hermitian matrix and an anti-Hermitian matrix.[23] However, the two indices on our matrix $v^{\dot{a}b}$ transform under representations of $SU(2)$. Notice that in the generators of these copies of $SU(2)$, both sets of generators N^+ and N^- are Hermitian (cf. (3.229)). So, we'll limit our discussion to the case where $v^{\dot{a}b}$ is a Hermitian 2×2 matrix.

So let's write down the most general Hermitian matrix we can. It will be

$$v^{\dot{a}b} = \begin{pmatrix} c & d - ie \\ d + ie & f \end{pmatrix}, \tag{3.259}$$

where c, d, e, and f are all real. But without loss of generality, we can make a slight change and write this instead as

$$v^{\dot{a}b} = \begin{pmatrix} c + f & d - ie \\ d + ie & c - f \end{pmatrix} \tag{3.260}$$

(this is just another way of writing the most general Hermitian matrix). Notice that we can now write this as

$$v^{\dot{a}b} = c \begin{pmatrix} 1 & 0 \\ 0 & 1 \end{pmatrix} + d \begin{pmatrix} 0 & 1 \\ 1 & 0 \end{pmatrix} + e \begin{pmatrix} 0 & -i \\ i & 0 \end{pmatrix} + f \begin{pmatrix} 1 & 0 \\ 0 & -1 \end{pmatrix}. \tag{3.261}$$

Or if we replace the c, d, e, f with vector indices $(c, d, e, f) = (v^0, v^1, v^2, v^3)$, then

$$v^{\dot{a}b} = v^\mu \sigma_\mu^{\dot{a}b}, \tag{3.262}$$

where σ_0 is the identity matrix and $\sigma_i^{\dot{a}b}$ for $i = 1, 2, 3$ are just the normal Pauli matrices. In other words, we have

$$v^{\dot{a}b} = \begin{pmatrix} v^0 & v^1 & v^2 & v^3 \end{pmatrix} \begin{pmatrix} \sigma_0 \\ \sigma_1 \\ \sigma_2 \\ \sigma_3 \end{pmatrix}. \tag{3.263}$$

[23] A Hermitian matrix is a matrix A satisfying $A = A^\dagger$ where $A^\dagger = A^{\star T}$, or the complex conjugate of the transpose. An anti-Hermitian matrix satisfies $A = -A^\dagger$.

Notice that this is nothing more than a particular choice of basis, like our arbitrary choice in (3.258). Writing out the terms gives (using (3.260) and again using (3.257))

$$\begin{pmatrix}1\\0\\0\\0\end{pmatrix} = \frac{1}{2}(|\uparrow\uparrow\rangle + |\downarrow\downarrow\rangle) \longleftrightarrow \frac{1}{2}(c + f + c - f) = c = v^0,$$

$$\begin{pmatrix}0\\1\\0\\0\end{pmatrix} = \frac{1}{2}(|\uparrow\downarrow\rangle + |\downarrow\uparrow\rangle) \longleftrightarrow \frac{1}{2}(d - ie + d + ie) = d = v^1,$$

$$\begin{pmatrix}0\\0\\1\\0\end{pmatrix} = \frac{i}{2}(|\uparrow\downarrow\rangle - |\downarrow\uparrow\rangle) \longleftrightarrow \frac{i}{2}(d - ie - d - ie) = e = v^2,$$

$$\begin{pmatrix}0\\0\\0\\1\end{pmatrix} = \frac{1}{2}(|\uparrow\uparrow\rangle - |\downarrow\downarrow\rangle) \longleftrightarrow \frac{1}{2}(c + f - c + f) = f = v^3. \quad (3.264)$$

As we said, the choice of basis is arbitrary, but using the Pauli matrices when talking about $SU(2)$ certainly seems like a good choice.

Now we are finally able to look at the meaning of the $(\frac{1}{2}, \frac{1}{2})$ representation of the Lorentz group. We'll do this by considering various transformations on $v^{\dot{a}b}$ and looking at the transformation that is applied to v^μ as a result.

For the sake of being concise, let's consider an arbitrary Lorentz transformation on $v^{\dot{a}b}$. Going only to first order (as usual), we'll combine (3.235)–(3.240) for the right-handed (dotted) transformations, and (3.245)–(3.250) for the left-handed (undotted) transformations. This will give

$$\begin{aligned}
v^{\dot{a}b} \longrightarrow v'^{\dot{c}d} &= \left(e^{\frac{i}{2}\boldsymbol{\theta}\cdot\boldsymbol{\sigma} - \frac{1}{2}\boldsymbol{\phi}\cdot\boldsymbol{\sigma}}\right)^{\dot{c}}_{\ \dot{a}} \left(e^{\frac{i}{2}\boldsymbol{\theta}\cdot\boldsymbol{\sigma} + \frac{1}{2}\boldsymbol{\phi}\cdot\boldsymbol{\sigma}}\right)^{d}_{\ b} v^{\dot{a}b} \\
&= \left(1 + \frac{i}{2}\boldsymbol{\theta}\cdot\boldsymbol{\sigma} - \frac{1}{2}\boldsymbol{\phi}\cdot\boldsymbol{\sigma}\right)^{\dot{c}}_{\ \dot{a}} \left(1 + \frac{i}{2}\boldsymbol{\theta}\cdot\boldsymbol{\sigma} + \frac{1}{2}\boldsymbol{\phi}\cdot\boldsymbol{\sigma}\right)^{d}_{\ b} v^{\dot{a}b} \\
&= \begin{pmatrix} 1 + \frac{1}{2}(i\theta_z - \phi_z) & \frac{1}{2}(i\theta_x + \theta_y - \phi_x + i\phi_y) \\ \frac{1}{2}(i\theta_x - \theta_y - \phi_x - i\phi_y) & 1 - \frac{1}{2}(i\theta_z - \phi_z) \end{pmatrix} \\
&\quad \times \begin{pmatrix} v^0 + v^3 & v^1 - iv^2 \\ v^1 + iv^2 & v^0 - v^3 \end{pmatrix}
\end{aligned}$$

$$\times\begin{pmatrix} 1+\frac{1}{2}(i\theta_z+\phi_z) & \frac{1}{2}(i\theta_x+\theta_y+\phi_x-i\phi_y) \\ \frac{1}{2}(i\theta_x-\theta_y+\phi_x+i\phi_y) & 1-\frac{1}{2}(i\theta_z+\phi_z) \end{pmatrix}$$

$$= \cdots$$

$$= \begin{pmatrix} A & B \\ C & D \end{pmatrix}, \tag{3.265}$$

where

$$\begin{aligned}
A &= v^0 + i\theta_z v^0 + i\phi_y v^1 + i\theta_x v^1 - i\phi_x v^2 + i\theta_y v^2 + v^3 + i\theta_z v^3, \\
B &= i\theta_x v^0 + \theta_y v^0 + v^1 - \phi_z v^1 - iv^2 + i\phi_z v^2 + \phi_x v^3 - i\phi_y v^3, \\
C &= i\theta_x v^0 - \theta_y v^0 + v^1 + \phi_z v^1 + iv^2 + i\phi_z v^2 - \phi_x v^3 - i\phi_y v^3, \\
D &= v^0 - i\theta_z v^0 - i\phi_y v^1 + i\theta_x v^1 + i\phi_x v^2 + i\theta_y v^2 - v^3 + i\theta_z v^3.
\end{aligned} \tag{3.266}$$

Then we can use (3.264) to find the new components. This gives

$$\begin{aligned}
v^0 \longrightarrow v'^0 &= \frac{1}{2}(A+D) \\
&= v^0 + i\theta_x v^1 + i\theta_y v^2 + i\theta_z v^3, \\
v^1 \longrightarrow v'^1 &= \frac{1}{2}(B+C) \\
&= i\theta_x v^0 + v^1 + i\phi_z v^2 - i\phi_y v^3, \\
v^2 \longrightarrow v'^2 &= \frac{i}{2}(B-C) \\
&= i\theta_y v^0 - i\phi_z v^1 + v^2 + i\phi_x v^3, \\
v^3 \longrightarrow v'^3 &= \frac{1}{2}(A-D) \\
&= i\theta_z v^0 + i\phi_y v^1 - i\phi_x v^2 + v^3.
\end{aligned} \tag{3.267}$$

You can work out the matrix representation of this and you will find that the transformation law here is

$$\begin{pmatrix} v^0 \\ v^1 \\ v^2 \\ v^3 \end{pmatrix} \longrightarrow \begin{pmatrix} v'^0 \\ v'^1 \\ v'^2 \\ v'^3 \end{pmatrix} = \begin{pmatrix} v^0 \\ v^1 \\ v^2 \\ v^3 \end{pmatrix} + \begin{pmatrix} 0 & i\theta_x & i\theta_y & i\theta_z \\ i\theta_x & 0 & i\phi_z & -i\phi_y \\ i\theta_y & -i\phi_z & 0 & i\phi_x \\ i\theta_z & i\phi_y & -i\phi_x & 0 \end{pmatrix} \begin{pmatrix} v^0 \\ v^1 \\ v^2 \\ v^3 \end{pmatrix} \tag{3.268}$$

Comparing this to equations (3.222)–(3.224) you can see that this is exactly the form of a general Lorentz transformation in the fundamental defining vector representation of $SO(1,3)$.

We therefore recognize this $(\frac{1}{2}, \frac{1}{2})$ representation *to be* the vector representation that we started with!

Finally, to give a bit more insight into what is going on here, consider again the expression $v^{\dot{a}b}$ as written out in (3.261). Taking the determinant, we get

$$\begin{aligned}\det(v^{\dot{a}b}) &= \det\begin{pmatrix} v^0 + v^3 & v^1 - iv^2 \\ v^1 + iv^2 & v^0 - v^3 \end{pmatrix} \\ &= (v^0 + v^3)(v^0 - v^3) - (v^1 - iv^2)(v^1 + iv^2) \\ &= (v^0)^2 - (v^1)^2 - (v^2)^2 - (v^3)^2, \end{aligned} \tag{3.269}$$

which is simply the invariant spacetime element that defined special relativity in the first place (cf. Sect. 1.4.2). Because we are operating on $v^{\dot{a}b}$ with two elements of $SU(2)$, which have determinant equal to one by definition, transformations of the form (3.265) will leave the determinant unchanged (cf. equation (3.68)), and therefore will leave the spacetime interval unchanged, as expected (and demanded) by special relativity.

So, in summary, we have four representations of the Lorentz group (in four dimensions): the $(0, 0)$ scalar representation where things don't transform, the $(\frac{1}{2}, 0)$ left-handed and the $(0, \frac{1}{2})$ right-handed representation which act on things called spinors (that have yet to be discussed in any detail), and the $(\frac{1}{2}, \frac{1}{2})$ representation that is the fundamental, defining vector representation of the Lorentz group in $1+3$ spacetime dimensions. You could of course choose any other (j, j') representation you want and work out the details.

3.3.4 The Vector Representation in Arbitrary Dimension

Everything we discussed in the previous sections was unique to *four* spacetime dimensions. We'll take a moment now to outline the case for the representation that acts on spacetime vectors of arbitrary dimension. We'll discuss spinor representations in arbitrary dimensions in Sect. 3.3.6.

Looking back at the forms of the rotations and boosts in (3.220) and (3.221), the pattern is fairly obvious. For rotations mixing spatial dimensions, you just put the sin's and cos's in the appropriate places, and for rotations mixing a space and a time dimension (boosts), you put the sinh's and cosh's in the appropriate places. The convention for the sign of each term is fairly straightforward as well. Then, you can use (3.86) to take the derivative, set the parameter equal to 0, and then multiply by $-i$. Then, if you lower all the indices like we did in (3.227) and (3.228), you will simply find that the generators are the basis set of antisymmetric matrices with two imaginary components. That is, they are the basis set of all Hermitian matrices with two components. In $n+1$ spacetime dimensions, there will be n boost generators and $\frac{n(n-1)}{2}$ rotation generators, for a total of $n + \frac{n(n-1)}{2} = \frac{n(n+1)}{2}$ generators.

We can generalize this very easily. We know that there will be $\frac{n(n+1)}{2}$ generators total. We know that $\frac{n(n+1)}{2}$ is the number of independent values in an $(n+1)\times(n+1)$ antisymmetric matrix. So rather than using a single index that runs from 1 to $\frac{n(n+1)}{2}$, let's use *two* indices that we take to be totally antisymmetric. In other words, one of our generators will be labeled $J^{\mu\nu} = -J^{\nu\mu}$, and it will have matrix components $(J^{\mu\nu})^{ab} = -(J^{\mu\nu})^{ba}$. We see that $J^{\mu\mu} = 0$, and $(J^{\mu\nu})^{aa} = 0$. Also note that μ, ν, a, and b all run from 0 to $n-1$ for n spacetime dimensions. We'll interpret $J^{\mu\nu}$ as the generator for rotations between the μ and ν dimensions. So J^{10} is a boost in the x direction, J^{23} is a rotation mixing y and z, etc.

We can now write out the general antisymmetric/Hermitian basis as

$$(J^{\mu\nu})^{ab} = -i(\eta^{\mu a}\eta^{\nu b} - \eta^{\nu a}\eta^{\mu b}) \tag{3.270}$$

You can work out the details to show that this recovers the familiar generators (3.220) and (3.221) in 4 dimensions.

The nice thing about the form (3.270) of the generators is that it is true for arbitrary dimension. So, while we already found the generators for $1+3$ dimensions, we can use (3.270) to find them for arbitrary dimension. We can work out the details as follows:

$$\begin{aligned}
[J^{\mu\nu}, J^{\rho\lambda}] &= (J^{\mu\nu})^a_b (J^{\rho\lambda})^{bc} - (J^{\rho\lambda})^a_b (J^{\mu\nu})^{bc} \\
&= \eta_{db}\big((J^{\mu\nu})^{ad}(J^{\rho\lambda})^{bc} - (J^{\rho\lambda})^{ad}(J^{\mu\nu})^{bc}\big) \\
&= \eta_{db}\big((\eta^{\rho a}\eta^{\lambda d} - \eta^{\lambda a}\eta^{\rho d})(\eta^{\mu b}\eta^{\nu c} - \eta^{\nu b}\eta^{\mu c}) \\
&\qquad -(\eta^{\mu a}\eta^{\rho d} - \eta^{\nu a}\eta^{\mu d})(\eta^{\rho b}\eta^{\lambda c} - \eta^{\lambda b}\eta^{\rho c})\big) \\
&= \eta_{db}(\eta^{\rho a}\eta^{\lambda d}\eta^{\mu b}\eta^{\nu c} - \eta^{\rho a}\eta^{\lambda d}\eta^{\nu b}\eta^{\mu c} - \eta^{\lambda a}\eta^{\rho d}\eta^{\mu b}\eta^{\nu c} + \eta^{\lambda a}\eta^{\rho d}\eta^{\nu b}\eta^{\mu c} \\
&\qquad -\eta^{\mu a}\eta^{\nu d}\eta^{\rho b}\eta^{\lambda c} + \eta^{\mu a}\eta^{\nu d}\eta^{\lambda b}\eta^{\rho c} + \eta^{\nu a}\eta^{\mu d}\eta^{\rho b}\eta^{\lambda c} - \eta^{\nu a}\eta^{\mu d}\eta^{\lambda b}\eta^{\rho c}) \\
&= \eta^{\rho a}\eta^{\nu c}\eta^{\lambda}_b\eta^{\nu b} - \eta^{\rho a}\eta^{\mu c}\eta^{\lambda}_b\eta^{\nu b} - \eta^{\lambda a}\eta^{\nu c}\eta^{\rho}_b\eta^{\mu b} + \eta^{\lambda a}\eta^{\mu c}\eta^{\rho}_b\eta^{\nu b} \\
&\qquad -\eta^{\mu a}\eta^{\lambda c}\eta^{\nu}_b\eta^{\rho b} + \eta^{\mu a}\eta^{\rho c}\eta^{\nu}_b\eta^{\lambda b} + \eta^{\nu a}\eta^{\lambda c}\eta^{\mu}_b\eta^{\rho b} - \eta^{\nu a}\eta^{\rho c}\eta^{\mu}_b\eta^{\lambda b} \\
&= \eta^{\rho a}\eta^{\nu c}\eta^{\lambda\mu} - \eta^{\rho a}\eta^{\mu c}\eta^{\lambda\nu} - \eta^{\lambda a}\eta^{\nu c}\eta^{\rho\mu} + \eta^{\lambda a}\eta^{\mu c}\eta^{\rho\nu} \\
&\qquad -\eta^{\mu a}\eta^{\lambda c}\eta^{\nu\rho} + \eta^{\mu a}\eta^{\rho c}\eta^{\nu\lambda} + \eta^{\nu a}\eta^{\lambda c}\eta^{\mu\rho} - \eta^{\nu a}\eta^{\rho c}\eta^{\mu\lambda} \\
&= \eta^{\lambda\mu}(\eta^{\rho a}\eta^{\nu c} - \eta^{\nu a}\eta^{\rho c}) - \eta^{\lambda\nu}(\eta^{\rho a}\eta^{\mu c} - \eta^{\mu a}\eta^{\rho c}) \\
&\qquad -\eta^{\rho\mu}(\eta^{\lambda a}\eta^{\nu c} - \eta^{\nu a}\eta^{\lambda c}) + \eta^{\rho\nu}(\eta^{\lambda a}\eta^{\mu c} - \eta^{\mu a}\eta^{\lambda c}) \\
&= i\eta^{\lambda\mu}(J^{\rho\nu})^{ac} - i\eta^{\lambda\nu}(J^{\rho\mu})^{ac} - i\eta^{\rho\mu}(J^{\lambda\nu})^{ac} + i\eta^{\rho\nu}(J^{\lambda\mu})^{ac} \\
&= i(\eta^{\lambda\mu}J^{\rho\nu} - \eta^{\lambda\nu}J^{\rho\mu} - \eta^{\rho\mu}J^{\lambda\nu} + \eta^{\rho\nu}J^{\lambda\mu}).
\end{aligned} \tag{3.271}$$

We can easily relate this to our previous form of the commutation relations (3.225). For example,

$$\begin{aligned}[J_x, J_y] &= [J^{23}, J^{31}] \\ &= i(\eta^{12}J^{33} - \eta^{13}J^{32} - \eta^{32}J^{13} + \eta^{33}J^{12}) \\ &= iJ^{12} \\ &= iJ_z. \end{aligned} \tag{3.272}$$

You can go on to confirm the remaining relations in (3.225) and see that it is consistent with (3.271). We will take (3.271) to be the *defining* commutation relations of the Lorentz group.

We could move on at this point, but before we do we want to provide one final calculation that isn't necessarily vital to what is to come later, but should provide slightly more insight into what is going on with the algebra of the Lorentz group. In order to do so, we want to write the generators in a different way – namely, as differential operators. Recall that in Sect. 1.4 we emphasized several times that we can think of all of the Lorentz transformations as rotations. It is the minus sign in the Minkowski metric that makes the rotations involving time so radically different than the rotations involving only space. So let's take this idea of Lorentz transformations being rotations more seriously and see what results we can get.

In order to do this, we need to make a brief comment. Consider an arbitrary function $f(x)$. Let's take the partial derivative with respect to x to be a generator (with the usual $-i$ factor) and see what happens when we exponentiate it and act on $f(x)$. This gives

$$\begin{aligned} e^{ia(-i)\frac{\partial}{\partial x}} f(x) &= e^{a\frac{\partial}{\partial x}} f(x) \\ &= \left(1 + a\frac{\partial}{\partial x} + \frac{1}{2}(a)^2\frac{\partial^2}{\partial x^2} + \frac{1}{3!}(a)^3\frac{\partial^3}{\partial x^3} + \cdots\right) f(x) \\ &= f(x+a) \end{aligned} \tag{3.273}$$

So the differential operator acts as a generator of spatial translation.[24]

This indicates that we can take the generator of rotation to be the cross product[25]

$$\mathbf{L} = -i\mathbf{r} \times \frac{\partial}{\partial \mathbf{x}}. \tag{3.274}$$

[24]Incidentally, recall from Sect. 1.2 that spatial translation symmetry results in the conservation of momentum. Recall from an introductory quantum mechanics course that momentum is identified with the partial derivative operator – $p_i \longleftrightarrow i\frac{\partial}{\partial x_i}$. This is the reason. A symmetry gives rise to a conserved quantity, and that conserved quantity is the generator of the symmetry. We'll be discussing this in much greater detail later in this series. We just wanted to point it out for now.

[25]Again, notice that quantum mechanically this is $\mathbf{L} = \mathbf{r} \times \mathbf{p} \longleftrightarrow i\mathbf{r} \times \frac{\partial}{\partial \mathbf{x}}$, the standard expression for angular momentum.

Using a more concise notation,

$$L^i = -i\epsilon^{ijk} x^j \partial^k. \tag{3.275}$$

Adopting the antisymmetric index convention from above, we have

$$L^{ij} = -i(x^i \partial^j - x^j \partial^i). \tag{3.276}$$

Notice that this is antisymmetric, as was (3.270). We can generalize this to include boosts by simply adding a time index and assuming that the indices have a Minkowski signature,

$$L^{ij} \longrightarrow L^{\mu\nu} = -i(x^\mu \partial^\nu - x^\nu \partial^\mu).$$

So let's work out the commutation relations of these generators.

$$\begin{aligned}
[L^{\mu\nu}, L^{\rho\lambda}] &= -(x^\mu \partial^\nu - x^\nu \partial^\mu)(x^\rho \partial^\lambda - x^\lambda \partial^\rho) + (x^\rho \partial^\lambda - x^\lambda \partial^\rho)(x^\mu \partial^\nu - x^\nu \partial^\mu) \\
&= -x^\mu \partial^\nu x^\rho \partial^\lambda + x^\mu \partial^\nu x^\lambda \partial^\rho + x^\nu \partial^\mu x^\rho \partial^\lambda - x^\nu \partial^\mu x^\lambda \partial^\rho \\
&\quad + x^\rho \partial^\lambda x^\mu \partial^\nu - x^\rho \partial^\lambda x^\nu \partial^\mu - x^\lambda \partial^\rho x^\mu \partial^\nu + x^\lambda \partial^\rho x^\nu \partial^\mu \\
&= -x^\mu \eta^{\nu\rho} \partial^\lambda - x^\mu x^\rho \partial^\nu \partial^\lambda + x^\mu \eta^{\nu\lambda} \partial^\rho + x^\mu x^\lambda \partial^\nu \partial^\rho \\
&\quad + x^\nu \eta^{\mu\rho} \partial^\lambda + x^\nu x^\rho \partial^\mu \partial^\lambda - x^\nu \eta^{\mu\lambda} \partial^\rho - x^\nu x^\lambda \partial^\mu \partial^\rho \\
&\quad + x^\rho \eta^{\lambda\mu} \partial^\nu + x^\rho x^\mu \partial^\lambda \partial^\nu - x^\rho \eta^{\lambda\nu} \partial^\mu - x^\rho x^\nu \partial^\lambda \partial^\mu \\
&\quad - x^\lambda \eta^{\rho\mu} \partial^\nu - x^\lambda x^\mu \partial^\rho \partial^\nu + x^\lambda \eta^{\rho\nu} \partial^\mu + x^\lambda x^\nu \partial^\rho \partial^\mu \\
&= -x^\mu \eta^{\nu\rho} \partial^\lambda + x^\mu \eta^{\nu\lambda} \partial^\rho + x^\nu \eta^{\mu\rho} \partial^\lambda - x^\nu \eta^{\mu\lambda} \partial^\rho \\
&\quad + x^\rho \eta^{\lambda\mu} \partial^\nu - x^\rho \eta^{\lambda\nu} \partial^\mu - x^\lambda \eta^{\rho\mu} \partial^\nu + x^\lambda \eta^{\rho\nu} \partial^\mu \\
&= \eta^{\lambda\mu}(x^\rho \partial^\nu - x^\nu \partial^\rho) - \eta^{\lambda\nu}(x^\rho \partial^\mu - x^\mu \partial^\rho) \\
&\quad - \eta^{\rho\mu}(x^\lambda \partial^\nu - x^\nu \partial^\lambda) + \eta^{\rho\nu}(x^\lambda \partial^\mu - x^\mu \partial^\lambda) \\
&= i(\eta^{\lambda\mu} L^{\rho\nu} - \eta^{\lambda\nu} L^{\rho\mu} - \eta^{\rho\mu} L^{\lambda\nu} + \eta^{\rho\nu} L^{\lambda\mu}).
\end{aligned} \tag{3.277}$$

This is the same as (3.271). So, taking the structure of the Lorentz transforms as rotations seriously and writing the generators as differential operators gives us the correct form of the algebra. This gives us yet another way of understanding the structure of the Lorentz group.

3.3.5 Spinor Indices

We now come to a topic that students typically find very frustrating and tedious. In Sect. 3.3.3 we derived the transformation laws for spinors of both handedness.

But for the spinor representations we only looked at the matrix forms of the transformations; we didn't consider their indices and we didn't look at the spinors themselves. With the vector representation, however, we introduced indices – dotted and undotted indices. We'll now go back and talk about the indices on the spinors and spinor representations.

To recap, we took the right-handed indices to be dotted and the left-handed indices to be undotted. We saw both indices in the expression (3.261), which we rewrite as

$$v^{\dot{a}b} = \begin{pmatrix} t+z & x-iy \\ x+iy & t-z \end{pmatrix}. \tag{3.278}$$

Notice the following identity:

$$(v^{\dot{a}b})^T = v^{b\dot{a}} = \begin{pmatrix} t+z & x+iy \\ x-iy & t-z \end{pmatrix} = \begin{pmatrix} t+z & x-iy \\ x+iy & t-z \end{pmatrix}^{\star} = (v^{\dot{a}b})^{\star} \tag{3.279}$$

Because the indices in this context are arbitrary labels, we have shown that complex conjugation has the affect of swapping dotted and undotted indices.

It turns out that this is correct; dotted and undotted indices are interchanged by taking a complex conjugate. However, we know from equation (3.253) that this isn't all that's involved in going from a left-handed to a right-handed spinor; we also need the $i\sigma^2$ factor. To understand why, we need to look more closely at $i\sigma^2$. We'll write its indices as raised undotted (left-handed) indices for now,

$$i\sigma^2 = (i\sigma^2)^{ab} = \begin{pmatrix} (i\sigma^2)^{11} & (i\sigma^2)^{12} \\ (i\sigma^2)^{21} & (i\sigma^2)^{22} \end{pmatrix} = \begin{pmatrix} 0 & 1 \\ -1 & 0 \end{pmatrix}. \tag{3.280}$$

Let's act on this with the Lorentz transformation for a simple rotation around, say the x axis (3.235). We need to act on both of the $i\sigma^2$ indices so that the transformed matrix is

$$\begin{aligned}
(i\sigma^2)'^{ac} &= (R_x)^a_b (R_x)^c_d (i\sigma^2)^{bd} \\
&= \begin{pmatrix} \cos\frac{\theta}{2} & i\sin\frac{\theta}{2} \\ i\sin\frac{\theta}{2} & \cos\frac{\theta}{2} \end{pmatrix} \begin{pmatrix} 0 & 1 \\ -1 & 0 \end{pmatrix} \begin{pmatrix} \cos\frac{\theta}{2} & i\sin\frac{\theta}{2} \\ i\sin\frac{\theta}{2} & \cos\frac{\theta}{2} \end{pmatrix} \\
&= \begin{pmatrix} i\cos\frac{\theta}{2}\sin\frac{\theta}{2} - i\cos\frac{\theta}{2}\sin\frac{\theta}{2} & \cos^2\frac{\theta}{2} + \sin^2\frac{\theta}{2} \\ -\sin^2\frac{\theta}{2} - \cos^2\frac{\theta}{2} & i\cos\frac{\theta}{2}\sin\frac{\theta}{2} - i\cos\frac{\theta}{2}\sin\frac{\theta}{2} \end{pmatrix} \\
&= \begin{pmatrix} 0 & 1 \\ -1 & 0 \end{pmatrix} \\
&= (i\sigma^2)^{ac}.
\end{aligned} \tag{3.281}$$

Or we can consider a boost,[26]

$$
\begin{aligned}
(i\sigma^2)'^{ac} &= (B_x)^a_b (B_x)^c_d (i\sigma^2)^{bd} \\
&= \begin{pmatrix} \cosh\frac{\phi}{2} & -\sinh\frac{\phi}{2} \\ -\sinh\frac{\phi}{2} & \cosh\frac{\phi}{2} \end{pmatrix} \begin{pmatrix} 0 & 1 \\ -1 & 0 \end{pmatrix} \begin{pmatrix} \cosh\frac{\phi}{2} & -\sinh\frac{\phi}{2} \\ -\sinh\frac{\phi}{2} & \cosh\frac{\phi}{2} \end{pmatrix} \\
&= \begin{pmatrix} -\cosh\frac{\phi}{2}\sinh\frac{\phi}{2} + \cosh\frac{\phi}{2}\sinh\frac{\phi}{2} & \cosh^2\frac{\phi}{2} - \sinh^2\frac{\phi}{2} \\ \sinh^2\frac{\phi}{2} - \cosh^2\frac{\phi}{2} & -\sinh\frac{\phi}{2}\cosh\frac{\phi}{2} + \cosh\frac{\phi}{2}\sinh\frac{\phi}{2} \end{pmatrix} \\
&= \begin{pmatrix} 0 & 1 \\ -1 & 0 \end{pmatrix} \\
&= (i\sigma^2)^{ac}.
\end{aligned}
\tag{3.282}
$$

You can work out these details for the rest of the boosts and show that $(i\sigma^2)^{ab}$ is invariant under all of them.

So we have a symbol that is unchanged under Lorentz transformations in the spinor representation. Recall that this idea – of something that is unchanged under Lorentz transformations (in the vector representation) – was the original definition of the Lorentz group (cf. equation (1.72)). There, it was the metric $\eta^{\mu\nu}$ that was unchanged by Lorentz transformations. Here, we see that it is $(i\sigma^2)^{ab}$. For this reason it is often said that $i\sigma^2$ is the "spinor metric". Because the metric was an "invariant symbol" under spacetime vector Lorentz transformations, it could be used to raise and lower the spacetime vector indices that *did* transform under Lorentz transformations. The same is true here. Because $(i\sigma^2)$ is invariant under Lorentz transformations in the *spinor* representation, we can use $(i\sigma^2)$ to raise and lower *spinor* indices that transform under the Lorentz spinor transformations.

Some textbooks use the notation for the antisymmetric tensor ϵ^{ab} instead because

$$
(i\sigma^2)^{ab} = \begin{pmatrix} 0 & 1 \\ -1 & 0 \end{pmatrix} = \epsilon^{ab}. \tag{3.283}
$$

Like we pointed out in (3.256), there is a sign difference on $i\sigma^2$ in going from left to right versus right to left (which one we put the minus sign on is arbitrary; we're just choosing to use this convention). Also, because all of the elements are real, $i\sigma^2$ is unchanged by complex conjugation and therefore the dotted values are the same as the undotted values. So to summarize, we have (as our convention in this book)

$$
(i\sigma^2)^{ab} = (i\sigma^2)^{\dot{a}\dot{b}} = -(i\sigma^2)_{ab} = -(i\sigma^2)_{\dot{a}\dot{b}} = \begin{pmatrix} 0 & 1 \\ -1 & 0 \end{pmatrix}. \tag{3.284}
$$

---

[26] We'll do a left-handed boost here, although it doesn't matter in this case.

Using these matrices we can raise and lower spinor indices at will. If we follow (3.284) exactly and are careful with our indices, we don't need to worry about remembering whether we use $i\sigma^2$ or $-i\sigma^2$ when going from left to right or right to left. The convention defined here takes care of it for us.

So, there are two differences between right-handed spinors and left-handed spinors. First you take a complex conjugate to interchange dotted and undotted indices, then you multiply by $i\sigma^2$ (or $-i\sigma^2$ depending on which way we're going) to raise or lower the index. This means that we have *four* different types of indices: raised and lowered, and dotted and undotted. By convention we'll choose our left-handed spinor to have a *lowered, undotted* index. Thus,

$$\psi_L = \psi_a = \begin{pmatrix} \psi_{L1} \\ \psi_{L2} \end{pmatrix}.$$

Then we can write down all four index types from here using complex conjugation and (3.284).

$$\begin{aligned}
\psi_L &= \psi_a = \begin{pmatrix} \psi_{L1} \\ \psi_{L2} \end{pmatrix}, \\
\psi_L^\star &= \psi_{\dot{a}} = \begin{pmatrix} \psi_{L1}^\star \\ \psi_{L2}^\star \end{pmatrix}, \\
\psi_R &= \psi^{\dot{a}} = (i\sigma^2)^{\dot{a}\dot{b}} \psi_{\dot{b}} = \begin{pmatrix} 0 & 1 \\ -1 & 0 \end{pmatrix} \begin{pmatrix} \psi_{L1}^\star \\ \psi_{L2}^\star \end{pmatrix} = \begin{pmatrix} \psi_{2L}^\star \\ -\psi_{L1}^\star \end{pmatrix} = \begin{pmatrix} \psi_R^1 \\ \psi_R^2 \end{pmatrix}, \\
\psi_R^\star &= \psi^a = \begin{pmatrix} \psi_{L2} \\ -\psi_{L1} \end{pmatrix} = \begin{pmatrix} \psi_R^{1\star} \\ \psi_R^{2\star} \end{pmatrix}.
\end{aligned} \tag{3.285}$$

Or, writing this out as a diagram,

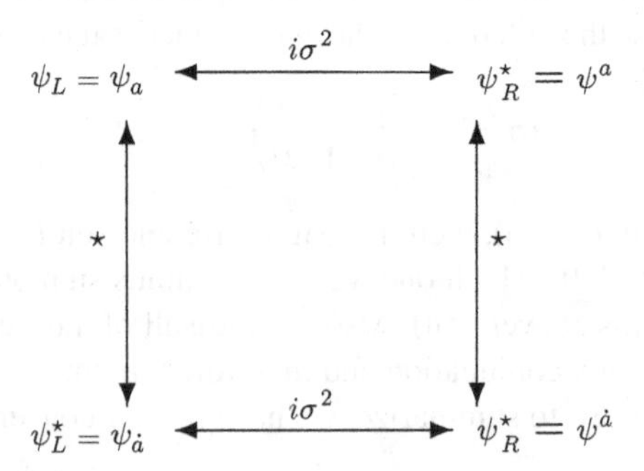

As indicated, the upper left and lower right are the left- and right-handed fields (respectively) which transform according to the rules we've outlined already (3.235)–(3.240) and (3.245)–(3.250). The other two corners are simply the complex conjugates of these states.

The details of this section are somewhat tedious, and students often have a difficult time keeping it all straight when it is first encountered. We will refer back to this section many times throughout the remainder of this book and we hope that its contents have been clear. The thing to take away is that there are two different types of spinors – left-handed and right-handed – and the difference is that while they transform the same under rotations, they transform differently under boosts. These details, including the exact ways they transform, are outlined in Sect. 3.3.3. We describe left-handed spinors (by convention) by lowered, undotted indices, and right-handed spinors (by convention) with raised, dotted spinors. We swap between dotted and undotted indices by complex conjugation and we raise and lower the spinor indices using the "spinor metric" $i\sigma^2$ according to the convention in (3.284).

Also, in Sect. 4.3.2 we'll discuss a geometrical interpretation of all of this, including not only the geometric meaning of the components of a spinor, but also the geometric meaning of the relationships between the four spinors on the above diagram.

3.3.6 Clifford Algebras

In the previous sections we discussed the vector representation of the Lorentz group in $4 = 1 + 3$ dimensions. It is fairly easy to derive this representation for any dimension by building the group elements and then using (3.86) to get the generators. Alternately, you can just use (3.270) and be careful with the signs from the metric. However, as we have said, we will find that the spinor representation also has an extremely important physical meaning and it will be useful to understand it in arbitrary dimensions.

However, the trick we did with the generators (3.229) only works in $1 + 3$ spacetime dimensions, so we can't blindly proceed as we did there. It turns out that there is an easy way to get the spinor representation in any dimension, and we'll discuss it later in this book. For now we simply introduce the underlying idea, restricting our discussion to the 4-dimensional case.

The framework we'll build in this section is, as with so many of the topics we're covering in this book, enormously deep. The underlying ideas largely have their origin in geometric algebra, which is fascinating, but the details aren't necessary to understand particle physics. You are encouraged to read up on it if you're interested, because it is a fascinating topic. It's just not necessary for the physical ideas we're considering.

Moving on – we have discussed the $(\frac{1}{2}, 0)$ and $(0, \frac{1}{2})$ spinor representations of the Lorentz group in some detail. We didn't discuss spinors (the object the spinor representations act on) very much, but we will later. What we want now is a more mathematically general way of looking at the spinor representations. We do so by using something called a **Clifford Algebra**. A Clifford algebra in n dimensions is a set of n matrices γ^μ for $i = 0, \ldots, n-1$ with components $(\gamma^\mu)^{ab}$ that satisfy

$$\{\gamma^\mu, \gamma^\nu\} = -2\eta^{\mu\nu}\mathbb{I}, \tag{3.286}$$

where $\{\gamma^\mu, \gamma^\nu\} = \gamma^\mu\gamma^\nu + \gamma^\nu\gamma^\mu$ is the anti-commutator of γ^μ and γ^ν.

So how does (3.286) relate to the Lorentz group? Consider the matrix defined by

$$S^{\mu\nu} = \frac{i}{4}[\gamma^\mu, \gamma^\nu]. \tag{3.287}$$

Obviously the indices of $S^{\mu\nu}$ are antisymmetric, and each $S^{\mu\nu}$ will be an $n \times n$ matrix with components $(S^{\mu\nu})^{ab}$.

We can rewrite $S^{\mu\nu}$ using (3.286). This gives

$$\begin{aligned}
S^{\mu\nu} &= \frac{i}{4}[\gamma^\mu, \gamma^\nu] \\
&= \frac{i}{4}(\gamma^\mu\gamma^\nu - \gamma^\nu\gamma^\mu) \\
&= \frac{i}{4}(\gamma^\mu\gamma^\nu - (-2\eta^{\mu\nu} - \gamma^\mu\gamma^\nu)) \\
&= \frac{i}{2}(\gamma^\mu\gamma^\nu + \eta^{\mu\nu}).
\end{aligned} \tag{3.288}$$

We want to look at the commutation of $S^{\mu\nu}$ with $S^{\rho\lambda}$, but it will be easier to first look at the following commutator:

$$\begin{aligned}
[S^{\mu\nu}, \gamma^\rho] &= \left[\frac{i}{4}[\gamma^\mu, \gamma^\nu], \gamma^\rho\right] \\
&= \left[\frac{i}{2}(\gamma^\mu\gamma^\nu + \eta^{\mu\nu}), \gamma^\rho\right] \\
&= \frac{i}{2}[\gamma^\mu\gamma^\nu, \gamma^\rho] \\
&= \frac{i}{2}(\gamma^\mu\gamma^\nu\gamma^\rho - \gamma^\rho\gamma^\mu\gamma^\nu) \\
&= \frac{i}{2}(\gamma^\mu\gamma^\nu\gamma^\rho + (\gamma^\mu\gamma^\rho\gamma^\nu - \gamma^\mu\gamma^\rho\gamma^\nu) \\
&\qquad - \gamma^\rho\gamma^\mu\gamma^\nu - (\gamma^\mu\gamma^\rho\gamma^\nu - \gamma^\mu\gamma^\rho\gamma^\nu)) \\
&= \frac{i}{2}(\gamma^\mu\{\gamma^\nu, \gamma^\rho\} - \gamma^\mu\gamma^\rho\gamma^\nu - \{\gamma^\rho, \gamma^\mu\}\gamma^\nu + \gamma^\mu\gamma^\rho\gamma^\nu) \\
&= \frac{i}{2}(-2\gamma^\mu\eta^{\nu\rho} + 2\gamma^\nu\eta^{\rho\mu}) \\
&= i(\gamma^\nu\eta^{\rho\mu} - \gamma^\mu\eta^{\nu\rho}).
\end{aligned} \tag{3.289}$$

We can now use this to determine the commutation relations among the $S^{\mu\nu}$'s.

$$\begin{aligned}
[S^{\mu\nu}, S^{\rho\lambda}] &= \left[S^{\mu\nu}, \frac{i}{4}[\gamma^\rho, \gamma^\lambda]\right] \\
&= \left[S^{\mu\nu}, \frac{i}{2}(\gamma^\rho\gamma^\lambda + \eta^{\rho\lambda})\right] \\
&= \frac{i}{2}[S^{\mu\nu}, \gamma^\rho\gamma^\lambda] \\
&= \frac{i}{2}(S^{\mu\nu}\gamma^\rho\gamma^\lambda - \gamma^\rho\gamma^\lambda S^{\mu\nu}) \\
&= \frac{i}{2}(S^{\mu\nu}\gamma^\rho\gamma^\lambda - \gamma^\rho S^{\mu\nu}\gamma^\lambda - \gamma^\rho\gamma^\lambda S^{\mu\nu} + \gamma^\rho S^{\mu\nu}\gamma^\lambda) \\
&= \frac{i}{2}([S^{\mu\nu}, \gamma^\rho]\gamma^\lambda + \gamma^\rho[S^{\mu\nu}, \gamma^\lambda]) \\
&= \frac{i}{2}\big(i(\gamma^\nu\eta^{\rho\mu} - \gamma^\mu\eta^{\nu\rho})\gamma^\lambda + i\gamma^\rho(\gamma^\nu\eta^{\lambda\nu} - \gamma^\mu\eta^{\nu\lambda})\big) \\
&= -\frac{1}{2}(\gamma^\nu\gamma^\lambda\eta^{\rho\mu} - \gamma^\mu\gamma^\lambda\eta^{\nu\rho} + \gamma^\rho\gamma^\nu\eta^{\lambda\nu} - \gamma^\rho\gamma^\nu\eta^{\nu\lambda}).
\end{aligned}$$

Then, inverting (3.288) to give $\gamma^\mu\gamma^\nu = -2iS^{\mu\nu} - \eta^{\mu\nu}$, we get

$$\begin{aligned}
[S^{\mu\nu}, S^{\rho\lambda}] &= -\frac{1}{2}\big((-2iS^{\nu\lambda} - \eta^{\nu\lambda})\eta^{\rho\mu} - (-2iS^{\mu\lambda} - \eta^{\nu\lambda})\eta^{\nu\rho} \\
&\quad + (-2iS^{\rho\nu} - \eta^{\rho\nu})\eta^{\lambda\mu} - (-2iS^{\rho\mu} - \eta^{\rho\mu})\eta^{\nu\lambda}\big) \\
&= i(\eta^{\lambda\mu}S^{\rho\nu} - \eta^{\nu\lambda}S^{\rho\mu} - \eta^{\rho\mu}S^{\lambda\nu} + \eta^{\nu\rho}S^{\lambda\mu}). \qquad (3.290)
\end{aligned}$$

Comparison with (3.271) shows that this is exactly the Lorentz algebra.

So, as long as the γ^μ matrices satisfy the Clifford algebra (3.286), we can build Lorentz generators $S^{\mu\nu}$ with (3.287). This reduces finding representations of the Lorentz group in any dimension to finding solutions of (3.286). We'll discuss how this comes up in physics later.

At this point, we've covered the "pure math" of the Lorentz group enough. We'll return to it when it comes up again in the context of physics.

3.4 References and Further Reading

The material in Sect. 3.1 came primarily from [27, 79]. The material in Sect. 3.2 came from [27, 29, 36]. The sections on $SU(2)$ also came from [80]. For further reading, we recommend [15, 23, 46, 47, 74].

Chapter 4
First Principles of Particle Physics and the Standard Model

4.1 Quantum Fields

Now that we've gone through quite a bit of mathematical formalism we're finally ready to start talking about physics. So that it's clear what we're doing, we'll state our goal up front: we want to formulate a *relativistic quantum mechanical* theory of *interactions*. If any of those three components (relativity, quantum, interaction) is missing we'll need to keep working. Our approach will be to start with what we know, see what is lacking, and then try to incorporate whatever is needed.

So, we'll start with the fundamental equation of quantum mechanics, Schroedinger's equation,[1]

$$H\Psi = i\hbar\frac{\partial\Psi}{\partial t}. \tag{4.1}$$

We know that for a non-interacting, non-relativistic particle, $H = \frac{\mathbf{p}^2}{2m} = -\frac{\hbar^2}{2m}\nabla^2$, so

$$-\frac{\hbar^2}{2m}\nabla^2\Psi = i\hbar\frac{\partial\Psi}{\partial t}. \tag{4.2}$$

For our purposes, we want to emphasize something that is usually not emphasized in introductory quantum courses – that Ψ is a **Scalar Field**. The physical meaning of this is that it only describes a particle with a single state. The way we interpret this is that it describes a particle with spin 0. Or in the language we learned about in the previous sections, it sits in a $j = 0$ representation of $SU(2)$, which is the trivial representation. A particle in a larger representation (i.e. a $j = \frac{1}{2}$

[1]If you're rusty on quantum mechanics at this point don't worry. We will talk more about what quantization is in Sect. 4.6. Now we're just providing motivation for the field equations we'll be using in this section.

M. Robinson, *Symmetry and the Standard Model: Mathematics and Particle Physics*,
DOI 10.1007/978-1-4419-8267-4_4,

representation) can be in multiple states (i.e. $+\frac{1}{2}$ or $-\frac{1}{2}$) and therefore requires multiple components. A scalar obviously can't have multiple states.

Furthermore, Ψ doesn't have any spacetime indices, so it also transforms trivially under the Lorentz group $SO(1, 3)$. However, despite this lack of difficulty with Lorentz transformations, we do have a fundamental problem in making this a truly relativistic theory. Namely, the spatial derivative in (4.2) acts quadratically (∇^2), whereas the time derivative is linear. Treating space and time differently in this way is unacceptable for a relativistic theory, because relativity requires that they be treated the same. The reason for this is that if space and time are to be seen simply as different components of a single geometry, we can no more treat them differently than the x and y dimensions can be treated differently in Euclidian space.

This is an indication of a much deeper problem with quantum mechanics: quantization involves "promoting" space from a parameter to an operator, whereas time is treated the same as in classical physics – as a parameter. This fundamental asymmetry is what ultimately prevents a straightforward generalization to a relativistic quantum theory. So to fix this problem, we have two choices: either promote time to an operator along with space, or demote space back to a parameter and quantize in a new way.

The first option would result in the Hermitian operators $\hat{X}, \hat{Y}, \hat{Z}$, and $\hat{T}$. It turns out that this approach is very difficult and less useful as far as building a relativistic quantum theory. So, physics has taken the second option.

In demoting position to a parameter along with time, we obviously have sacrificed the operators which we imposed commutation relations on to get a "quantum" theory in the first place. Because we obviously can't impose commutation relations on parameters (because they are scalars), quantization appears impossible. So, we are going to have to make a fairly radical reinterpretation.

Rather than letting the coordinates be Hermitian operators that act on the state in the Hilbert space representing a particle, *we now interpret the particle as the Hermitian operator*, and *this* operator (or particle) will be parameterized by the spacetime coordinates. The physical state that the particle operators act on is then the vacuum itself, $|0\rangle$. So, whereas before you acted on the "electron" $|\Psi\rangle$ with the operator $\hat{x}$, now the "electron" (parameterized by x) $\hat{\Psi}(x^\mu)$ acts on the vacuum $|0\rangle$, creating the state $\big(\hat{\Psi}(x^\mu)|0\rangle\big)$. In other words, the operator representing an electron excites the vacuum (empty space) resulting in an electron. We will see that all quantum fields contain appropriate raising and lowering operators to do just this. This approach, where the quantum mechanical entities are no longer the coordinates acting on the fields, but the fields themselves, is called **Quantum Field Theory** (QFT). Some books also refer to this process as **Second Quantization**.

Thus, whereas before, quantization was defined by imposing commutation relations on the coordinate operators $[x, p] \neq 0$, we now quantize by imposing commutation relations on the field operators, $[\Psi, \Pi] \neq 0$, where Π is the canonical momentum associated with the field Ψ (cf. equation (1.115)). Also, whereas before the non-zero relation $[x, p] \neq 0$ meant that we couldn't measure both the position and momentum of a particle simultaneously, now the non-zero relation $[\Psi, \Pi] \neq 0$ means that we can't measure both the field and its time rate of change simultaneously.

So, that's how quantization will work. We'll find that this allows us to get to a relativistic theory in a much nicer way. However, as with any theory, before we can quantize we need something *to* quantize. In other words, we need some equations of motion that govern the dynamics of these fields. Therefore, before coming back to the quantization process we'll spend the rest of this section coming up with the classical equations governing the fields we want to work with.

4.2 Spin-0 Fields

4.2.1 Equation of Motion for Scalar Fields

As we said above, Schroedinger's equation (4.2) describes the time evolution of a spin-0 field, or a scalar field. Generalizing to higher spins will come later. Now we see how to make this description relativistic.

The most obvious guess for a relativistic form is to simply plug in the standard relativistic Hamiltonian, which we derived in (1.98),

$$H = \sqrt{\mathbf{p}^2c^2 + m^2c^4}. \tag{4.3}$$

Note that, using the standard Taylor expansion $\sqrt{1+x^2} \approx 1 + \frac{1}{2}x$ for $x \ll 1$ gives $H \approx mc^2 + \frac{\mathbf{p}^2}{2m}$, for $\mathbf{p}^2 \ll c^2$, which is the standard non-relativistic form (plus a constant, the rest energy) we'd expect from a low speed limit. So, plugging (4.3) into (4.2), we have

$$i\hbar\frac{\partial\phi}{\partial t} = \sqrt{-\hbar^2c^2\nabla^2 + m^2c^4}\phi. \tag{4.4}$$

But there are two problems with this:

1. The space and time derivatives are still treated differently, so this is inadequate as a relativistic equation, and
2. Taylor expanding the square root will give an infinite number of derivatives acting on ϕ, making this theory non-local.

One solution is to square the operator on both sides, giving

$$-\hbar^2\frac{\partial^2\phi}{\partial t^2} = \left(-\hbar^2c^2\nabla^2 + m^2c^4\right)\phi,$$

$$\Rightarrow \left(-\partial^0\partial_0 + \nabla^2 - \frac{m^2c^2}{\hbar^2}\right)\phi = 0. \tag{4.5}$$

If we choose the so called "natural units" or "God units", where $c = \hbar = 1$, we have

$$(\partial^2 - m^2)\phi = 0 \tag{4.6}$$

(the expression ∂^2 here is assuming the "metric" form of the dot product $\partial^2 = \partial^\mu \partial_\mu$ with the Minkowski metric – cf. Sects. 1.4.1 and 1.4.2). Equation (4.6) is called the **Klein-Gordon** equation. It is nothing more than an operator version of the standard relativistic relation from (1.98),

$$E^2 = m^2c^4 + \mathbf{p}^2c^2. \tag{4.7}$$

Note that because we will be quantizing fields and not coordinates, there is absolutely nothing "quantum" about the Klein-Gordon equation. It is, at this point, merely a relativistic wave equation for a classical, spinless, non-interacting field.

Finally, we note one major problem with the Klein-Gordon equation. When we squared the Hamiltonian

$$H = \sqrt{m^2c^4 + \mathbf{p}^2c^2} \tag{4.8}$$

to get

$$H^2 = m^2c^4 + \mathbf{p}^2c^2, \tag{4.9}$$

the energy eigenvalues became

$$E = \pm\sqrt{m^2c^4 + \mathbf{p}^2c^2}. \tag{4.10}$$

It appears that we have a negative energy eigenvalue! Obviously this is unacceptable in a physically meaningful theory, because negative energy means that we don't have a true vacuum, and therefore a particle can cascade down forever, giving off an infinite amount of radiation. We will see that this problem plagues the spin-1/2 particles as well, so we wait to talk about the solution until then.

4.2.2 Lagrangian for Scalar Fields

We now want to find a Lagrangian for the scalar (spin $j = 0$) Klein-Gordon fields. We know that the terms in any Lagrangian are all Lorentz scalars (in order for the Lagrangian to be relativistically invariant). This doesn't really pose a problem for the case of scalar fields because they're scalars to begin with. The simplest scalar term (the mass term) will be proportional to ϕ^2.[2] Another scalar term (the kinetic energy term) is proportional to $\partial^\mu\phi\partial_\mu\phi$. It turns out that these are the only terms we need. The Lagrangian for a scalar Klein-Gordon field is

$$\mathcal{L}_{KG} = -\frac{1}{2}\partial^\mu\phi\partial_\mu\phi - \frac{1}{2}m^2\phi^2. \tag{4.11}$$

[2] A term proportional to ϕ would simply become a constant in the equation of motion, so we don't include such terms.

It is then straightforward to take the variation of this to show that it gives (4.6). Using (1.110) we have

$$\frac{\partial \mathcal{L}}{\partial \phi} = -\frac{1}{2}m^2\frac{\partial}{\partial \phi}\phi^2 = -m^2\phi,$$
$$\partial_\mu\left(\frac{\partial \mathcal{L}}{\partial(\partial_\mu \phi)}\right) = \partial_\mu\big(-\partial^\mu \phi\big) = -\partial^2\phi. \tag{4.12}$$

So, the equation of motion is

$$-\partial^2\phi + m^2\phi = 0 \Longrightarrow \left(\partial^2 - m^2\right)\phi = 0, \tag{4.13}$$

which is exactly (4.6), as we'd expect.

Before moving on, let's consider a slight modification of the scalar Lagrangian we just considered. We were actually considering *real* scalar fields. We could also consider *complex* scalar fields. In this case the Lagrangian is

$$\mathcal{L}_{KG} = -\partial^\mu\phi^\dagger\partial_\mu\phi - m^2\phi^\dagger\phi. \tag{4.14}$$

In the case of the real field, ϕ had only a single (real) degree of freedom. Now ϕ has *two* real degrees of freedom ($\phi = \phi_1 + i\phi_2$). Rather than treating ϕ_1 and ϕ_2 as the independent fields, we'll treat ϕ and $\phi^\dagger$ as the independent fields (this is simply a rotation/change of basis in the complex plane). So (4.14) describes two independent fields, ϕ and $\phi^\dagger$ (this is why we exclude the factors of $1/2$ – because there are two fields here). We can take the variation with respect to $\phi^\dagger$ to get

$$\frac{\partial \mathcal{L}}{\partial \phi^\dagger} = -m^2\phi,$$
$$\partial_\mu\left(\frac{\partial \mathcal{L}}{\partial(\partial_\mu \phi^\dagger)}\right) = -\partial_\mu(\partial^\mu\phi), \tag{4.15}$$

so the equation of motion is

$$(\partial^2 - m^2)\phi = 0. \tag{4.16}$$

Then we can take a variation with respect to ϕ and we find

$$\frac{\partial \mathcal{L}}{\partial \phi} = -m^2\phi^\dagger,$$
$$\partial_\mu\left(\frac{\partial \mathcal{L}}{\partial(\partial_\mu \phi)}\right) = -\partial_\mu(\partial^\mu\phi^\dagger), \tag{4.17}$$

so the equation of motion is

$$(\partial^2 - m^2)\phi^\dagger = 0. \tag{4.18}$$

Notice that the equations of motion for ϕ and $\phi^\dagger$ are both simply the Klein-Gordon equation (both fields satisfy the same equation of motion). Because scalar $j = 0$ fields have no indices and therefore transform trivially under any type of transformation, there isn't much more to say about this other than to just solve the Klein-Gordon equation.

4.2.3 Solutions to the Klein-Gordon Equation

In this section we'll focus on the real scalar field. The complex version is easy to work out. The Klein-Gordon equation (4.6),

$$\left(\partial^2 - m^2\right)\phi = 0, \tag{4.19}$$

can be solved by inspection. The solutions are simply plane waves subject to the usual relativistic constraint (4.7),[3]

$$\phi(x) = e^{\pm ip\cdot x} = e^{\pm ip_\mu x^\mu} = e^{\pm i(\mathbf{p}\cdot\mathbf{x}-Et)}, \tag{4.20}$$

where $\mathbf{p}$ is the momentum of the particle and E is the energy. To confirm that this is a solution we simply plug it in directly,

$$\begin{aligned}
\partial^2\phi &= \partial_\mu\partial^\mu\left(e^{\pm ip_\nu x^\nu}\right) \\
&= (\pm ip^\mu)\,\partial_\mu e^{\pm ip_\nu x^\nu} \\
&= (\pm ip^\mu)\left(\pm ip_\mu\right)e^{\pm ip_\nu x^\nu} \\
&= -p_\mu p^\mu e^{\pm ip\cdot x} \\
&= -\left(-E^2 + \mathbf{p}^2\right)e^{\pm ip\cdot x} \\
&= -\left(-m^2\right)e^{\pm ip\cdot x} \\
&= m^2 e^{\pm ip\cdot x}.
\end{aligned} \tag{4.21}$$

So,

$$\left(\partial^2 - m^2\right)\phi = \left(m^2 - m^2\right)\phi = 0. \tag{4.22}$$

We generally break the solutions up based on the sign of Et. The solutions with $+Et$ in the exponent are called **Negative Frequency Solutions**, and the solutions with $-Et$ in the exponent are called **Positive Frequency Solutions**. So, we have

$$\phi(x) = \phi_{pos}(x) + \phi_{neg}(x) = ae^{ip\cdot x} + be^{-ip\cdot x}. \tag{4.23}$$

where a and b are arbitrary constants of integration.

---

[3]The argument x by itself (with no indices) should be understood to mean all spacetime components (i.e. $x = x^\mu$). So $p \cdot x = p_\mu x^\mu$.

4.3 Spin-1/2 Fields

Spin-0 fields were, as we saw, extremely straightforward. Because they're scalar fields they transform trivially under everything, and therefore there isn't much to say. As we will see, however, spin-1/2 fields are quite different. There are an enormous number of very subtle details that make them very, very frustrating when they are first encountered.

One of the biggest difficulties is that the sheer number of subtle details makes gaining any holistic picture of what is going on a daunting task. So much time gets spent working out annoying calculations (that usually provide little to no intuition for what is going on) that it is easy to lose sight of what's actually happening. It's not necessarily the case that the material is particularly difficult; it's just a lot of detail to keep track of. Once you get all the details in your head you'll find that it's actually a very elegant framework.

We can't promise that our introduction in the following sections will be any different (though we hope it is). We'll try to walk through the main ideas in as logical a progression as we can. The first time you read through these sections, we encourage you not to get bogged down in the details. If it all makes perfect sense the first time you see it, great. Otherwise, just keep reading, trying to get the big picture. Then, after you've gone through the whole section (Sects. 4.3.1–4.3.13), if you feel like you missed something *then* go back and start thinking through the details. Furthermore, the last part (Sect. 4.3.13) is a summary of everything we'll talk about to help reinforce the big picture.

With that said, let's dive in.

4.3.1 A Brief Review of Spin

In the last section we discussed the physical equations of motion for particles with spin 0, or scalar particles. The next step is to discuss particles with spin. Before diving into the mathematical details of how spin shows up in quantum field theory, we'll review the idea in more basic terms. We'll also attempt to give some sort of geometric picture of what is going on with spin.

With classical particles there are two quantities contributing to angular momentum: the orbital angular momentum and the rotational angular momentum. The orbital angular momentum is the angular momentum of the object as it rotates around some fixed point. The earth's rotation *around* the sun contributes to its orbital angular momentum (and the four seasons). Rotational angular momentum is the momentum an object has as it rotates around its own center of mass. This is the rotation of the earth around its own axis (which causes us to have days and nights). The subtlety is that rotational angular momentum is the sum of the orbital angular momentum of the individual parts around the axis of rotation. In other words, a rock on the surface of the earth has orbital angular momentum as it spins around the earth's axis of rotation, and therefore contributes to the earth's rotational angular momentum.

With fundamental particles the concept of orbital angular momentum makes sense. Electrons can rotate[4] around nucleons. However, particles are generally taken to be fundamental and point-like,[5] meaning that they have no internal structure. So the concept of rotational angular momentum doesn't make sense as in the classical case.

However, nature didn't accept that answer. In the "early days" of quantum mechanics, it was noticed in various experiments[6] that the results were not consistent with particles being merely non-rotating point particles – the point particle part was okay, but the non-rotating part was not. The problem was that while the particles did indeed seem to be rotating, it wasn't in the $SO(3)$ sense we're used to thinking of rotation in. With classical rotation, the "spin" of a particle can be defined as a (typically normalized) vector and an angular velocity, or $\boldsymbol{\omega} = \omega\hat{\mathbf{n}}$. In other words, there were three values (each component of $\hat{\mathbf{n}}$ times ω) in the state, and it would make sense to ask what the rotational value is along each axis. For example, a particle with rotation $\boldsymbol{\omega} = \omega(1, 0, 0)^T$ would have rotation ω around the x axis and rotation 0 around the y and z axes.

With the quantum systems the situation seemed different. In any direction, there appeared to be exactly two states possible, which came to be called "up" and "down" and which seemed to be opposite of each other. Furthermore, as with all quantum systems, the general state of a particle should be described by complex numbers, not real numbers. So, what we need for a quantum system is something that describes rotation, has two components that are "opposite" of each other, and includes complex coefficients. At this point in our notes the choice is obvious: we should use $SU(2)$. Historically it was a more difficult conclusion. After all, using $SU(2)$ to describe a property of a particle was a large addition to the already disturbingly large list of quantum phenomena. This paradigm shift raised many questions – especially about what spin "is". It's obviously not a rotation through spacetime like the $SO(3)$ rotations that we're used to, but it's still a rotation. But then, if it's not rotating through spacetime, what space is it rotating through? The obvious answer is that it's rotating through "$SU(2)$" space, but that's merely a mathematical statement that doesn't really illuminate the physical meaning.

There were many attempts to "explain" spin in terms of more familiar concepts, but they seem to all ultimately fall short. It appears that spin is simply a fundamental fact of quantum mechanical life. Particles do actually rotate through an "internal" space that isn't spacetime, and the angular momentum it has as it rotates through "spinor space" (as it has come to be called) is described by $SU(2)$ in much the same way as $SO(3)$ describes classical rotation. The two initially striking differences are that $SU(2)$ spin is described by two complex components (not three real ones

[4]Of course the notion of an electron around a nucleus is modified in quantum mechanics, but its orbital angular momentum is well defined and can be easily calculated.

[5]Note that modifications to these assumptions will be considered in the next chapter.

[6]See almost any introductory text on quantum mechanics for a discussion of this history.

like $SO(3)$ spin), and that because there are exactly two states and it is the $j = \frac{1}{2}$ representation of $SU(2)$ that has two states, $SU(2)$ spin is quantized to be either $+\frac{1}{2}$ or $-\frac{1}{2}$ (cf. Sect. 3.2.9).[7]

At this point, most introductory quantum texts take the following approach. A physical state is described by the eigenvectors of the Hermitian operators that correspond to physically observable quantities. If spin is a physically measurable quantity and it is described by $SU(2)$, then the (Hermitian) generators of $SU(2)$ are those operators. We know that those operators are the Pauli matrices (3.144), so we pick one of them to get our eigenvectors. Of course different books choose different ways of saying all of that, with varying degrees of mathematical detail. We hope that after reading Sect. 3.2, especially Sect. 3.2.12, this has much deeper mathematical meaning. Quantum mechanical states are described by points in a vector (Hilbert) space, and to describe a point in a vector space we need a basis.

As we've learned about nature, the different physical states in this vector space are related to each other through some representation of some Lie group. Further, as we saw in Sect. 3.2 it is convenient to choose as our basis the eigenvectors of the Cartan generators, which is the z component Pauli matrix. So, our three generators are the Pauli matrices

$$J^x = \frac{1}{2}\sigma^x = \frac{1}{2}\begin{pmatrix} 0 & 1 \\ 1 & 0 \end{pmatrix}, \quad J^y = \frac{1}{2}\sigma^y = \frac{1}{2}\begin{pmatrix} 0 & -i \\ i & 0 \end{pmatrix}, \quad J^z = \frac{1}{2}\sigma^z = \frac{1}{2}\begin{pmatrix} 1 & 0 \\ 0 & -1 \end{pmatrix}. \tag{4.24}$$

As we know, the only Cartan generator is J^z, which has eigenvectors

$$|\uparrow\rangle_z = \begin{pmatrix} 1 \\ 0 \end{pmatrix}, \qquad |\downarrow\rangle_z = \begin{pmatrix} 0 \\ 1 \end{pmatrix}, \tag{4.25}$$

with eigenvalues

$$\begin{aligned} J^z|\uparrow\rangle_z &= +\frac{1}{2}|\uparrow\rangle_z, \\ J^z|\downarrow\rangle_z &= -\frac{1}{2}|\downarrow\rangle_z, \end{aligned} \tag{4.26}$$

where we have included the z subscript to remind ourselves that these are the eigenvectors of the z generator J^z.

Thus, any arbitrary physical state can be written in terms of $|\uparrow\rangle_z$ and $|\downarrow\rangle_z$ as a basis. Such a state may be written as

$$|\psi\rangle_z = \begin{pmatrix} \alpha \\ \beta \end{pmatrix} = \alpha|\uparrow\rangle_z + \beta|\downarrow\rangle_z, \tag{4.27}$$

[7] We're choosing units so things like c and $\hbar$ are 1.

where α and β are complex numbers subject to the constraint

$$|\alpha|^2 + |\beta|^2 = \alpha^\star \alpha + \beta^\star \beta = 1, \tag{4.28}$$

where the * denotes complex conjugate. The meaning of this is that if we measure the spin in the z direction we will find it "up" (having the $+\frac{1}{2}$ eigenvalue) with probability $|\alpha|^2$ and "down" (having the $-\frac{1}{2}$ eigenvalue) with probability $|\beta|^2$. If the state is $|\psi\rangle = |\uparrow\rangle_z$, measuring the z component will give $+\frac{1}{2}$ every time because that is a pure state.

But what happens if we measure the rotation in the x direction? We can determine this by finding the eigenvectors and eigenvalues of J^x. You can work out the details of finding them. They are

$$|\uparrow\rangle_x = \frac{1}{\sqrt{2}}\begin{pmatrix}1\\1\end{pmatrix}, \qquad |\downarrow\rangle_x = \frac{1}{\sqrt{2}}\begin{pmatrix}1\\-1\end{pmatrix}, \tag{4.29}$$

with eigenvalues

$$\begin{aligned} J^x|\uparrow\rangle_x &= +\frac{1}{2}|\uparrow\rangle_x, \\ J^x|\downarrow\rangle_x &= -\frac{1}{2}|\downarrow\rangle_x. \end{aligned} \tag{4.30}$$

We can write each of these in terms of the J^z eigenvectors,

$$\begin{aligned} |\uparrow\rangle_x &= \frac{1}{\sqrt{2}}(|\uparrow\rangle_z + |\downarrow\rangle_z), \\ |\downarrow\rangle_x &= \frac{1}{\sqrt{2}}(|\uparrow\rangle_z - |\downarrow\rangle_z). \end{aligned} \tag{4.31}$$

In other words, we can describe the same state in either of the following ways:

$$|\psi\rangle = \alpha|\uparrow\rangle_z + \beta|\downarrow\rangle_z = \left(\frac{\alpha+\beta}{\sqrt{2}}\right)|\uparrow\rangle_x + \left(\frac{\alpha-\beta}{\sqrt{2}}\right)|\downarrow\rangle_x. \tag{4.32}$$

Finally, we can do the same thing with J^y and we get

$$\begin{aligned} &|\uparrow\rangle_y = \frac{1}{\sqrt{2}}\begin{pmatrix}1\\i\end{pmatrix}, \qquad |\downarrow\rangle_y = \frac{1}{\sqrt{2}}\begin{pmatrix}1\\-i\end{pmatrix}, \\ &J^y|\uparrow\rangle_y = +\frac{1}{2}|\uparrow\rangle_y, \qquad J^y|\downarrow\rangle_y = -\frac{1}{2}|\downarrow\rangle_y. \end{aligned} \tag{4.33}$$

We can write these in terms of the z eigenvectors as well,

$$|\uparrow\rangle_y = \frac{1}{\sqrt{2}}(|\uparrow\rangle_z + i|\downarrow\rangle_z),$$
$$|\downarrow\rangle_y = \frac{1}{\sqrt{2}}(|\uparrow\rangle_z - i|\downarrow\rangle_z). \tag{4.34}$$

This gives us a third equivalent description of $|\psi\rangle$,

$$\begin{aligned}|\psi\rangle &= \alpha|\uparrow\rangle_z + \beta|\downarrow\rangle_z \\ &= \left(\frac{\alpha+\beta}{\sqrt{2}}\right)|\uparrow\rangle_x + \left(\frac{\alpha-\beta}{\sqrt{2}}\right)|\downarrow\rangle_x \\ &= \left(\frac{\alpha+i\beta}{\sqrt{2}}\right)|\uparrow\rangle_y + \left(\frac{\alpha-i\beta}{\sqrt{2}}\right)|\downarrow\rangle_y. \end{aligned} \tag{4.35}$$

So, for our arbitrary state $|\psi\rangle = \alpha|\uparrow\rangle_z + \beta|\downarrow\rangle_z$, if we measure the spin along the x axis we find it up with probability $|\frac{\alpha+\beta}{\sqrt{2}}|^2$ and down with probability $|\frac{\alpha-\beta}{\sqrt{2}}|^2$, or if we measure the spin along the y axis we find it up with probability $|\frac{\alpha+i\beta}{\sqrt{2}}|^2$ and down with probability $|\frac{\alpha-i\beta}{\sqrt{2}}|^2$.

Of course, the quantum mechanical peculiarity is made obvious by these probabilities. Consider a particle in the pure state $|\psi\rangle = |\uparrow\rangle_z$. If we measure the spin along the z direction we will get $+\frac{1}{2}$ every time. But even though it is in a pure z component spin state, the probability of finding $+\frac{1}{2}$ if we measure the x or y component spin is also $\frac{1}{2}$! Additionally, if it starts out in the pure z state and we measure its x component spin and get $+\frac{1}{2}$, the particle is now in a pure x state. If we measure the z component at this point we'll get $+\frac{1}{2}$ with probability $\frac{1}{2}$ and $-\frac{1}{2}$ with probability $\frac{1}{2}$!

Another peculiarity is the form of the state vector, or spinor. Students taking introductory quantum will often realize that the spinor relates to spin components in the usual x, y, and z directions and wonder about the exact relationship between "vectors" and "spinors" (the fact that spinors are vectors in the abstract sense, but not in the spacetime sense compounds this confusion). In the context of introductory quantum it's difficult to precisely explain this distinction. The relationship between them, however, is exactly what we discussed in Sect. 3.3.3, namely the way the objects transform. A "vector" in spacetime transforms under rotations according to (3.220), whereas a spinor transforms according to (3.235)–(3.237). It is precisely these transformation rules that differentiate vectors and spinors.

This difference in transformation rules can be seen very strikingly in the general form of $SO(3)$ vs. $SU(2)$ transformations. A generic $SO(3)$ transformation on a vector is given by (using ϕ, ψ, and θ as the x, y, and z angles respectively),

$$R_v = e^{i(\phi J^x + \psi J^y + \theta J^z)}. \tag{4.36}$$

On the other hand, a generic $SU(2)$ transformation is given by

$$R_s = e^{i\left(\phi \frac{\sigma^x}{2} + \psi \frac{\sigma^x}{2} + \theta \frac{\sigma^z}{2}\right)}. \tag{4.37}$$

So consider a rotation of 2π around the z axis. For $SO(3)$ this will be

$$R_v = e^{i2\pi J^z}. \tag{4.38}$$

In comparison, for $SU(2)$ this will be

$$R_s = e^{i\pi\sigma^z}. \tag{4.39}$$

Notice the factor of $1/2$ difference. It means that the $SU(2)$ spinor space rotates through only half of the angle that the $SO(3)$ vector space does. So, with a 2π rotation, a vector $\mathbf{v}$ goes to $\mathbf{v}$, and a spinor ψ goes to $-\psi$. So, both ψ and $-\psi$ correspond to $\mathbf{v}$. This isn't a problem because the only thing that is physically measurable is $|\psi|^2$ and the factor of -1 squares away. While this is mathematically straightforward, it isn't particularly easy to intuitively picture something that is facing the opposite direction when you rotate it 360°.

Students typically learn the mathematical tools necessary to do calculations regarding spin, but as is so often the case, being able to churn out calculations doesn't always provide intuition. Unfortunately, because spin is a purely quantum mechanical phenomenon there is no meaningful classical analogue. However, it would still be nice to have a better picture of how spinors (two component complex objects) relate to directions in space (three component real objects). We now try to provide this picture.

4.3.2 A Geometric Picture of Spin

In this section we'll try to provide a way of "picturing" spin. Specifically, we would like a way of picturing how a spinor relates to an axis of rotation in space, and how we get a minus sign when we do a full spatial rotation.

So we want to relate directions in space to spinors. We'll begin with directions in space. Because quantum mechanical spin is quantized to be either $\pm\frac{1}{2}$ and because the spinors end up normalized anyway, we don't need to bother with length in spacetime, just direction. A particular direction will be specified by a point on the unit sphere. It will be convenient to think of the unit sphere as a null cone of light rays emitted from the origin. In other words we're considering the set of points satisfying

$$-t^2 + x^2 + y^2 + z^2 = 0 \Longrightarrow x^2 + y^2 + z^2 = t^2. \tag{4.40}$$

This will be a sphere in space with radius t. We'll keep this in mind, but we can stick with the unit sphere and take $t = 1$ without loss of generality (for now).

Taking the usual spherical coordinates θ and ϕ, we will use

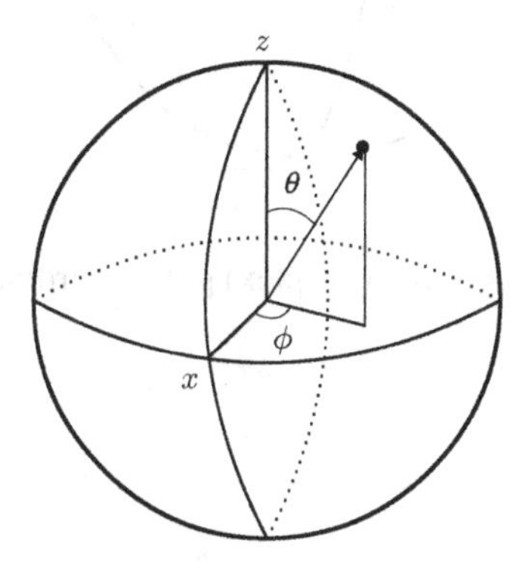

Next comes the first trick: imagine the sphere above superimposed with the complex plane through the $z = 0$ axis as shown below;

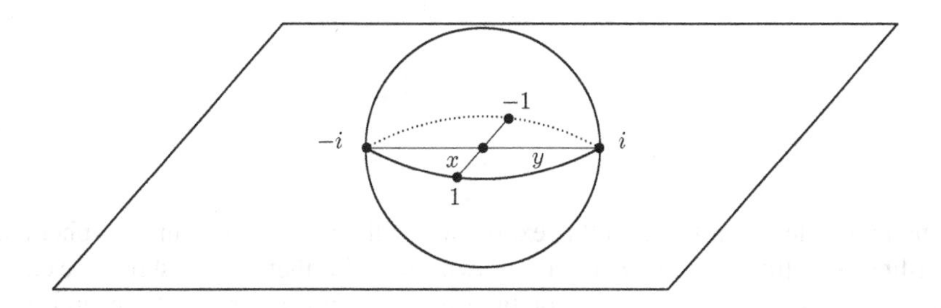

Hence, the positive x axis, or the axis corresponding to $\theta = \frac{\pi}{2}$ and $\phi = 0$ in the spherical coordinates, is aligned with 1, and the positive y axis, or the axis corresponding to $\theta = \frac{\pi}{2}$ and $\phi = \frac{\pi}{2}$, is aligned with i.

Next, we define a map called a **Stereographic Projection**. We start with the "North Pole" of the unit sphere – the point at $(x, y, z) = (0, 0, 1)$, or in spherical coordinates, the $\theta = 0$ point. As we said, a given spin direction is defined by a point on the unit sphere. So, pick any point on the sphere and connect the point at the north pole to this point, extending it on forever. This line will intersect the north pole (of course), the point on the sphere we're interested in, and the complex plane at some value $z = a + ib$, where a and b are real numbers. The picture here is

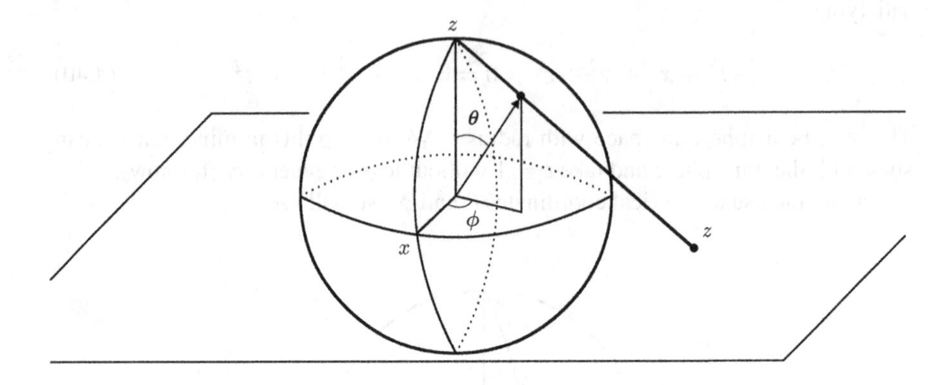

If we represent z in terms of its (complex) polar coordinates, it is

$$z = re^{i\phi}. \tag{4.41}$$

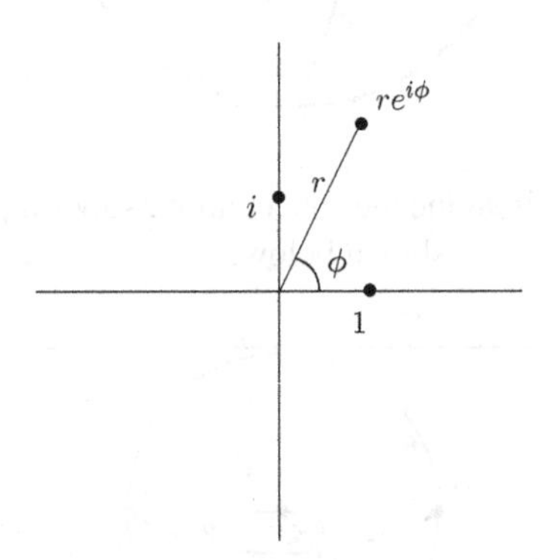

It should be clear that the ϕ in this expression will be the same ϕ in our spherical coordinates on the unit sphere in space drawn above. Furthermore, it should be clear that the radius should depend only on the value of θ. It is a very simple exercise to work out that r is

$$r = r(\theta) = \cot \frac{\theta}{2}. \tag{4.42}$$

Thus, the point (θ, ϕ) on the unit sphere in space is mapped to the point

$$z = e^{i\phi} \cot \frac{\theta}{2} \tag{4.43}$$

on the complex plane.

Now comes the next trick. We're going to take the complex number z and write it as the ratio of two other complex numbers α and β,

$$z = \frac{\alpha}{\beta}. \tag{4.44}$$

However, because we can take α and β to be literally any two complex numbers that have z as their ratio, there is a huge redundancy in what these can be. Therefore, we can make certain assumptions about them without losing generality. For example, we can assume that α is always real, putting the complex part of z entirely in β. We can also assume that α is always non-negative, putting any negative sign of z entirely on β. It then makes sense to write this as

$$z = e^{i\phi} \cot\frac{\theta}{2} = \frac{\cos\frac{\theta}{2}}{e^{-i\phi}\sin\frac{\theta}{2}} = \frac{\alpha}{\beta}, \tag{4.45}$$

from which we can read off the obvious choices

$$\begin{aligned} \alpha &= \cos\frac{\theta}{2}, \\ \beta &= e^{-i\phi}\sin\frac{\theta}{2}. \end{aligned} \tag{4.46}$$

Next comes the third trick. Just as a point on the unit sphere uniquely specifies a point z on the complex plane, the above mentioned constraints on α allows a complex point z to uniquely specify *two* complex points α and β. So, we define the map that takes

$$z \in \mathbb{C} \longrightarrow \begin{pmatrix} \alpha \\ \beta \end{pmatrix} \in \mathbb{C} \times \mathbb{C}. \tag{4.47}$$

In other words, we have the map

$$(\theta, \phi) \longrightarrow \psi_R^\star = \begin{pmatrix} \alpha \\ \beta \end{pmatrix} = \begin{pmatrix} \cos\frac{\theta}{2} \\ e^{-i\phi}\sin\frac{\theta}{2} \end{pmatrix}. \tag{4.48}$$

Note that we're calling this state $\psi_R^\star$. This is because (as we will soon show explicitly) this is actually the complex conjugate of a right-handed spinor in a state corresponding to the direction in spacetime as specified by the point on the unit sphere that produced this spinor.

By trusting us for now that this state is $\psi_R^\star = \psi^a$, we can easily turn it into a right-handed spinor by taking the complex conjugate. This gives

$$\psi_R = \psi^{\dot{a}} = \begin{pmatrix} \cos\frac{\theta}{2} \\ e^{i\phi}\sin\frac{\theta}{2} \end{pmatrix} = \begin{pmatrix} \alpha \\ \beta \end{pmatrix}, \tag{4.49}$$

where we have relabeled α and β as indicated (α didn't change and we've replaced β by its complex conjugate).

We can justify our calling ψ_R a spinor as follows. Let's consider a few points in space and see where they are mapped. First of all, let's look at the point corresponding to the positive z direction; this is given by $(\theta, \phi) = (0, 0)$, which yields

$$\psi_R = \begin{pmatrix} 1 \\ 0 \end{pmatrix}. \tag{4.50}$$

This is the spinor for spin "up" in the z direction, so it makes sense that the positive z direction maps to it.[8] Then the point corresponding to the negative z direction will be $(\theta, \phi) = (\pi, 0)$, which becomes

$$\psi_R = \begin{pmatrix} 0 \\ 1 \end{pmatrix}, \tag{4.51}$$

which is of course the spinor corresponding to spin down in the z direction. You can easily work out the details to show that the x and y directions give

$$\begin{aligned}
\text{positive } x \text{ direction} &\Leftrightarrow (\theta, \phi) = \left(\frac{\pi}{2}, 0\right) \longrightarrow \frac{1}{\sqrt{2}} \begin{pmatrix} 1 \\ 1 \end{pmatrix}, \\
\text{negative } x \text{ direction} &\Leftrightarrow (\theta, \phi) = \left(\frac{\pi}{2}, \pi\right) \longrightarrow \frac{1}{\sqrt{2}} \begin{pmatrix} 1 \\ -1 \end{pmatrix}, \\
\text{positive } y \text{ direction} &\Leftrightarrow (\theta, \phi) = \left(\frac{\pi}{2}, \frac{\pi}{2}\right) \longrightarrow \frac{1}{\sqrt{2}} \begin{pmatrix} 1 \\ i \end{pmatrix}, \\
\text{negative } y \text{ direction} &\Leftrightarrow (\theta, \phi) = \left(\frac{\pi}{2}, -\frac{\pi}{2}\right) \longrightarrow \frac{1}{\sqrt{2}} \begin{pmatrix} 1 \\ -i \end{pmatrix}.
\end{aligned} \tag{4.52}$$

You can glance back at Sect. 4.3.1 and see that this picture of spin – of deriving (4.49) through the stereographic projection with the Riemann sphere (knowing that we were actually getting $z^\star$ instead of z) – has given us an exact geometric visualization of spin along the appropriate axes.

To use more familiar notation, we typically write (4.49) as

$$\psi_R = \begin{pmatrix} \cos\frac{\theta}{2} \\ e^{i\phi} \sin\frac{\theta}{2} \end{pmatrix} = \cos\frac{\theta}{2} | \uparrow\rangle_z + e^{i\phi} \sin\frac{\theta}{2} | \downarrow\rangle_z. \tag{4.53}$$

This construction we've done, of mapping the unit sphere in space to the space of all spinors, is called the **Bloch Sphere** construction of spinor space. The

[8] We're using z here to refer to the third coordinate in three-dimensional Cartesian space, whereas we're also using z as a complex number. Sorry. The context should make clear which we mean.

stereographic projection is a nice geometrical way of deriving it, but in what sense is it right-handed (other than the fact that it came from the right-handed unit sphere)?

Because we have an explicit relationship between the components of this spinor and points in spacetime, we can invert the map (and swap from spherical to Cartesian coordinates) as follows to get[9]

$$\begin{aligned}
t &= \alpha\alpha^\star + \beta\beta^\star \\
&= \cos^2\frac{\theta}{2} + \sin^2\frac{\theta}{2} \\
&= 1, \\
x &= \alpha\beta^\star + \alpha^\star\beta \\
&= e^{-i\phi}\cos\frac{\theta}{2}\sin\frac{\theta}{2} + e^{i\phi}\cos\frac{\theta}{2}\sin\frac{\theta}{2} \\
&= \cos\phi\sin\theta \\
&= x, \\
y &= i(\alpha\beta^\star - \alpha^\star\beta) \\
&= i\left(e^{-i\phi}\cos\frac{\theta}{2}\sin\frac{\theta}{2} - e^{i\phi}\cos\frac{\theta}{2}\sin\frac{\theta}{2}\right) \\
&= \sin\phi\sin\theta \\
&= y, \\
z &= \alpha\alpha^\star - \beta\beta^\star \\
&= \cos^2\frac{\theta}{2} - \sin^2\frac{\theta}{2} \\
&= \cos\theta \\
&= z,
\end{aligned} \tag{4.54}$$

where we used the standard relationships between spherical coordinates on the unit sphere and Cartesian coordinates.

The expressions in (4.54) should look familiar; recall the relationship we found between spacetime and spinor space in Sect. 3.3.3 (for the $(1/2, 1/2)$ representation), namely in equations (3.257) and (3.262). Here we can form the matrix

$$A = \begin{pmatrix} \alpha\alpha^\star & \alpha^\star\beta \\ \alpha\beta^\star & \beta\beta^\star \end{pmatrix} = \begin{pmatrix} t+z & x-iy \\ x+iy & t-z \end{pmatrix} = x^\mu\sigma_\mu. \tag{4.55}$$

So the Pauli matrices have emerged from the stereographic projection.

---

[9] We're using a number of trig identities in these calculations. We leave it to you to look them up.

Next we can act on ψ_R with the (right-handed) Lorentz transformations and then use (4.54) to see what spacetime transformation this corresponds to. Starting with the spinor rotation (of angle γ) around the x axis we have (from (3.235))

$$\psi_R = \begin{pmatrix}\alpha\\ \beta\end{pmatrix} \longrightarrow \begin{pmatrix}\cos\frac{\gamma}{2} & i\sin\frac{\gamma}{2}\\ i\sin\frac{\gamma}{2} & \cos\frac{\gamma}{2}\end{pmatrix}\begin{pmatrix}\alpha\\ \beta\end{pmatrix} = \begin{pmatrix}\alpha\cos\frac{\gamma}{2} + i\beta\sin\frac{\gamma}{2}\\ i\alpha\sin\frac{\gamma}{2} + \beta\cos\frac{\gamma}{2}\end{pmatrix} = \begin{pmatrix}\alpha'\\ \beta'\end{pmatrix}. \quad (4.56)$$

We can now calculate the corresponding transformation in spacetime as follows.

$$\begin{aligned}
t' &= \alpha'\alpha'^{\star} + \beta'\beta'^{\star}\\
&= \left(\alpha\cos\frac{\gamma}{2} + i\beta\sin\frac{\gamma}{2}\right)\left(\alpha^{\star}\cos\frac{\gamma}{2} - i\beta^{\star}\sin\frac{\gamma}{2}\right)\\
&\quad + \left(i\alpha\sin\frac{\gamma}{2} + \beta\cos\frac{\gamma}{2}\right)\left(-i\alpha^{\star}\sin\frac{\gamma}{2} + \beta^{\star}\cos\frac{\gamma}{2}\right)\\
&= \alpha\alpha^{\star}\cos^2\frac{\gamma}{2} + \beta\beta^{\star}\sin^2\frac{\gamma}{2} + \alpha\alpha^{\star}\sin^2\frac{\gamma}{2} + \beta\beta^{\star}\cos^2\frac{\gamma}{2}\\
&= \alpha\alpha^{\star} + \beta\beta^{\star}\\
&= t,\\
x' &= \alpha'\beta'^{\star} + \alpha'^{\star}\beta'\\
&= \left(\alpha\cos\frac{\gamma}{2} + i\beta\sin\frac{\gamma}{2}\right)\left(-i\alpha^{\star}\sin\frac{\gamma}{2} + \beta^{\star}\cos\frac{\gamma}{2}\right)\\
&\quad + \left(\alpha^{\star}\cos\frac{\gamma}{2} - i\beta^{\star}\sin\frac{\gamma}{2}\right)\left(i\alpha\sin\frac{\gamma}{2} + \beta\cos\frac{\gamma}{2}\right)\\
&= (\alpha\beta^{\star} + \alpha^{\star}\beta)\cos^2\frac{\gamma}{2} + (\alpha\beta^{\star} + \alpha^{\star}\beta)\sin^2\frac{\gamma}{2}\\
&= x,\\
y' &= i\left(\alpha'\beta'^{\star} - \alpha'^{\star}\beta'\right)\\
&= i\left[\left(\alpha\cos\frac{\gamma}{2} + i\beta\sin\frac{\gamma}{2}\right)\left(-i\alpha^{\star}\sin\frac{\gamma}{2} + \beta^{\star}\cos\frac{\gamma}{2}\right)\right.\\
&\qquad \left. - \left(\alpha^{\star}\cos\frac{\gamma}{2} - i\beta^{\star}\sin\frac{\gamma}{2}\right)\left(i\alpha\sin\frac{\gamma}{2} + \beta\cos\frac{\gamma}{2}\right)\right]\\
&= i\left(\alpha\beta^{\star} - \alpha^{\star}\beta\right)\left(\cos^2\frac{\gamma}{2} - \sin^2\frac{\gamma}{2}\right) + (\alpha\alpha^{\star} - \beta\beta^{\star})\,2\sin\frac{\gamma}{2}\cos\frac{\gamma}{2}\\
&= y\cos\gamma + z\sin\gamma,\\
z' &= \alpha'\alpha'^{\star} - \beta'\beta'^{\star}\\
&= \left(\alpha\cos\frac{\gamma}{2} + i\beta\sin\frac{\gamma}{2}\right)\left(\alpha^{\star}\cos\frac{\gamma}{2} - i\beta^{\star}\sin\frac{\gamma}{2}\right)\\
&\quad - \left(i\alpha\sin\frac{\gamma}{2} + \beta\cos\frac{\gamma}{2}\right)\left(-i\alpha^{\star}\sin\frac{\gamma}{2} + \beta^{\star}\cos\frac{\gamma}{2}\right)\\
&= (\alpha\alpha^{\star} - \beta\beta^{\star})\left(\cos^2\frac{\gamma}{2} - \sin^2\frac{\gamma}{2}\right) - i\left(\alpha\beta^{\star} - \alpha^{\star}\beta\right)2\sin\frac{\gamma}{2}\cos\frac{\gamma}{2}\\
&= z\cos\gamma - y\sin\gamma.
\end{aligned} \quad (4.57)$$

Therefore, we have that the spinor transformation

$$R_x(\gamma) = \begin{pmatrix} \cos\frac{\gamma}{2} & i\sin\frac{\gamma}{2} \\ i\sin\frac{\gamma}{2} & \cos\frac{\gamma}{2} \end{pmatrix} \tag{4.58}$$

on a spinor state is equivalent to

$$R_x(\gamma) = \begin{pmatrix} 1 & 0 & 0 & 0 \\ 0 & 1 & 0 & 0 \\ 0 & 0 & \cos\gamma & \sin\gamma \\ 0 & 0 & -\sin\gamma & \cos\gamma \end{pmatrix} \tag{4.59}$$

on a spacetime vector, all through the stereographic projection.

We strongly encourage you to work out the other Lorentz transformations and confirm that they all produce the correct results. You will notice that if you try the *left* handed spinor boost transformations, you will get the wrong results for the spacetime spinor boost transformations. That is why we called the above spinor a *right* handed $\psi_R = \psi^{\dot{a}}$.

Now we want to understand *left* handed spinors. We're also ready to make sense of why we call then "right-handed" and "left-handed". Recall from Sect. 1.4.6 that transformations that change the handedness of a system are called parity transformations (cf. the matrix in equation (1.102)). Therefore, if we have a right-handed spinor $\psi_R = \psi^{\dot{a}}$ and we want a left-handed spinor, we should do a parity transformation.

Starting with the point in the complex plane that corresponded to ψ_R (the complex conjugate of (4.45)), a parity transformation will be given by

$$(x, y, z) \longrightarrow (-x, -y, -z), \tag{4.60}$$

or

$$\theta \longrightarrow \pi - \theta, \qquad \phi \longrightarrow \phi + \pi. \tag{4.61}$$

Applying these to the point on the complex plane

$$z = e^{-i\phi}\cot\frac{\theta}{2} \longrightarrow z' = e^{-i(\phi+\pi)}\cot\left(\frac{\pi-\theta}{2}\right) = -e^{-i\phi}\tan\frac{\theta}{2}. \tag{4.62}$$

Notice that this transformation can be written

$$z \longrightarrow -\frac{1}{z^\star} = -\frac{1}{e^{i\phi}\cot\frac{\theta}{2}} = -e^{-i\phi}\tan\frac{\theta}{2} = z'. \tag{4.63}$$

So, under the parity transformation $z \to -\frac{1}{z^\star}$, our spinors can now be written

$$-e^{-i\phi}\tan\frac{\theta}{2} = \frac{e^{-i\phi}\sin\frac{\theta}{2}}{-\cos\frac{\theta}{2}} = \frac{\alpha}{\beta}, \tag{4.64}$$

from which we write the new state

$$\psi_L = \begin{pmatrix} e^{-i\phi} \sin\frac{\theta}{2} \\ -\cos\frac{\theta}{2} \end{pmatrix} = \begin{pmatrix} \alpha \\ \beta \end{pmatrix} = e^{-i\phi} \sin\frac{\theta}{2} |\uparrow\rangle_z - \cos\frac{\theta}{2} |\downarrow\rangle_z. \tag{4.65}$$

Now (you are encouraged to work these out explicitly) we have

$$\begin{aligned} t &= \alpha\alpha^\star + \beta\beta^\star, \\ x &= -(\alpha\beta^\star + \alpha^\star\beta), \\ y &= i(\alpha^\star\beta - \alpha\beta^\star), \\ z &= -(\alpha\alpha^\star - \beta\beta^\star). \end{aligned} \tag{4.66}$$

Notice that these are the same as (4.54) except the x, y, and z parts have a minus sign; this is consistent with the fact that we have changed the handedness of the system (cf. (1.102)).

You can apply the *left-handed* spinor Lorentz transformations on (4.66) and you will find that the correct vector transformations are induced on the spacetime coordinates. The right-handed unit sphere gets mapped to (the complex conjugate of) a right-handed spinor, and the left-handed unit sphere (the right-handed sphere after a parity operation) gets mapped to (the complex conjugate of) a left-handed spinor.

Incidentally, let's take a closer look at the transformation (4.63),

$$z \longrightarrow -\frac{1}{z^\star}.$$

This transformation can be equivalently written

$$z = \frac{\alpha}{\beta} \longrightarrow -\frac{1}{z^\star} = -\frac{\beta^\star}{\alpha^\star}. \tag{4.67}$$

If we follow our usual procedure for creating the spinor state from this, we have that the new spinor is

$$-\frac{\beta^\star}{\alpha^\star} \longrightarrow \begin{pmatrix} \beta^\star \\ -\alpha^\star \end{pmatrix}. \tag{4.68}$$

Note that this is exactly what we have in (4.65). However, notice that this transformation can be written as

$$-\frac{\beta^\star}{\alpha^\star} = \begin{pmatrix} 0 & 1 \\ -1 & 0 \end{pmatrix} \begin{pmatrix} \alpha \\ \beta \end{pmatrix}^\star = i\sigma^2 \begin{pmatrix} \alpha \\ \beta \end{pmatrix}^\star. \tag{4.69}$$

Therefore, as promised, we've found a geometric derivation of the results of Sect. 3.3.5.

So we've managed to do what we set out to do: finding a geometric picture of spin. We've shown specifically how a direction in space relates to a direction in "spinor space". The fact that a full rotation in space results in an overall minus sign in spinor space comes from the fact that the two states $|\uparrow\rangle_z$ and $|\downarrow\rangle_z$, which are perpendicular in spinor space, correspond to the $+z$ and $-z$ directions in space (through the Bloch sphere), which are π radians apart.

Now, as we said above, spin is a purely quantum mechanical effect. We know experimentally that it only allows two values, but we saw that $SO(3)$ doesn't have the appropriate structure. However, the $j = \frac{1}{2}$ representation of $SU(2)$ does have the appropriate structure. So we use $SU(2)$. There initially seems to be a problem with this because $SU(2)$ is $2 \to 1$ with $SO(3)$, so rotations through space don't seem to line up with quantum mechanical spin rotations (both S and $-S$ correspond to the same R in $SO(3)$). However, this isn't actually a problem because, once again, we are in a quantum mechanical system where only $|\psi|^2$ is measurable, and the minus sign difference doesn't matter. So, using $SU(2)$ instead of $SO(3)$ gives us a physically meaningful and mathematically consistent *quantum mechanical* theory of spin. As a comment, the fact that there is a $2 \to 1$ relationship between $SU(2)$ and $SO(3)$ leads mathematicians to say that $SU(2)$ is the **Non-Trivial Double Cover** of $SO(3)$. It turns out that all of the rotation groups $SO(n)$ have such non-trivial double covers, called the **Spin Groups**, denoted $Spin(n)$. Don't worry about this for now (we won't need to talk about general spin groups in this books) – we're just mentioned it so you're aware of it. We'll talk about $Spin(n)$ much more in later volumes.

An important thing to understand is that "spin" is not a rotation through spacetime in any meaningful way. It is a rotation in another space called "spinor space", which is an *internal* degree of freedom. Like many things in quantum mechanics, spinor space is a mathematical structure. All we can say for certain is what we can measure or know ($|\psi|^2$), not what ψ "is".

As one final comment, it is often difficult to make intuitive sense of quantum mechanical spin when it is first encountered. Classical spin in spacetime is, of course, easy to visualize because we are familiar with things rotating through spacetime. Students are often left with the impression that quantum mechanical spin isn't actually a "rotation", but rather some mathematical device. However, we will see later in this series that quantum mechanical spin is very much a rotation through a space – just not spacetime. As we will see, a great, great deal of theoretical physics involves the assumption of various "geometries" that are associated with spacetime. They aren't spaces we can "see", but their effects are very real. Spin is one example of this. It is indeed believed that particles are genuinely *spinning*, but through a space that is "attached" to spacetime at every point. If this isn't clear now, that isn't a problem; we'll talk much more about it later. In fact, the idea of various geometries "attached" to spacetime and the physically observable effects they have on what we can see is in many ways the dominant theme of particle physics. We will actually spend the next several books in this series developing exactly that theme.

Now that we've discussed what spin "looks like" (as best as we can), we're ready to talk about how to incorporate it into quantum field theory.

4.3.3 *Spin-1/2 Fields*

As we've seen in the past two sections, any arbitrary spin state is described by two complex parameters (subject to the condition of being normalized). We'll change the notation we've been using so far and describe a particle of spin-1/2 with the two-component **Spinor**

$$\psi = \begin{pmatrix} \psi_1 \\ \psi_2 \end{pmatrix}, \tag{4.70}$$

where ψ_1 and ψ_2 are both in $\mathbb{C}$. We now want an equation to govern their behavior.

In what follows we'll take Dirac's approach to finding the equation of motion that governs ψ. While he wanted such an equation to stand on its own, he reasoned that it should also be consistent with, or somehow imply, the Klein-Gordon equation (which merely makes the theory relativistic). Furthermore he wanted it to contain only first-order derivatives. However, this is problematic. If we simply act on ψ with ∂_μ, we get

$$\partial_\mu \psi = \begin{pmatrix} \partial_\mu \psi_1 \\ \partial_\mu \psi_2 \end{pmatrix}, \tag{4.71}$$

which has an uncontracted spacetime index. This has to be contracted with something else that has a spacetime index. A first guess may be some vector v^μ, but the expression $v^\mu \partial_\mu \psi$ has a "preferred" direction in spacetime (the direction of v^μ) and is therefore not acceptable.

Instead of acting on ψ with $v^\mu \partial_\mu$, Dirac sought a differential operator in the form of a 2×2 matrix (to act on the 2 component spinor). In other words, he wanted to find an operator of the form

$$\gamma^\mu \partial_\mu = \gamma^0 \partial_0 + \gamma^1 \partial_1 + \gamma^2 \partial_2 + \gamma^3 \partial_3, \tag{4.72}$$

where the γ^μ's are 2×2 matrices. The equation of motion would then be

$$\gamma^\mu \partial_\mu \psi = -im\psi, \tag{4.73}$$

or writing it in a more standard way,

$$(i\gamma^\mu \partial_\mu - m)\psi = 0. \tag{4.74}$$

Dirac's reasoning was that given this equation, the Klein-Gordon equation should come about by acting twice with the "Dirac operator" $(i\gamma^\mu \partial_\mu = m)$. Writing this out gives

$$\begin{aligned}(i\gamma^\nu \partial_\nu - m)(i\gamma^\mu \partial_\mu - m)\psi = 0 &\Longrightarrow (-\gamma^\nu\gamma^\mu \partial_\nu \partial_\mu - 2mi\gamma^\mu \partial_\mu + m^2)\psi = 0,\\ &\Longrightarrow (-\gamma^\nu\gamma^\mu \partial_\nu \partial_\mu - 2m(m) + m^2)\psi = 0,\\ &\Longrightarrow (\gamma^\nu\gamma^\mu \partial_\nu \partial_\mu + m^2)\psi = 0. \end{aligned} \tag{4.75}$$

This will give the Klein-Gordon equation if the symmetric part of $\gamma^\mu\gamma^\nu$ satisfies

$$\gamma^\mu\gamma^\nu = -\eta^{\mu\nu}\mathbb{I}. \tag{4.76}$$

Using the symmetry of the sum in (4.75) to ensure that we're working with the symmetric part, we require

$$\frac{1}{2}(\gamma^\mu\gamma^\nu + \gamma^\nu\gamma^\mu) = -\eta^{\mu\nu}\mathbb{I}. \tag{4.77}$$

This is great news! You can recognize this as the Clifford algebra from equation (3.286), which we already know a little about. Specifically, we showed how to get the Lorentz generators from solutions to (3.286). It is encouraging that in trying to find a relativistically acceptable theory of spinors we've come up with the Clifford algebra – something we know to be related to the Lorentz group.

Let's see how this equation implies the Klein-Gordon equation. Consider

$$\{\gamma^\mu, \gamma^\nu\} = \gamma^\mu\gamma^\nu + \gamma^\nu\gamma^\mu = -2\eta^{\mu\nu}\mathbb{I}. \tag{4.78}$$

If the γ^μ matrices satisfy (4.78), then (4.75) gives

$$(\gamma^\mu\gamma^\nu \partial_\mu \partial_\nu + m^2)\psi = 0 \Rightarrow (-\eta^{\mu\nu}\partial_\mu\partial_\nu + m^2)\psi = 0 \Rightarrow (\partial^2 - m^2)\psi = 0,$$

which is exactly the Klein-Gordon equation (4.6). Thus, we have that the Dirac equation is the "square root" of the Klein-Gordon equation,

$$\left(i\gamma^\mu \partial_\mu - m\right)^2 = (\partial^2 - m^2), \tag{4.79}$$

whenever the Clifford algebra is satisfied.

Armed with the knowledge that as long as we choose γ^μ matrices so that (4.78) holds we have a relativistically consistent theory, we can start with the Dirac equation

$$(i\gamma^\mu \partial_\mu - m)\psi = 0, \tag{4.80}$$

and try to find the proper γ^μ matrices.

To begin we want to find a solution to (4.78). A few moments thought should make it clear that there are no 2×2 or 3×3 matrix solutions.[10] The smallest possible representation consists of 4×4 matrices. If you've been reading carefully, this should trouble you for two reasons. First of all, because we're trying to write down an equation for a spin-$1/2$ particle, which has exactly *two* states, not four. Second, we know from Sect. 3.3.3 that there are *two* different 2×2 representations of the Lorentz group (the $(\frac{1}{2},0)$ and $(0,\frac{1}{2})$ representation). Why would the Clifford algebra (which we know to give rise to the Lorentz group) seem to rule these out?

For now we'll put these questions aside and accept that ψ must actually have *four* components, not *two*. We'll work with whatever solutions to (4.78) we can find and press on. The (extremely elegant) physical meaning will come shortly.

4.3.4 Solutions to the Clifford Algebra

We introduced the Clifford algebra in Sect. 3.3.6, but we said nothing about how to solve it. We're now ready to do that. It turns out that there is a nice "trick" we can use to find solutions to (4.78). Consider the following 2×2 matrices,

$$A = \begin{pmatrix} a_{11} & a_{12} \\ a_{21} & a_{22} \end{pmatrix}, \qquad B = \begin{pmatrix} b_{11} & b_{12} \\ b_{21} & b_{22} \end{pmatrix}. \tag{4.81}$$

Then, consider the following expression for a type of "product" between them,

$$\begin{aligned} A \otimes B &= \begin{pmatrix} Ab_{11} & Ab_{12} \\ Ab_{21} & Ab_{22} \end{pmatrix} \\ &= \begin{pmatrix} \begin{pmatrix} a_{11} & a_{12} \\ a_{21} & a_{22} \end{pmatrix} b_{11} & \begin{pmatrix} a_{11} & a_{12} \\ a_{21} & a_{22} \end{pmatrix} b_{12} \\ \begin{pmatrix} a_{11} & a_{12} \\ a_{21} & a_{22} \end{pmatrix} b_{21} & \begin{pmatrix} a_{11} & a_{12} \\ a_{21} & a_{22} \end{pmatrix} b_{22} \end{pmatrix} \\ &= \begin{pmatrix} a_{11}b_{11} & a_{12}b_{11} & a_{11}b_{12} & a_{12}b_{12} \\ a_{21}b_{11} & a_{22}b_{11} & a_{21}b_{12} & a_{22}b_{12} \\ a_{11}b_{21} & a_{12}b_{21} & a_{11}b_{22} & a_{12}b_{22} \\ a_{21}b_{21} & a_{22}b_{21} & a_{21}b_{22} & a_{22}b_{22} \end{pmatrix}. \end{aligned} \tag{4.82}$$

[10] We encourage you to spend a few minutes trying to find such a solution if this isn't clear. A brief attempt should make it clear that it can't be done.

This notation is convenient because it allows us to write the following identity:

$$(A \otimes B)(C \otimes D) = (AC \otimes BD), \tag{4.83}$$

where AC and BD are simply the usual 2×2 matrix products (working out the details to prove this is extremely easy, but we encourage you to do it if you're not convinced).

So how is this useful? We want to satisfy (4.78). Hence, the γ^μ matrices should satisfy

$$\begin{aligned}(\gamma^0)^2 &= (-2)\eta^{00} = 2,\\ (\gamma^i)^2 &= (-2)\eta^{ii} = -2,\end{aligned} \tag{4.84}$$

while otherwise they should anticommute. As a way of building this, consider the matrices σ^μ used in equation (3.261) (the Pauli matrices plus the identity matrix). We can work out the anticommutation relations for the Pauli matrices easily,

$$\{\sigma^i, \sigma^j\} = 2\delta^{ij}, \tag{4.85}$$

and

$$\{\sigma^0, \sigma^\mu\} = 2\sigma^\mu. \tag{4.86}$$

Now, let's consider the 4×4 matrices $(\sigma^\mu \otimes \sigma^\nu)$. We can generate various products of this form and easily calculate their anticommutation relations using (4.83). Consider for example the set

$$\begin{aligned}\gamma^0 &= (\sigma^0 \otimes \sigma^1),\\ \gamma^i &= i(\sigma^i \otimes \sigma^2).\end{aligned} \tag{4.87}$$

The anticommutation relations are

$$\begin{aligned}\{\gamma^0, \gamma^0\} &= 2(\sigma^0 \otimes \sigma^1)(\sigma^0 \otimes \sigma^1)\\ &= 2(\sigma^0\sigma^0 \otimes \sigma^1\sigma^1)\\ &= 2,\\ \{\gamma^i, \gamma^j\} &= i^2(\sigma^i \otimes \sigma^2)(\sigma^j \otimes \sigma^2) + i^2(\sigma^j \otimes \sigma^2)(\sigma^i \otimes \sigma^2)\\ &= -(\sigma^i\sigma^j \otimes \sigma^2\sigma^2) - (\sigma^j\sigma^i \otimes \sigma^2\sigma^2)\\ &= -(\{\sigma^i, \sigma^j\} \otimes 1)\\ &= -2\delta^{ij},\end{aligned}$$

$$
\begin{aligned}
\{\gamma^0, \gamma^i\} &= i(\sigma^0 \otimes \sigma^1)(\sigma^i \otimes \sigma^2) + i(\sigma^i \otimes \sigma^2)(\sigma^0 \otimes \sigma^1) \\
&= i(\sigma^i \otimes \sigma^1\sigma^2) + i(\sigma^i \otimes \sigma^2\sigma^1) \\
&= i(\sigma^i \otimes \{\sigma^1, \sigma^2\}) \\
&= 0, \qquad (4.88)
\end{aligned}
$$

which is exactly what we want.

The expressions (4.87) are a representation of the Clifford algebra. Writing them out explicitly we have

$$
\begin{aligned}
\gamma^0 &= \begin{pmatrix} 0 & \sigma^0 \\ \sigma^0 & 0 \end{pmatrix} = \begin{pmatrix} 0 & 1 \\ 1 & 0 \end{pmatrix}, \\
\gamma^i &= i \begin{pmatrix} 0 & -i\sigma^i \\ i\sigma^i & 0 \end{pmatrix} = \begin{pmatrix} 0 & \sigma^i \\ -\sigma^i & 0 \end{pmatrix}. \qquad (4.89)
\end{aligned}
$$

For reasons that will be clear later, this is called the **Chiral Representation** of the Clifford algebra. It is sometimes also called the **Weyl Representation**.

There are of course many other solutions to the Clifford algebra, but it turns out that *any* two solutions will be related by a similarity transformation. Therefore, up to similarity transformation, there is only one solution. Because the similarity transformation won't change the fact that the γ^μ's solve (4.78), they are all perfectly good representations. They are just different bases to represent the same things. Furthermore, because physics shouldn't depend on a basis, any of them are as good as any other. A few moments studying the form of the γ^μ matrices defined in (4.89) should make it fairly easy to write down many more representations (once you get the hang of what can go where in the $(\sigma^\mu \otimes \sigma^\nu)$ product).

It may of course be the case, however, that a particular basis makes calculations easier. This is indeed the case with the γ^μ matrices. There are three bases that we'll make use of. The first is the chiral representation we've already seen. Another representation that will be useful is the **Dirac Representation**, where the γ^i matrices are defined the same as in the chiral representation, but γ^0 changes. The Dirac representation is

$$
\begin{aligned}
\gamma^0 &= (\sigma^0 \otimes \sigma^3) = \begin{pmatrix} 1 & 0 \\ 0 & -1 \end{pmatrix}, \\
\gamma^i &= i(\sigma^i \otimes \sigma^2) = \begin{pmatrix} 0 & \sigma^i \\ -\sigma^i & 0 \end{pmatrix}. \qquad (4.90)
\end{aligned}
$$

You can work out the details yourself to show that this does indeed satisfy (4.78).

Finally, there is the **Majorana Representation**, which is defined as

$$\gamma^0 = (\sigma^2 \otimes \sigma^1) = \begin{pmatrix} 0 & \sigma^2 \\ \sigma^2 & 0 \end{pmatrix},$$

$$\gamma^1 = i(\sigma^3 \otimes \sigma^0) = \begin{pmatrix} i\sigma^3 & 0 \\ 0 & i\sigma^3 \end{pmatrix},$$

$$\gamma^2 = -i(\sigma^2 \otimes \sigma^2) = \begin{pmatrix} 0 & -\sigma^2 \\ \sigma^2 & 0 \end{pmatrix},$$

$$\gamma^3 = -i(\sigma^1 \otimes \sigma^0) = \begin{pmatrix} -i\sigma^1 & 0 \\ 0 & -i\sigma^1 \end{pmatrix}. \tag{4.91}$$

You can work out these details as well to show that this satisfies (4.78). The interesting thing about the Majorana representation is that every nonzero element is imaginary. We'll discuss the "meaning" of each of these representations later, as well as what kind of physical information they each give.

We've made progress, but we still have quite a bit of work to do. We have an equation that we assume governs the behavior of a spin $j = 1/2$ particle, the Dirac equation (4.80). However, when we require that it "implies" the Klein-Gordon equation (or in other words, when we require that it satisfy special relativity), we see the Clifford algebra emerge. While on one hand this is encouraging because we know the Clifford algebra is related to the Lorentz group, it is also troubling because it doesn't allow our particles to have the number of spin states we know they have. Two extra states seem to have been forced into our theory.

As we said, the physical meaning of all of this will be made clear shortly. It turns out that we're seeing a very deep and very profound aspect of spinor fields, and in a few pages these peculiarities will actually make perfect sense (hopefully). But before we get there, let's press on developing a theory of the *four* component ψ. The next step is to find the Lagrangian.

4.3.5 The Action for a Spin-1/2 Field

We know the Dirac equation (4.80). We could work backwards from it and come up with a Lagrangian that produces it. But doing so could obscure some of the details of how spinors work. We'll therefore seek to build a Dirac field Lagrangian "from scratch" and then check to make sure that we can recover the Dirac equation.

So we need to build a Lagrangian. Once again, we know that the terms in a Lagrangian are all Lorentz scalars. Typically this means terms with no spacetime indices or terms with a spacetime index contracted with the differential operator ∂_μ. This was easy for the Klein-Gordon field ϕ because it was a scalar to begin with. Now, however, we need to try to find a pure scalar from ψ, which has four components and therefore a spinor index (that we haven't been writing explicitly).

We'll begin by trying something that seems like it should work (but won't). We want to build a scalar from ψ. We know that ψ is a four component object with complex entries, which we'll write in terms of *two* 2 component quantities ψ_1 and ψ_2 (in other words ψ_1 has two complex components and ψ_2 has two complex components). So it seems natural to guess that a scalar can be formed from the Hermitian conjugate

$$\psi^\dagger = \begin{pmatrix} \psi_1 \\ \psi_2 \end{pmatrix}^{\star T} = \left(\psi_1^{\star T} \;\; \psi_2^{\star T} \right). \tag{4.92}$$

The scalar would then be $\psi^\dagger \psi$.

Let's look at how $\psi^\dagger \psi$ transforms. We'll write a generic Lorentz transformation in the spinor representation as $\mathcal{S}$, and the generators as S. So we write this transformation as

$$\psi^\dagger \psi \longrightarrow \psi'^\dagger \psi' = \psi^\dagger \mathcal{S}^\dagger \mathcal{S} \psi, \tag{4.93}$$

This transformation will indicate that $\psi^\dagger \psi$ is indeed a Lorentz scalar as long as $\mathcal{S}^\dagger \mathcal{S} = 1$. But we know exactly what this representation looks like from (3.287); the generators are given by

$$S^{\mu\nu} = \frac{i}{4}[\gamma^\mu, \gamma^\nu] \tag{4.94}$$

So a general finite transformation will be

$$\mathcal{S} = e^{\frac{i}{2}\omega_{\mu\nu} S^{\mu\nu}}, \tag{4.95}$$

where $\omega_{\mu\nu}$ is an antisymmetric real matrix set of parameters (angles of rotation and boost parameters) specifying *which* Lorentz transformation we are doing. For example $\omega_{01} = -\omega_{10}$ is ϕ_x, the boost parameter in the x direction, $\omega_{31} = -\omega_{13}$ is θ_y, the angle of rotation mixing x and z, etc. Because the sum will go over both $\omega_{\mu\nu}$ and $\omega_{\nu\mu}$, the factor of $1/2$ is included to keep from double counting. Writing this out fully, we have

$$\mathcal{S} = e^{\frac{i}{2}\omega_{\mu\nu} S^{\mu\nu}} = e^{-\frac{1}{8}\omega_{\mu\nu}[\gamma^\mu, \gamma^\nu]}. \tag{4.96}$$

Taking the Hermitian conjugate of the exponent will give

$$\begin{aligned} \left(\frac{i}{2}\omega_{\mu\nu} S^{\mu\nu} \right)^\dagger &= \left(-\frac{1}{8}\omega_{\mu\nu}[\gamma^\mu, \gamma^\nu] \right)^\dagger \\ &= \left(-\frac{1}{8}\omega_{\mu\nu}(\gamma^\mu\gamma^\nu - \gamma^\nu\gamma^\mu) \right)^\dagger \end{aligned}$$

$$
\begin{aligned}
&= -\frac{1}{8}\omega^\dagger_{\mu\nu}(\gamma^{\nu\dagger}\gamma^{\mu\dagger} - \gamma^{\mu\dagger}\gamma^{\nu\dagger}) \\
&= -\frac{1}{8}\omega_{\mu\nu}(\gamma^{\mu\dagger}\gamma^{\nu\dagger} - \gamma^{\nu\dagger}\gamma^{\mu\dagger}) \\
&= -\frac{1}{8}\omega_{\mu\nu}[\gamma^{\mu\dagger}, \gamma^{\nu\dagger}] \\
&= -\frac{i}{2}\omega_{\mu\nu}\left(\frac{-i}{4}[\gamma^{\mu\dagger}, \gamma^{\nu\dagger}]\right) \\
&= -\frac{i}{2}\omega_{\mu\nu}S^{\mu\nu\dagger}, \qquad (4.97)
\end{aligned}
$$

where

$$
S^{\mu\nu\dagger} = -\frac{i}{4}[\gamma^{\mu\dagger}, \gamma^{\nu\dagger}]. \qquad (4.98)
$$

We have used the fact that $\omega^\dagger_{\mu\nu} = -\omega_{\mu\nu}$ (because $\omega_{\mu\nu}$ is real and antisymmetric). Now, if all of the γ^μ matrices are either Hermitian ($\gamma^{\mu\dagger} = \gamma^\mu$) or all of them are anti-Hermitian ($\gamma^{\mu\dagger} = -\gamma^\mu$), then $S^{\mu\nu}$ is Hermitian ($S^{\mu\nu\dagger} = S^{\mu\nu}$) and we have

$$
\begin{aligned}
\mathcal{S}^\dagger\mathcal{S} &= e^{-\frac{i}{2}\omega_{\mu\nu}S^{\mu\nu\dagger}} e^{\frac{i}{2}\omega_{\mu\nu}S^{\mu\nu}} \\
&= e^{-\frac{i}{2}\omega_{\mu\nu}S^{\mu\nu}} e^{\frac{i}{2}\omega_{\mu\nu}S^{\mu\nu}} \\
&= 1, \qquad (4.99)
\end{aligned}
$$

as expected to make (4.93) a scalar transformation. However, this isn't the case; for example, look at the definitions of the γ^μ matrices in (4.89). The γ^0 matrix is Hermitian, but because the Pauli matrices are all Hermitian, $\gamma^{i\dagger} = -\gamma^i$. In other words, we have

$$
\begin{aligned}
\gamma^{0\dagger} &= \gamma^0, \\
\gamma^{i\dagger} &= -\gamma^i. \qquad (4.100)
\end{aligned}
$$

So, we've arrived at the rather startling fact that the Lorentz group does not produce unitary transformations on states! It turns out that there are *no* finite dimensional unitary representations of the Lorentz group. In other words, there are no generators $S^{\mu\nu}$ satisfying the Lorentz algebra that are also Hermitian, satisfying

$$
S^{\mu\nu\dagger} = S^{\mu\nu}. \qquad (4.101)
$$

As a consequence, $\psi^\dagger\psi$ is not a scalar. All is not lost though; our failure to create a scalar has given us a clue as to how to build a genuine scalar term. We know from the Clifford algebra (4.78) that γ^0 satisfies

$$
\begin{aligned}
\{\gamma^0, \gamma^0\} &= -2\eta^{00} = -2(-1) = 2 \Longrightarrow (\gamma^0)^2 = 1 \\
\{\gamma^0, \gamma^i\} &= -2\eta^{0i} = 0 \Longrightarrow \gamma^0\gamma^i = -\gamma^i\gamma^0. \qquad (4.102)
\end{aligned}
$$

Thus, consider the expression $\gamma^0\gamma^\mu\gamma^0$. It is either

$$\gamma^0\gamma^0\gamma^0 = \gamma^0,$$

or

$$\gamma^0\gamma^i\gamma^0 = -\gamma^0\gamma^0\gamma^i = -\gamma^i. \tag{4.103}$$

Comparing this to (4.100), we have

$$\gamma^{\mu\dagger} = \gamma^0\gamma^\mu\gamma^0 \tag{4.104}$$

for all spacetime indices μ. Hence, the reason $\psi^\dagger\psi$ is not a scalar is that in (4.99), the generators $S^{\mu\nu}$ weren't Hermitian. Now we can write

$$\begin{aligned} S^{\mu\nu\dagger} &= -\frac{i}{4}[\gamma^{\mu\dagger}, \gamma^{\nu\dagger}] \\ &= -\frac{i}{4}[\gamma^0\gamma^\mu\gamma^0, \gamma^0\gamma^\nu\gamma^0] \\ &= -\frac{i}{4}\gamma^0[\gamma^\mu, \gamma^\nu]\gamma^0 \\ &= -S^{\mu\nu}. \end{aligned} \tag{4.105}$$

Therefore we can write

$$\begin{aligned} S^\dagger &= e^{-\frac{i}{2}\omega_{\mu\nu}S^{\mu\nu\dagger}} \\ &= \gamma^0 e^{-\frac{i}{2}\omega_{\mu\nu}S^{\mu\nu}}\gamma^0 \\ &= \gamma^0 S^{-1}\gamma^0. \end{aligned} \tag{4.106}$$

This is the key; we can now easily write down the Lorentz scalar using ψ. We first define the expression

$$\bar{\psi} = \psi^\dagger\gamma^0. \tag{4.107}$$

Then, the scalar will be $\bar{\psi}\psi$. We can write the transformation out easily; it is

$$\begin{aligned} \bar{\psi}\psi \longrightarrow \bar{\psi}'\psi' &= \psi'^\dagger\gamma^0\psi' \\ &= \psi^\dagger S^\dagger\gamma^0 S\psi \\ &= \psi^\dagger\gamma^0 S^{-1}\gamma^0\gamma^0 S\psi \\ &= \psi^\dagger\gamma^0 S^{-1} S\psi \\ &= \psi^\dagger\gamma^0\psi \\ &= \bar{\psi}\psi. \end{aligned} \tag{4.108}$$

So with some difficulty we've managed to build a Lorentz scalar out of the spinor field ψ. Our next goal, as we said at the beginning of this section, is to build a Lorentz vector. We'll skip all of the "motivation" discussion we included in finding $\bar{\psi}\psi$ and simply state the answer. The expression $\bar{\psi}\gamma^\mu\psi$ is a Lorentz vector. To show this we need one preliminary result. Consider again the commutator in (3.289). From our result there and from (3.270), we can show that

$$\begin{aligned}[\gamma^\rho, S^{\mu\nu}] &= -i(\eta^{\rho\mu}\gamma^\nu - \eta^{\nu\rho}\gamma^\mu) \\ &= -i(\eta^{\rho\mu}\eta^{\nu c} - \eta^{\nu\rho}\eta^{\mu c})\eta_{bc}\gamma^b \\ &= (J^{\mu\nu})^\rho_b\gamma^b,\end{aligned} \tag{4.109}$$

where $J^{\mu\nu}$ is the Lorentz generator in the vector representation. Then, to see that $\bar{\psi}\gamma^\mu\psi$ is a vector, we simply work out the transformation, using (4.109). This gives

$$\begin{aligned}\bar{\psi}\gamma^\mu\psi \longrightarrow \bar{\psi}'\gamma^\mu\psi' &= \psi'^\dagger\gamma^0\gamma^\mu\psi' \\ &= \psi^\dagger S^\dagger\gamma^0\gamma^\mu S\psi \\ &= \psi^\dagger\gamma^0 S^{-1}\gamma^0\gamma^0\gamma^\mu S\psi \\ &= \psi^\dagger\gamma^0 S^{-1}\gamma^\mu S\psi \\ &= \bar{\psi}(S^{-1}\gamma^\mu S)\psi.\end{aligned} \tag{4.110}$$

We now need to look only at the part in parentheses. This is (to first order only, as usual)

$$\begin{aligned}S^{-1}\gamma^\mu S &= \left(1 - \frac{i}{2}\omega_{\alpha\beta}S^{\alpha\beta}\right)\gamma^\mu\left(1 + \frac{i}{2}\omega_{\alpha\beta}S^{\alpha\beta}\right) \\ &= \gamma^\mu + \frac{i}{2}\omega_{\alpha\beta}\gamma^\mu S^{\alpha\beta} - \frac{i}{2}\omega_{\alpha\beta}S^{\alpha\beta}\gamma^\mu \\ &= \gamma^\mu + \frac{i}{2}\omega_{\alpha\beta}[\gamma^\mu, S^{\alpha\beta}] \\ &= \gamma^\mu + \frac{i}{2}\omega_{\alpha\beta}(J^{\alpha\beta})^\mu_\sigma\gamma^\sigma.\end{aligned} \tag{4.111}$$

We can recognize the last line as the first-order term in $e^{\frac{i}{2}\omega_{\mu\nu}J^{\mu\nu}}$, the vector transformation on γ^μ (cf. equation (3.268)). Hence, $\bar{\psi}\gamma^\mu\psi$ is indeed a spacetime vector, which we can contract with the differentiation operator ∂_μ to form a scalar. Thus, our two Lorentz scalars are $\bar{\psi}\psi$ and $\bar{\psi}\gamma^\mu\partial_\mu\psi$. Writing down our Lagrangian is now quite easy. It is

$$\mathcal{L}_D = \bar{\psi}(i\gamma^\mu\partial_\mu - m)\psi. \tag{4.112}$$

(the i is there with foresight). Let's see if this reproduces the Dirac equation.

We can take variations with respect to either $\bar{\psi}$ or ψ (the results will be conjugates of each other). Taking the conjugate with respect to $\bar{\psi}$ is easy:

$$\begin{aligned}
\frac{\partial \mathcal{L}_D}{\partial \bar{\psi}} &= (i\gamma^\mu \partial_\mu - m)\psi, \\
\partial_\mu \left(\frac{\partial \mathcal{L}_D}{\partial(\partial_\mu \bar{\psi})} \right) &= 0, \\
&\Longrightarrow \partial_\mu \left(\frac{\partial \mathcal{L}_D}{\partial(\partial_\mu \bar{\psi})} \right) - \frac{\partial \mathcal{L}_D}{\partial \bar{\psi}} = 0, \\
&\Longrightarrow (i\gamma^\mu \partial_\mu - m)\psi = 0,
\end{aligned} \tag{4.113}$$

which is the correct form of the Dirac equation. Taking the variation with respect to ψ is a bit trickier:

$$\begin{aligned}
\frac{\partial \mathcal{L}_D}{\partial \psi} &= -m\bar{\psi}, \\
\partial_\mu \left(\frac{\partial \mathcal{L}_D}{\partial(\partial_\mu \psi)} \right) &= \partial_\mu (i\bar{\psi}\gamma^\mu), \\
&\Longrightarrow \partial_\mu \left(\frac{\partial \mathcal{L}_D}{\partial(\partial_\mu \psi)} \right) - \frac{\partial \mathcal{L}_D}{\partial \psi} = 0, \\
&\Longrightarrow i\partial_\mu(\bar{\psi}\gamma^\mu) + m\bar{\psi} = 0, \\
&\Longrightarrow i\partial_\mu(\psi^\dagger\gamma^0\gamma^\mu) + m(\psi^\dagger\gamma^0) = 0, \\
&\Longrightarrow i\partial_\mu(\gamma^{\mu\dagger}\gamma^{0\dagger}\psi)^\dagger + m(\gamma^{0\dagger}\psi)^\dagger = 0, \\
&\Longrightarrow i\partial_\mu(\gamma^0\gamma^\mu\gamma^0\gamma^0\gamma^0\gamma^0\psi)^\dagger + m(\gamma^0\gamma^0\gamma^0\psi)^\dagger = 0, \\
&\Longrightarrow i\partial_\mu(\gamma^0\gamma^\mu\psi)^\dagger + m(\gamma^0\psi)^\dagger = 0, \\
&\Longrightarrow \big(-i\partial_\mu(\gamma^0\gamma^\mu\psi) + m\gamma^0\psi\big)^\dagger = 0, \\
&\Longrightarrow -i\gamma^0\gamma^\mu\partial_\mu\psi + \gamma^0 m\psi = 0, \\
&\Longrightarrow -\gamma^0(i\gamma^\mu\partial_\mu - m)\psi = 0, \\
&\Longrightarrow (i\gamma^\mu\partial_\mu - m)\psi = 0.
\end{aligned} \tag{4.114}$$

So taking the variation of (4.112) with respect to either $\bar{\psi}$ or ψ gives us the Dirac equation, as expected. Before moving on, let's pause for a moment and consider what we've found. We wanted an equation for the two component spinor ψ that would be relativistically invariant. We were able to do this, but at the price of doubling the number of states in ψ, from 2 to 4. We accepted this and moved on. We were able to find the relevant equation, called the Dirac equation (4.80), which was both first order in the differential operator and still relativistically invariant. In other words, we solved the problem that contracting ∂_μ with something caused – there is no "preferred" direction in spacetime as with a term like $v^\mu \partial_\mu \psi$. The reason for this is the algebraic structure of the γ^μ Clifford algebra. Now we want to understand why we were forced from 2 to 4 spin components.

4.3.6 Parity and Handedness

As we know, we are able to create a field equation for a spinor field, but we did so at the price of the number of spin components we expected. Namely, we found that we can't satisfy special relativity with only *two* components – we need $four$. This is very strange because we *know* that there are *two* component representations of the Lorentz group (the $(1/2, 0)$ and $(0, 1/2)$ representations) from Sect. 3.3.3. What has happened?

It is now time to bring together a few ideas that have been scattered along the way so far. We first have to go back to Sect. 1.4.6 where we discussed the improper (but still important) Lorentz transformations like parity and time reversal. Recall we said that, while it is not possible to do a Lorentz transformation from an object's rest frame to a frame described by an improper transformation, improper transformations are still valid Lorentz transform because they satisfy (1.72). Further, as we said there, this means that we still need to factor them into a theory in order for it to be a relativistically invariant theory. This hasn't really mattered up until now, but now we must deal with it. Specifically, recall the improper transformation Λ_P, the parity transformation. The idea of parity is that the handedness of a coordinate frame is switched; right-handed frames become left-handed frames and vice versa.

Now consider Sect. 3.3.3. In addition to the trivial scalar representation and the familiar vector representation of the Lorentz group, we also discussed the $(1/2, 0)$ and $(0, 1/2)$ representations, which we called (for reasons that probably weren't clear at the time) "left-handed" and "right-handed", respectively. We also pointed out in that section that we can relate left- and right-handed spinors; if ψ_L is a left-handed spinor, then $i\sigma^2\psi_L^\star$ is a right-handed spinor.

So, let's consider the two sets of transformation laws for the left- and right-handed spinor representations. They are given by equations (3.235)–(3.240) and (3.245)–(3.250). Recall that in Sect. 3.3.3 we pointed out that the rotation transformations were the same regardless of handedness; the only differences were a few minus signs in the boost equations. We can now discuss the origin of those minus signs.

Consider first the left-handed boosts. They were

$$B_x(\phi_x) = \begin{pmatrix} \cosh\frac{\phi_x}{2} & \sinh\frac{\phi_x}{2} \\ \sinh\frac{\phi_x}{2} & \cosh\frac{\phi_x}{2} \end{pmatrix},$$

$$B_y(\phi_y) = \begin{pmatrix} \cosh\frac{\phi_y}{2} & -i\sinh\frac{\phi_y}{2} \\ i\sinh\frac{\phi_y}{2} & \cosh\frac{\phi_y}{2} \end{pmatrix}, \text{ and}$$

$$B_z(\phi_z) = \begin{pmatrix} e^{\frac{\phi_z}{2}} & 0 \\ 0 & e^{-\frac{\phi_z}{2}} \end{pmatrix}.$$

Now recall from Sect. 1.4.3 that the "angle" in the Lorentz boosts is given by the ratio of the speed of a frame to the speed of light – $\beta = \frac{v}{c}$. If we consider a parity transformation, the x, y and z axes are all reversed, picking up a minus sign. This means that the velocity components will also pick up a minus sign (obviously if you were moving in the positive x direction and then that direction becomes the minus x direction, the $v_x \to -v_x$). Thus, under parity, the Lorentz boost "angles" all pick up a minus sign. Applying this to ϕ_x, ϕ_y, and ϕ_z above, we have

$$B_x(-\phi_x) = \begin{pmatrix} \cosh\frac{\phi_x}{2} & -\sinh\frac{\phi_x}{2} \\ -\sinh\frac{\phi_x}{2} & \cosh\frac{\phi_x}{2} \end{pmatrix},$$

$$B_y(-\phi_y) = \begin{pmatrix} \cosh\frac{\phi_y}{2} & i\sinh\frac{\phi_y}{2} \\ -i\sinh\frac{\phi_y}{2} & \cosh\frac{\phi_y}{2} \end{pmatrix}, \quad \text{and}$$

$$B_z(-\phi_z) = \begin{pmatrix} e^{-\frac{\phi_z}{2}} & 0 \\ 0 & e^{\frac{\phi_z}{2}} \end{pmatrix}.$$

Comparing these with (3.248)–(3.250), we see that they are identical. Thus, the parity transformation transforms a left-handed spinor into a right-handed spinor.[11]

We have now solved the "mystery" of why we were forced to have 4 components in our spinor ψ. As we said in Sect. 1.4.6, if a theory is to be relativistically invariant it must be invariant under *all* Lorentz transforms, even the improper ones. Because this includes parity, a theory that has only, say, left-handed spinors will not be invariant. It will be changed under parity from left- to right-handed. Keep in mind that this doesn't mean that there is an actual physical transformation that can transform a right-handed particle into a left-handed particle or vice versa. The point is that because such a (non-physical) transformation does in fact preserve the metric, our theory must still incorporate them. So, our theory *must* have both left- and right-handed spinors – exactly what the Clifford algebra "told us"! So, we have our 4 component spinor ψ defined by

$$\psi = \begin{pmatrix} \psi_1 \\ \psi_2 \end{pmatrix}, \tag{4.115}$$

[11] This shouldn't really be surprising at this point, especially after Sect. 4.3.2 where we saw that it is exactly parity that swaps between left- and right-handedness.

where ψ_1 and ψ_2 are both two component spinors (of which the two components correspond to $\pm\frac{1}{2}$, as we'd expect). We'll see shortly how to determine which of ψ_1 and ψ_2 is left-handed and which is right-handed.

This is a very profound point. We wanted to make a relativistically invariant theory of spinors. However, there are two types of spinors – left- and right-handed. We found that a nice way of generating spinor representation Lorentz generators was to use the Clifford algebra (4.78), which we saw showed up almost automatically in trying to build an equation for spinors. Because of the deep algebraic and geometric structure of the Clifford algebra, placing this constraint on the Dirac equation automatically made it relativistically invariant – even under the improper Lorentz transforms. Had we tried to force our spinor ψ to have only two components (like we initially wanted), it wouldn't have been invariant under parity and the theory wouldn't have been relativistically invariant. However, by accepting the need for all 4 components, we have a theory that is both unchanged under proper Lorentz transformations, and under the improper transformations. In other words, special relativity has forced us into having *both* representations.

So to rephrase the results of Sect. 3.3.3, we actually have only three physically acceptable irreducible representations of the Lorentz group,

$$\begin{aligned}(0,0) &: \text{ Scalar,}\\ (1/2,1/2) &: \text{ Vector,}\\ (1/2,0)\oplus(0,1/2) &: \text{ Spinor.}\end{aligned}$$

4.3.7 Weyl Spinors in the Chiral Representation

We now understand why our theory was required to go from a *two* component ψ to a *four* component ψ. We have chosen to break ψ up into *two* component spinors

$$\psi = \begin{pmatrix}\psi_1\\ \psi_2\end{pmatrix} = \begin{pmatrix}\psi_{11}\\ \psi_{12}\\ \psi_{21}\\ \psi_{22}\end{pmatrix}. \tag{4.116}$$

What is the meaning of these four components? We know from relativity that parity requires our theory to have left- and right-handed spinors, so which is which? It may be tempting to guess that ψ_1 is left-handed and ψ_2 is right-handed (or vice versa), but it is not this simple. It turns out that the decision depends on our choice of the Clifford algebra γ^μ matrices.

So how can we break up the components of the Weyl fields into left- and right-handed parts? In order to proceed, let's choose (for clarity) a specific representation of the γ^μ matrices. Namely, let's choose the chiral representation (4.89). We can write this in a more compact way by using notation similar to that used in (3.261). We define

$$\sigma^\mu = (1, \sigma^1, \sigma^2, \sigma^3)^T,$$
$$\bar{\sigma}^\mu = (1, -\sigma^1, -\sigma^2, -\sigma^3), \tag{4.117}$$

and then we have that the chiral representation is

$$\gamma^\mu = \begin{pmatrix} 0 & \sigma^\mu \\ \bar{\sigma}^\mu & 0 \end{pmatrix}. \tag{4.118}$$

Now it is straightforward to use (3.287) to determine the Lorentz transformations in this representation. These gives

$$\begin{aligned} S^{\mu\nu} &= \frac{i}{4}[\gamma^\mu, \gamma^\nu] \\ &= \frac{i}{4}\left[\begin{pmatrix} 0 & \sigma^\mu \\ \bar{\sigma}^\mu & 0 \end{pmatrix}\begin{pmatrix} 0 & \sigma^\nu \\ \bar{\sigma}^\nu & 0 \end{pmatrix} - \begin{pmatrix} 0 & \sigma^\nu \\ \bar{\sigma}^\nu & 0 \end{pmatrix}\begin{pmatrix} 0 & \sigma^\mu \\ \bar{\sigma}^\mu & 0 \end{pmatrix}\right] \\ &= \frac{i}{4}\begin{pmatrix} \sigma^\mu\bar{\sigma}^\nu - \sigma^\nu\bar{\sigma}^\mu & 0 \\ 0 & \bar{\sigma}^\mu\sigma^\nu - \bar{\sigma}^\nu\sigma^\mu \end{pmatrix}. \end{aligned} \tag{4.119}$$

From these it is straightforward to work out that the rotations are given by

$$\begin{aligned} S^{12} &= J^z = \frac{1}{2}\begin{pmatrix} \sigma^3 & 0 \\ 0 & \sigma^3 \end{pmatrix}, \\ S^{31} &= J^y = \frac{1}{2}\begin{pmatrix} \sigma^2 & 0 \\ 0 & \sigma^2 \end{pmatrix}, \\ S^{23} &= J^x = \frac{1}{2}\begin{pmatrix} \sigma^1 & 0 \\ 0 & \sigma^1 \end{pmatrix}, \end{aligned} \tag{4.120}$$

and the boosts are given by

$$\begin{aligned} S^{01} &= K^x = \frac{i}{2}\begin{pmatrix} -\sigma^1 & 0 \\ 0 & \sigma^1 \end{pmatrix}, \\ S^{02} &= K^y = \frac{i}{2}\begin{pmatrix} -\sigma^2 & 0 \\ 0 & \sigma^2 \end{pmatrix}, \\ S^{03} &= K^z = \frac{i}{2}\begin{pmatrix} -\sigma^3 & 0 \\ 0 & \sigma^3 \end{pmatrix}. \end{aligned} \tag{4.121}$$

We can then exponentiate these to get (combining them all together)

$$R(\boldsymbol{\theta}) = e^{i\boldsymbol{\theta}\cdot\mathbf{J}} = \begin{pmatrix} e^{i\boldsymbol{\theta}\cdot\frac{\boldsymbol{\sigma}}{2}} & 0 \\ 0 & e^{i\boldsymbol{\theta}\cdot\frac{\boldsymbol{\sigma}}{2}} \end{pmatrix},$$

$$B(\boldsymbol{\phi}) = e^{i\boldsymbol{\phi}\cdot\boldsymbol{K}} = \begin{pmatrix} e^{i\boldsymbol{\phi}\cdot\frac{\boldsymbol{\sigma}}{2}} & 0 \\ 0 & e^{-i\boldsymbol{\phi}\cdot\frac{\boldsymbol{\sigma}}{2}} \end{pmatrix}. \tag{4.122}$$

Thus, with the chiral representation of the γ^μ matrices, we have found that the *four* component field $\psi = \binom{\psi_1}{\psi_2}$ transforms under Lorentz rotations according to

$$\begin{pmatrix} \psi_1 \\ \psi_2 \end{pmatrix} \longrightarrow \begin{pmatrix} e^{i\boldsymbol{\theta}\cdot\frac{\boldsymbol{\sigma}}{2}} & 0 \\ 0 & e^{i\boldsymbol{\theta}\cdot\frac{\boldsymbol{\sigma}}{2}} \end{pmatrix} \begin{pmatrix} \psi_1 \\ \psi_2 \end{pmatrix}, \tag{4.123}$$

and under Lorentz boosts according to

$$\begin{pmatrix} \psi_1 \\ \psi_2 \end{pmatrix} \longrightarrow \begin{pmatrix} e^{\boldsymbol{\phi}\cdot\frac{\boldsymbol{\sigma}}{2}} & 0 \\ 0 & e^{-\boldsymbol{\phi}\cdot\frac{\boldsymbol{\sigma}}{2}} \end{pmatrix} \begin{pmatrix} \psi_1 \\ \psi_2 \end{pmatrix}. \tag{4.124}$$

It is clear that these are the exact transformation laws for left- and right-handed spinors from Sect. 3.3.3 *if* we take ψ_1 to be left-handed and ψ_2 to be right-handed (cf. equations (3.234) and (3.244)).

Hence, in the chiral representation of the Clifford algebra, we have that the Dirac field ψ is nicely split up into its left- and right-handed parts automatically. This is why this is called the "chiral" representation. Notice that with the fields written this way our γ^μ matrices in the chiral representation (4.118) has the *right-handed* Pauli matrices acting on the *right-handed* field, and the *left-handed* Pauli matrices (which differ from the right-handed ones by a minus sign on each of the spatial parts) acting on the *left-handed* field.

Also, notice that with γ^μ taking on the form (4.118), we can write the Dirac equation very nicely in terms of

$$\psi = \begin{pmatrix} \psi_1 \\ \psi_2 \end{pmatrix} = \begin{pmatrix} \psi_L \\ \psi_R \end{pmatrix}. \tag{4.125}$$

With the field written this way we have

$$(i\gamma^\mu\partial_\mu - m)\psi = 0 \Longrightarrow \begin{pmatrix} -m & i\sigma^\mu\partial_\mu \\ i\bar{\sigma}^\mu\partial_\mu & -m \end{pmatrix} \begin{pmatrix} \psi_L \\ \psi_R \end{pmatrix} = 0. \tag{4.126}$$

This breaks into two equations:

$$i\sigma^\mu\partial_\mu\psi_R - m\psi_L = 0,$$

$$i\bar{\sigma}^\mu\partial_\mu\psi_L - m\psi_R = 0. \tag{4.127}$$

This shows that when the mass m is not zero, the left- and right-handed parts are coupled together. However, as $m \to 0$, the left- and right-handed fields decouple and are not related to each other. We'll discuss this limit more later. Finally, note that we can also write the Dirac Lagrangian (4.112) in terms of these fields. We get (again, in the chiral representation)

$$\begin{aligned}\mathcal{L}_D &= \bar{\psi}(i\gamma^\mu\partial_\mu - m)\psi \\ &= \begin{pmatrix}\psi_L^\dagger & \psi_R^\dagger\end{pmatrix}\begin{pmatrix}-m & i\sigma^\mu\partial_\mu \\ i\bar{\sigma}^\mu\partial_\mu & -m\end{pmatrix}\begin{pmatrix}\psi_L \\ \psi_R\end{pmatrix} \\ &= i\psi_R^\dagger\sigma^\mu\partial_\mu\psi_R + i\psi_L^\dagger\bar{\sigma}^\mu\partial_\mu\psi_L - m(\psi_R^\dagger\psi_L + \psi_L^\dagger\psi_R). \qquad (4.128)\end{aligned}$$

4.3.8 Weyl Spinors in Any Representation

In the previous section we looked at the components of the four component spinor $\psi = \binom{\psi_1}{\psi_2}$ in the chiral representation and we found that this representation gives the components a very nice form: the two components simply are the left- and right-handed fields. We were able to determine this because we used (3.287) to get the exact form of the Lorentz transformations, and because we already knew from Sect. 3.3.3 what these looked like, it was clear what ψ_1 and ψ_2 were.

Nevertheless, as we said, this is unique to the chiral representation. What about in an arbitrary representation? We *could* try to find the Lorentz transformations (via (3.287)) for any set of γ^μ matrices, but this can be extremely difficult for non-chiral representations.

The question we'd now like to ask is 'is there a way of finding which part of ψ is left-handed and which part is right-handed?' It turns out that there is. Consider the matrix defined as

$$\gamma^5 = i\gamma^0\gamma^1\gamma^2\gamma^3. \qquad (4.129)$$

It is straightforward (though very tedious) to work out the details using the known commutation and anticommutation relations of the γ^μ matrices (4.78) to show that (in any representation) γ^5 commutes with the Lorentz generators,

$$\left[S^{\mu\nu}, \gamma^5\right] = 0. \qquad (4.130)$$

And therefore γ^5 is invariant under Lorentz transformations.

Next, we form the operators

$$P_\pm = \frac{1}{2}(1 \pm \gamma^5). \qquad (4.131)$$

Notice that because $(\gamma^5)^2 = 1$ we have

$$
\begin{aligned}
P_+P_+ &= \frac{1}{4}I(1+\gamma^5)(1+\gamma^5) = \frac{1}{4}(1+2\gamma^5+(\gamma^5)^2) = \frac{1}{2}(1+\gamma^5) = P_+,\\
P_-P_- &= \frac{1}{4}(1-\gamma^5)(1-\gamma^5) = \frac{1}{4}(1-2\gamma^5+(\gamma^5)^2) = \frac{1}{2}(1-\gamma^5) = P_-,\\
P_+P_- &= \frac{1}{4}(1+\gamma^5)(1-\gamma^5) = \frac{1}{4}(1-(\gamma^5)^2) = 0 = P_-P_+.
\end{aligned} \tag{4.132}
$$

This tells us that $P_\pm$ are projection operators. And it turns out that they are specifically the projection operators onto the left- and right-handed parts of the field ψ. We have

$$
\begin{aligned}
\psi_R &= P_+\psi,\\
\psi_L &= P_-\psi.
\end{aligned} \tag{4.133}
$$

These relationships will hold in *any* representation of the Clifford algebra γ^μ matrices.

To get a better feel for how this works we can work it out in the (already somewhat familiar) chiral representation. We get

$$
\begin{aligned}
\gamma^5 &= i\gamma^0\gamma^1\gamma^2\gamma^3\\
&= i(\sigma^0\otimes\sigma^1)i(\sigma^1\otimes\sigma^2)i(\sigma^2\otimes\sigma^2)i(\sigma^3\otimes\sigma^2)\\
&= (\sigma^0\sigma^1\sigma^2\sigma^3\otimes\sigma^1\sigma^2\sigma^2\sigma^2)\\
&= (i\sigma^0)\otimes(i\sigma^3)\\
&= \begin{pmatrix} -1 & 0\\ 0 & 1\end{pmatrix}.
\end{aligned} \tag{4.134}
$$

Then the projection operators give us

$$
\begin{aligned}
\psi_R = P_+\psi &= \frac{1}{2}(1+\gamma^5)\psi = \begin{pmatrix} 0 & 0\\ 0 & 1\end{pmatrix}\begin{pmatrix}\psi_1\\ \psi_2\end{pmatrix} = \begin{pmatrix}0\\ \psi_2\end{pmatrix},\\
\psi_L = P_-\psi &= \frac{1}{2}(1-\gamma^5)\psi = \begin{pmatrix} 1 & 0\\ 0 & 0\end{pmatrix}\begin{pmatrix}\psi_1\\ \psi_2\end{pmatrix} = \begin{pmatrix}\psi_1\\ 0\end{pmatrix}.
\end{aligned} \tag{4.135}
$$

This is exactly what we found before by looking at the explicit Lorentz transformations in the chiral representation.

Rather than work out the details of other representations now, we'll do so in the next section when we start to look at explicit solutions to the Dirac equation.

4.3.9 Solutions to the Dirac Equation

We're now ready to start talking about actual solutions to the Dirac equation. To begin with, we know that the Dirac equation "implies" the Klein-Gordon equation (cf. equation (4.79)), and therefore any solution of the Dirac equation must also be a solution to the Klein-Gordon equation. So, we simply assume a certain form for the solutions ψ of the Dirac equation. Namely, we assume the form of the plane wave solutions of Klein-Gordon with an additional four component spinor that we'll take to depend only on the momentum of the field. Separating the positive and negative frequency solutions, we have[12]

$$\psi(x) = av(\mathbf{p})e^{ip\cdot x} + bu(\mathbf{p})e^{-ip\cdot x}, \tag{4.136}$$

where $v(\mathbf{p})$ and $u(\mathbf{p})$ are four component objects with complex elements that depend on the momentum 3-vector $\mathbf{p}$, and a and b are arbitrary constants of integration (which we'll ignore for now). It should be clear that each component will solve the Klein-Gordon equation exactly.

4.3.9.1 Solutions to the Dirac Equation for Particles at Rest

To make sense of this, let's start with something easy: a particle in the rest frame with the chiral representation of the γ^μ matrices. Because it is in the rest frame, $\mathbf{p} = 0$, so $p_\mu = (m, 0, 0, 0)$. Focusing (for now) on only the positive frequency solution,

$$\begin{aligned}(i\gamma^\mu \partial_\mu - m)v(\mathbf{p})e^{ip\cdot x} = 0 &\Longrightarrow (i\gamma^\mu(ip_\mu) - m)v(0)e^{ip\cdot x} = 0,\\ &\Longrightarrow (-\gamma^\mu p_\mu - m)v(0) = 0,\\ &\Longrightarrow m(\gamma^0 + 1)v(0) = 0,\\ &\Longrightarrow \begin{pmatrix}1 & 1\\ 1 & 1\end{pmatrix}\begin{pmatrix}v_1(0)\\ v_2(0)\end{pmatrix} = 0.\end{aligned} \tag{4.137}$$

where we have written out the *two* component terms in v (both v_1 and v_2 are two component elements, for a total of four components in v). Suppressing the (0) argument, this equation gives

$$v_1 = -v_2. \tag{4.138}$$

[12] Again, we are writing x to represent all 4 spacetime components – i.e. $x = x^\mu$. So $p\cdot x = p_\mu x^\mu$.

We therefore take the general solution in this case to be

$$v(0) = A \begin{pmatrix} \xi \\ -\xi \end{pmatrix}, \tag{4.139}$$

where ξ is some constant *two* component spinor and A is some arbitrary constant. Recall from (4.125) that v_1 is the left-handed part and v_2 is the right-handed part.

$$v_L = -v_R \tag{4.140}$$

Thus, the Dirac equation requires that the right and left-handed fields be related by a minus sign; once you fix the left-handed field, the right-handed field is completely determined. In other words, you can't have one without the other (in the rest frame).

Had we looked at the negative frequency solutions we would have found

$$(i\gamma^\mu \partial_\mu - m)u(\mathbf{p})e^{-ip\cdot x} = 0 \Longrightarrow (\gamma^0 - 1)u(0) = 0$$
$$\Longrightarrow \begin{pmatrix} -1 & 1 \\ 1 & -1 \end{pmatrix} \begin{pmatrix} u_1 \\ u_2 \end{pmatrix} = 0, \tag{4.141}$$

which gives

$$u_1 = u_2, \tag{4.142}$$

or

$$u_L = u_R. \tag{4.143}$$

From this we conclude that the solution is

$$u(0) = B \begin{pmatrix} \chi \\ \chi \end{pmatrix}, \tag{4.144}$$

where again χ is a constant two component spinor and B is an arbitrary constant. Once again, the Dirac equation is putting constraints on the right-handed field in terms of the left-handed field – once one is chosen the other is fixed. You can't have one without the other. Thus, we see that the general solution to the Dirac equation for a particle at rest is (recalling that $p \cdot x = -Et + \mathbf{p} \cdot \mathbf{x}$ and in this case $\mathbf{p} = 0$ and $p_0 = m$)

$$\psi(x) = A \begin{pmatrix} \xi \\ -\xi \end{pmatrix} e^{-imt} + B \begin{pmatrix} \chi \\ \chi \end{pmatrix} e^{imt}. \tag{4.145}$$

To get a better feel for this let's look at the same thing in a different representation of the γ^μ matrices. We expect that because this is nothing but a change of basis, the physics should be the same. Let's confirm this. We'll look at the Dirac basis, given in (4.90). We'll start by finding γ^5. It is

$$\begin{aligned}\gamma^5 &= i\gamma^0\gamma^1\gamma^2\gamma^3 \\ &= i(\sigma^0 \otimes \sigma^3)i(\sigma^1 \otimes \sigma^2)i(\sigma^2 \otimes \sigma^2)i(\sigma^3 \otimes \sigma^2) \\ &= (\sigma^0\sigma^1\sigma^2\sigma^3 \otimes \sigma^3\sigma^2\sigma^2\sigma^2) \\ &= (i\sigma^0) \otimes (-i\sigma^1) \\ &= \begin{pmatrix} 0 & 1 \\ 1 & 0 \end{pmatrix}. \end{aligned} \tag{4.146}$$

In this representation the Weyl fields break apart as

$$\begin{aligned}\psi_R = P_+\psi = \frac{1}{2}(1+\gamma^5)\psi = \frac{1}{2}\begin{pmatrix} 1 & 1 \\ 1 & 1 \end{pmatrix}\begin{pmatrix} \psi_1 \\ \psi_2 \end{pmatrix} = \frac{1}{2}\begin{pmatrix} \psi_1 + \psi_2 \\ \psi_1 + \psi_2 \end{pmatrix}, \\ \psi_L = P_-\psi = \frac{1}{2}(1-\gamma^5)\psi = \frac{1}{2}\begin{pmatrix} 1 & -1 \\ -1 & 1 \end{pmatrix}\begin{pmatrix} \psi_1 \\ \psi_2 \end{pmatrix} = \frac{1}{2}\begin{pmatrix} \psi_1 - \psi_2 \\ \psi_1 - \psi_2 \end{pmatrix}. \end{aligned} \tag{4.147}$$

Now, for the particle in the rest frame, we can look at the Dirac equation for the positive frequency solution,

$$m(\gamma^0 + 1)v(0) = 0 \Longrightarrow \begin{pmatrix} 2 & 0 \\ 0 & 0 \end{pmatrix}\begin{pmatrix} v_1 \\ v_2 \end{pmatrix} = 0. \tag{4.148}$$

This tells us that

$$v_1 = 0, \tag{4.149}$$

but it leaves v_2 unconstrained. However, we know from (4.147) that

$$\begin{aligned} v_R &= \frac{1}{2}\begin{pmatrix} v_1 + v_2 \\ v_1 + v_2 \end{pmatrix}, \\ v_L &= \frac{1}{2}\begin{pmatrix} v_1 - v_2 \\ v_1 - v_2 \end{pmatrix}. \end{aligned} \tag{4.150}$$

So if $v_1 = 0$, we have

$$v_L = -v_R, \tag{4.151}$$

which is in agreement with (4.138).

For the negative frequency solutions we have

$$m(\gamma^0 - 1)v(0) = 0 \Longrightarrow \begin{pmatrix} 0 & 0 \\ 0 & 2 \end{pmatrix}\begin{pmatrix} v_1 \\ v_2 \end{pmatrix} = 0, \tag{4.152}$$

from which we get

$$v_L = v_R, \tag{4.153}$$

which is in agreement with (4.138). We do indeed get identical results in both representations. We'll leave it to you to confirm that the Majorana representation (4.91) once again gives the same result. The point is that for a particle sitting still, the left and right parts are not independent. They are exactly fixed by each other, and you can't have one without the other. We'll now move on to consider the general case.

4.3.9.2 Solutions to the Dirac Equation for Moving Particles

As we've seen, physics is independent of the representation of the γ^μ matrices we choose. Therefore, for the sake of simplicity we'll only look at the chiral representation in this section. Looking only at the positive frequency solutions (for now and for simplicity), the Dirac equation is

$$\begin{aligned}(i\gamma^\mu\partial_\mu - m)v(\mathbf{p})e^{ip\cdot x} = 0 &\Longrightarrow (i\gamma^\mu(ip_\mu) - m)v(\mathbf{p}) = 0,\\ &\Longrightarrow (\gamma^\mu p_\mu + m)v(\mathbf{p}) = 0,\\ &\Longrightarrow \begin{pmatrix} m & p_\mu\sigma^\mu \\ p_\mu\bar{\sigma}^\mu & m\end{pmatrix}\begin{pmatrix} v_1(\mathbf{p}) \\ v_2(\mathbf{p})\end{pmatrix} = 0, \end{aligned}\tag{4.154}$$

where we have used the notation from (4.118). Writing this out in terms of the Weyl spinor parts gives

$$\begin{aligned} mv_1(\mathbf{p}) + (p\cdot\sigma)v_2(\mathbf{p}) &= 0,\\ mv_2(\mathbf{p}) + (p\cdot\bar{\sigma})v_1(\mathbf{p}) &= 0. \end{aligned}\tag{4.155}$$

At this point we can be helped by a very useful identity. Consider the product

$$\begin{aligned}(p\cdot\sigma)(p\cdot\bar{\sigma}) &= (p_\mu\sigma^\mu)(p_\nu\bar{\sigma}^\nu)\\ &= (-p^0 + p_i\sigma^i)(-p^0 - p_j\sigma^j)\\ &= (p^0)^2 - p_i p_j\sigma^i\sigma^j\\ &= E^2 - p_i p_j\delta^{ij}\\ &= E^2 - \mathbf{p}^2\\ &= m^2. \end{aligned}\tag{4.156}$$

We can rewrite this as

$$m = \sqrt{p \cdot \sigma}\sqrt{p \cdot \bar{\sigma}}, \tag{4.157}$$

where it is understood that we're taking the square root of each component of the matrix. Plugging this into, say, the first of (4.155), we have (suppressing the $(\mathbf{p})$ argument for notational simplicity)

$$\begin{aligned} mv_1 + (p \cdot \sigma)v_2 = 0 &\Longrightarrow \sqrt{p \cdot \sigma}\sqrt{p \cdot \bar{\sigma}}v_1 + (p \cdot \sigma)v_2 = 0, \\ &\Longrightarrow \sqrt{p \cdot \bar{\sigma}}v_1 + \sqrt{p \cdot \sigma}v_2 = 0, \\ &\Longrightarrow \sqrt{p \cdot \bar{\sigma}}v_1 = -\sqrt{p \cdot \sigma}v_2, \\ &\Longrightarrow \frac{v_1}{v_2} = -\frac{\sqrt{p \cdot \sigma}}{\sqrt{p \cdot \bar{\sigma}}}. \end{aligned} \tag{4.158}$$

From this it makes sense to write (restoring the $(\mathbf{p})$ argument)

$$v(\mathbf{p}) = \begin{pmatrix} v_1(\mathbf{p}) \\ v_2(\mathbf{p}) \end{pmatrix} = \begin{pmatrix} \sqrt{p \cdot \sigma}\xi \\ -\sqrt{p \cdot \bar{\sigma}}\xi \end{pmatrix}. \tag{4.159}$$

Notice that when the particle is at rest, $\mathbf{p} = 0$, we have

$$v(0) = \sqrt{m}\begin{pmatrix} \xi \\ -\xi \end{pmatrix}, \tag{4.160}$$

which is in agreement with (4.139) as long as we take $A = \sqrt{m}$ (which we are of course free to do since A was arbitrary).

For the negative frequency solutions we have that the Dirac equation is (you should work this out yourself)

$$\begin{pmatrix} -m & p \cdot \sigma \\ p \cdot \bar{\sigma} & -m \end{pmatrix}\begin{pmatrix} u_1(\mathbf{p}) \\ u_2(\mathbf{p}) \end{pmatrix} = 0, \tag{4.161}$$

from which we get

$$\sqrt{p \cdot \bar{\sigma}}u_1 = \sqrt{p \cdot \sigma}u_2, \tag{4.162}$$

and therefore

$$u(\mathbf{p}) = \begin{pmatrix} \sqrt{p \cdot \sigma}\chi \\ \sqrt{p \cdot \bar{\sigma}}\chi \end{pmatrix}. \tag{4.163}$$

You can again notice that the $\mathbf{p} = 0$ case matches (4.144) when $B = \sqrt{m}$.

So finally we have that the general solution to the Dirac equation is

$$\psi(x) = av(\mathbf{p})e^{ip\cdot x} + bu(\mathbf{p})e^{-ip\cdot x} = a\begin{pmatrix}\sqrt{p\cdot\sigma}\xi\\ -\sqrt{p\cdot\bar{\sigma}}\xi\end{pmatrix}e^{ip\cdot x} + b\begin{pmatrix}\sqrt{p\cdot\sigma}\chi\\ \sqrt{p\cdot\bar{\sigma}}\chi\end{pmatrix}e^{-ip\cdot x}, \tag{4.164}$$

where once again a and b are arbitrary constants of integration.

The meaning of the ξ and χ spinors is exactly the usual meaning of the *two* component spinors we're used to dealing with. Their meaning is summarized in (4.25), (4.29), and (4.33). Thus, for example, a particle at rest with spin up in the z direction will be described by the field

$$\begin{aligned}\psi &= a\sqrt{m}\begin{pmatrix}|\uparrow\rangle_z\\ -|\uparrow\rangle_z\end{pmatrix}e^{-imt} + b\sqrt{m}\begin{pmatrix}|\uparrow\rangle_z\\ |\uparrow\rangle_z\end{pmatrix}e^{imt}.\\ &= a\sqrt{m}\begin{pmatrix}1\\0\\-1\\0\end{pmatrix}e^{-imt} + b\sqrt{m}\begin{pmatrix}1\\0\\1\\0\end{pmatrix}e^{imt}.\end{aligned} \tag{4.165}$$

You can work out the details for other spins and other momentums.

4.3.9.3 The Ultra-Relativistic Limit

Consider the case where the particle approaches the speed of light. This means that

$$E \approx |\mathbf{p}|. \tag{4.166}$$

And when this holds, the standard relativistic relationship (1.98)

$$E^2 - p^2 = m^2, \tag{4.167}$$

tells us that in this limit m (the rest mass) goes to 0 (compared to the momentum). Therefore, the Dirac equation in the limit that m goes to 0 is (remember that we're sticking with the chiral representation of γ^μ)

$$\begin{pmatrix}0 & p\cdot\sigma\\ p\cdot\bar{\sigma} & 0\end{pmatrix}\begin{pmatrix}v_1\\ v_2\end{pmatrix} = 0, \tag{4.168}$$

where we have suppressed the $(\mathbf{p})$ argument on the v's. The first thing to notice about this limit is that v_1 and v_2 have decoupled from each other. Our equations are

$$\begin{aligned}(p\cdot\sigma)v_2 &= 0,\\ (p\cdot\bar{\sigma})v_1 &= 0.\end{aligned} \tag{4.169}$$

These are called the **Weyl Equations**. In the non-relativistic limit (when the particle is at rest) we found that the left- and right-handed parts are completely related to each other – if you define one then the other is automatically defined. It was somewhat clear from sets of equations like (4.155) that it is exactly the mass that couples the left and right parts together. This confirms it; when $m = 0$ the two parts are independent. Note that this would be true even in a non-relativistic limit for particles that actually are massless.

Anyway, moving on – we are interested in the relativistic limit. In this limit, (4.166) says that we have

$$\begin{aligned}(p\cdot\sigma)v_2 = 0 &\Longrightarrow (-p^0+\mathbf{p}\cdot\boldsymbol{\sigma})v_2 = 0,\\ &\Longrightarrow (-E+\mathbf{p}\cdot\boldsymbol{\sigma})v_2 = 0,\\ &\Longrightarrow \mathbf{p}\cdot\boldsymbol{\sigma}v_2 = Ev_2,\\ &\Longrightarrow \hat{\mathbf{p}}\cdot\boldsymbol{\sigma}v_2 = v_2,\end{aligned} \tag{4.170}$$

where $\hat{\mathbf{p}}$ is the unit vector in the direction of the particle's momentum. Similarly we have

$$\hat{\mathbf{p}}\cdot\boldsymbol{\sigma}v_1 = -v_1. \tag{4.171}$$

Let's think about what these equations, (4.170) and (4.171), are telling us. In order to understand them we need to understand the operator $\hat{\mathbf{p}}\cdot\boldsymbol{\sigma}$. It will be helpful to think of it in terms of our discussion in Sect. 4.3.2. The idea was that a given spin state could be directly related to a particular point on the unit sphere (which related to a particular direction in space). The spinor $(\alpha,\beta)^T$ that we found through the stereographic projection was the exact eigenvector of the linear combination of the Pauli matrices "in that direction". By this we mean that given any unit vector $\hat{\mathbf{n}}$, the operator $\hat{\mathbf{n}}\cdot\boldsymbol{\sigma}$ would have eigenvectors corresponding to the $(\alpha,\beta)^T$ spinor that resulted from the stereographic projection. For example, if $\hat{\mathbf{n}} = (0,0,1)^T$, the unit vector in the z direction, then

$$\hat{\mathbf{n}}\cdot\boldsymbol{\sigma} = \sigma^3. \tag{4.172}$$

We saw that the point $(0,0,1)^T$ on the unit sphere was mapped to the spinor $(1,0)^T$, which is exactly the "positive" eigenvector of σ^3 (by "positive" we mean that it is the eigenvector with the positive eigenvalue). This was true for all 6 "unit" directions, as demonstrated in (4.50), (4.51), and (4.52). Of course, if we have a spinor with spin up in the z direction, it won't be a proper eigenvector of any operator other than σ^3.

Looking at $\hat{\mathbf{p}}\cdot\boldsymbol{\sigma}$, we see that we have the same situation. The vector $\hat{\mathbf{p}}$ is simply a unit vector in a particular direction and thus plays the same role as $\hat{\mathbf{n}}$ in the previous paragraph. So, equations (4.170) and (4.171) are actually very profound; they are telling us that the spinors v_1 and v_2 are actually eigenvectors of the linear combination of the Pauli matrices corresponding to the direction the particle is traveling! In other words, in the extreme relativistic limit, the particle is always spinning "around its direction of motion" – we can't talk about its spin around another axis. So as the particle moves faster and faster, its spin will be "forced" to line up with the direction it's moving in. For this reason, we call $\hat{\mathbf{p}}\cdot\boldsymbol{\sigma}$ the **Helicity Operator**. A spinor with $+1$ as its $\hat{\mathbf{p}}\cdot\boldsymbol{\sigma}$ eigenvalue is said to be "right-handed", and a spinor with -1 as its $\hat{\mathbf{p}}\cdot\boldsymbol{\sigma}$ eigenvalue is said to be "left-handed".[13]

We can look at a specific example of an ultra-relativistic limit. Consider the particle moving in the z direction very near the speed of light. In this case we have $p_\mu = (E, 0, 0, p) = (E, 0, 0, E)$. Let's consider a particle with spin up in the z direction, so $\xi = (1, 0)^T$, in this limit. The helicity operator is "in the positive z direction" because that is the direction of motion. Furthermore, the helicity operator requires that v_2 have a positive eigenvalue, and this is not a problem – we are taking it to have spin up in the positive z direction. However, the helicity equation says that v_1 must have a negative eigenvalue with the helicity operator. Further, we know that it is also in a spin up in the positive z direction state. Additionally, a state with spin up in the positive z direction can't have a negative eigenvalue with the helicity operator in the positive z direction. So we expect that the equation will allow v_2, but somehow get rid of the v_1 part.

We will have from (4.164)

$$v = \begin{pmatrix} \sqrt{p\cdot\sigma}\begin{pmatrix}1\\0\end{pmatrix} \\ \sqrt{p\cdot\bar{\sigma}}\begin{pmatrix}1\\0\end{pmatrix} \end{pmatrix}. \tag{4.173}$$

In this case, for p_μ

$$\begin{aligned} p\cdot\sigma &= -E + E\sigma^3 = -E(1-\sigma^3) = -2E\begin{pmatrix}0 & 0\\ 0 & 1\end{pmatrix}, \\ p\cdot\bar{\sigma} &= -E - \sigma^3 = -E(1+\sigma^3) = -2E\begin{pmatrix}1 & 0\\ 0 & 0\end{pmatrix}. \end{aligned} \tag{4.174}$$

Because the overall constants don't matter,

$$\begin{aligned} \sqrt{p\cdot\sigma}\begin{pmatrix}1\\0\end{pmatrix} &= 0, \\ \sqrt{p\cdot\bar{\sigma}}\begin{pmatrix}1\\0\end{pmatrix} &= \begin{pmatrix}1\\0\end{pmatrix}. \end{aligned} \tag{4.175}$$

[13] It is also very common to write the helicity operator as $\frac{1}{2}\hat{\mathbf{p}}\cdot\boldsymbol{\sigma}$ and then say that the eigenvector with $+\frac{1}{2}$ eigenvalue is right-handed and the eigenvector with $-\frac{1}{2}$ eigenvalue is left-handed. For our purposes it doesn't matter which is used; we just wanted to make you aware of the different convention.

So, in this limit,

$$v = \begin{pmatrix} 0 \\ 0 \\ 1 \\ 0 \end{pmatrix}. \tag{4.176}$$

This isn't really surprising. As we said, the helicity equations (4.170) and (4.171) tell us that v_2 must be a right-handed eigenvector of the Pauli matrices (in the direction of motion) and that v_1 must be a left-handed eigenvector of the Pauli matrices (in the direction of motion). However, both v_1 and v_2 are $(1, 0)^T$ (spin up in the z direction), and therefore both have $+1$ as their helicity operator eigenvalue. So, the left-handed field is forced to vanish. The only way they can both be non-zero would be if they had opposite handedness (i.e. $(1, 0)^T$ and $(0, 1)^T$), but the Dirac equation disallows this (because both the left- and right-handed parts are in terms of the same spinor ξ, cf. (4.159)). Had we started with a spin down spinor ($\xi = (0, 1)^T$), we would have found that only the left-handed field would be non-zero when the momentum is in the positive z direction.

We strongly encourage you to work out other examples of various spins with various boosts and rotations to see what they look like, so as to gain some familiarity with spinors.

4.3.10 The Dirac Sea Interpretation of Antiparticles

We now have a very interesting question to answer. We understand mathematically why nature only allows us to have 4 components (not just 2) in our theory. But what are these particles physically? We'll first look at Dirac's initial guess (which was brilliant, but incorrect). Also, recall from Sect. 4.2.1 that we still have to deal with negative energy states.

Dirac suggested that because spin-1/2 particles obey the **Pauli Exclusion Principle**,[14] there could be an infinite number of particles *already* in the negative energy levels, and so they are already occupied, preventing any more particles from falling down and giving off infinite energy. Thus, the negative energy problem was solved.

Furthermore, he said that it is possible for one of the particles in this infinite negative sea to be excited and jump up into a positive energy state, leaving behind a hole. This would appear to us, experimentally, as a particle with the same mass, but the opposite charge. He called such particles **Antiparticles**. For example, the antiparticle of the electron is the antielectron, or the positron (same mass,

---

[14]Which we'll talk about shortly, in Sect. 4.6.3. The point is that two particles obeying this principle can't be in the same state – i.e. can't occupy the same place.

opposite charge). The positron is not a particle in the same sense as the electron, but rather is a hole in an infinite sea of electrons. Where this negative charge is missing, all that is left is a hole which appears as a positively charged particle.

The point is that we do indeed need only two components to understand spin-1/2 particles, but the extra two components were forced on us by the holes in the "Dirac Sea". Everything about this worked out mathematically, and when antiparticles were detected about 5 years after Dirac's prediction of them, it appeared that Dirac's suggestion was correct.

However, there were two major problems with Dirac's idea, and they ultimately proved fatal to the "Dirac Sea" interpretation:

1. This theory, which was supposed to be a theory of single particles, now requires an infinite number of them.
2. Particles like photons, pions, mesons, or Klein-Gordon scalars don't obey the Pauli Exclusion Principle, but still have negative energy states, and therefore Dirac's argument doesn't work.

However, his labeling them "antiparticles" has stuck, and we therefore still refer to them this way. We will soon see exactly how particles and antiparticles come out of quantum field theory (it is not as simple as ψ_1 is the particle and ψ_2 is the antiparticle – both ψ_1 and ψ_2 contain terms involving both particles and antiparticles. We'll discuss this in more detail later.) In any case, we must have some other way of understanding the existence of the antiparticles.

4.3.11 The QFT Interpretation of Antiparticles

In presenting the problem of negative energy states, we have been somewhat intentionally sloppy. To take stock, we have two equations of motion: the Klein-Gordon (4.6) for scalar/spin-0 fields, and the Dirac equation (4.80) for spin-1/2 fields.

In our discussion of negative energy states, we were "pretending" that the ψ's and ϕ's are "states" with negative energy. But, as we said in Sect. 4.1, QFT offers a different interpretation of the fields. Namely, the fields are not states – they are operators. And consequently they can't have energy. A state is *made* by acting on the vacuum with either of the *operators* ϕ or ψ, and then the state $\phi|0\rangle$ or $\psi|0\rangle$ has some energy (we'll see exactly how this is done shortly).

QFT allows us to see the antiparticle as a real, actual particle, rather than the absence of a particle. We do not need the conceptually difficult idea of an infinite sea of negative energy particles. The vacuum $|0\rangle$, with no particles in it, is now our state with the lowest possible energy level. As we will see, there are never negative energy states with these particles.

How exactly $|0\rangle$ works will become clearer when we quantize. The point to be understood for now is that QFT solves the problem of negative energy by reinterpreting what is a state and what is an operator. The fields ϕ and ψ are operators, not states, and therefore they do not have energy associated with them (any more than the operator $\hat{x}$ or $\hat{p}_x$ did in non-relativistic quantum mechanics). So, without any problems of negative energy, we merely accept that nature, due to relativity, demands that particles come in particle/antiparticle pairs, and we move on.

However, as a warning before moving on, the necessity of having left-handed and right-handed representations is a *mathematical* necessity about group representations. The physical consequences are a bit more subtle. What we will see is that while all spin-$1/2$ particles will be described by a field sitting in either a left- or right-handed representation, there may or may not be a corresponding field with the opposite handedness in nature. It is in this sense that the Standard Model (which we will look at later) is a **Chiral** theory. In other words, nature is not symmetric regarding particles and antiparticles of given handedness.

4.3.12 Dirac and Majorana Fields

There is one final topic we need to discuss before moving on. We know that the spinor ψ has complex elements. It will be convenient later to try to impose a reality condition on these fields. We won't always need to do this, but it will be helpful to be able to talk about a "real" field and a "complex" field. We'll discuss why this is helpful later. Trusting for now that doing this is worthwhile, we could try to impose a reality condition,

$$\psi = \psi^{\star}, \tag{4.177}$$

but this wouldn't hold after doing a Lorentz transformation (cf. equation (4.56) – even if α and β start off real there, α' and β' won't in general be real). So how can we impose a reality condition on ψ?

It turns out that there is an easy way to impose a reality condition on ψ, but not using the chiral or Dirac representations of γ^{μ}. Recall when we introduced the solutions of the γ^{μ} matrices we introduced three solutions, but we've only really talked about two of them so far (the chiral and Dirac). It will now be useful to use the Majorana representation which was given in (4.91). Recall that we said the unique thing about the Majorana representation is that it is purely imaginary. This means that, via (3.287), the Lorentz group generators are purely imaginary, and so when we exponentiate them with the factor of i that goes with the exponential map, the actual Lorentz transformations will be purely real! We encourage you to work all of this out (explicitly find all 6 Lorentz transformations). So in the Majorana representation, if we impose (4.177), the field will remain real in any frame.

What about in another γ^{μ} representation? We know that we can't keep a field ψ real in general, because the Lorentz transformations aren't in general real. Is there a constraint that is analogous to $\psi = \psi^{\star}$ in general? The answer is yes. To understand what this constraint is, consider our discussion in Sect. 4.3.5. There, we found that the expression $\psi^\dagger\psi$ is *not* a scalar under the Lorentz group. However, the expression $\bar{\psi}\psi$, where $\bar{\psi} = \psi^\dagger\gamma^0$, is a Lorentz scalar (cf. (4.107)). So what does this tell us? It tells us that, in a sense, the quantity $\bar{\psi}$ is the "conjugate" of ψ. In a similar way that $z^\star$ is the conjugate of z (complex numbers), or $\mathbf{v}^T$ (transpose) is the conjugate of $\mathbf{v}$ for a real vector $\mathbf{v}$, or $\mathbf{v}^\dagger$ is the conjugate of a complex vector $\mathbf{v}$. In every case, the product is simply a (real) scalar.

If ψ is a field (in the chiral representation), we can think of $\bar{\psi}$ as its conjugate. But in what sense? Consider a four component spinor field ψ. The top components are left-handed and the bottom components are right-handed (cf. (4.125)). Also, recall from Sects. 3.3.3 and 3.3.5 that we can express right and left-handed fields in terms of each other. Let's take λ and ρ to both be two component left-handed fields. Then the spinor field ψ can be written in the chiral representation as

$$\psi = \begin{pmatrix} \psi_L \\ \psi_R \end{pmatrix} = \begin{pmatrix} \lambda \\ (i\sigma^2)\rho^\star \end{pmatrix}, \tag{4.178}$$

where our conventions for $i\sigma^2$ come from (3.284). For the sake of clarity we can write this out even more explicitly. We'll take

$$\begin{aligned}
\lambda = \lambda_a &= \begin{pmatrix} \lambda_1 \\ \lambda_2 \end{pmatrix}, \\
\rho = \rho_a &= \begin{pmatrix} \rho_1 \\ \rho_2 \end{pmatrix}, \\
(i\sigma^2)\rho^\star &= (i\sigma^2)^{\dot{a}\dot{b}}(\rho_b)^\star = \begin{pmatrix} 0 & 1 \\ -1 & 0 \end{pmatrix}\begin{pmatrix} \rho_1^\star \\ \rho_2^\star \end{pmatrix} = \begin{pmatrix} \rho_2^\star \\ -\rho_1^\star \end{pmatrix}, \\
\psi &= \begin{pmatrix} \lambda_1 \\ \lambda_2 \\ \rho_2^\star \\ -\rho_1^\star \end{pmatrix}.
\end{aligned} \tag{4.179}$$

Then we can write out $\bar{\psi}$. It will be

$$\begin{aligned}
\bar{\psi} &= \psi^\dagger\gamma^0 \\
&= \begin{pmatrix} \lambda_1^\star & \lambda_2^\star & \rho_2 & -\rho_1 \end{pmatrix}\begin{pmatrix} 0 & 0 & 1 & 0 \\ 0 & 0 & 0 & 1 \\ 1 & 0 & 0 & 0 \\ 0 & 1 & 0 & 0 \end{pmatrix} \\
&= \begin{pmatrix} \rho_2 & -\rho_1 & \lambda_1^\star & \lambda_2^\star \end{pmatrix} \\
&= \begin{pmatrix} ((i\sigma^2)\rho^\star)^{\star T} & \lambda^{\star T} \end{pmatrix}.
\end{aligned} \tag{4.180}$$

Now recall from Sect. 4.3.5 that $\bar{\psi}$ and ψ transform differently under Lorentz transformations (cf. equation (4.108)). Remember that our goal is to find a constraint analogous to the reality condition $\psi = \psi^\star$ in the Majorana representation. This means that we need something of the form $\psi = \psi^{(c)}$ where $\psi^{(c)}$ is some sort of conjugate of ψ. As we've said, we're taking $\bar{\psi}$ to be the conjugate of ψ, but these aren't the same type of object (one is a column spinor with a left- and a right-handed field, the other is a row object with the complex conjugates of left- and right-handed spinors). So, our idea is to turn $\bar{\psi}$ into something that we can compare to ψ. The obvious first step is to turn this back into a column object. So, we start with

$$\bar{\psi}^T = \begin{pmatrix} \rho_2 \\ -\rho_1 \\ \lambda_1^\star \\ \lambda_2^\star \end{pmatrix} = \begin{pmatrix} ((i\sigma^2)\rho^\star)^\star \\ \lambda^\star \end{pmatrix} = \begin{pmatrix} (i\sigma^2)\rho \\ \lambda^\star \end{pmatrix}. \tag{4.181}$$

Notice that this is *close* to being a spinor like ψ (with a left-handed field on top and a right-handed field on the bottom). In fact, both the bottom and the top are *only* off by a factor of $i\sigma^2$. So, if we define the matrix

$$C = \begin{pmatrix} 0 & -1 & 0 & 0 \\ 1 & 0 & 0 & 0 \\ 0 & 0 & 0 & 1 \\ 0 & 0 & -1 & 0 \end{pmatrix} = \begin{pmatrix} -(i\sigma^2) & 0 \\ 0 & (i\sigma^2) \end{pmatrix} = \begin{pmatrix} (i\sigma^2)_{ab} & 0 \\ 0 & (i\sigma^2)^{\dot{a}\dot{b}} \end{pmatrix}, \tag{4.182}$$

then we can write out the expression

$$C\bar{\psi}^T = \begin{pmatrix} -(i\sigma^2) & 0 \\ 0 & (i\sigma^2) \end{pmatrix} \begin{pmatrix} (i\sigma^2)\rho \\ \lambda^\star \end{pmatrix} = \begin{pmatrix} -(i\sigma^2)(i\sigma^2)\rho \\ (i\sigma^2)\lambda^\star \end{pmatrix} = \begin{pmatrix} \rho \\ (i\sigma^2)\lambda^\star \end{pmatrix}. \tag{4.183}$$

Notice that we can also write this as (in the chiral representation)

$$C\bar{\psi}^T = -i\gamma^2\psi^\star = -i\begin{pmatrix} 0 & \sigma^2 \\ -\sigma^2 & 0 \end{pmatrix}\begin{pmatrix} \lambda^\star \\ (i\sigma^2)\rho \end{pmatrix} = -i\begin{pmatrix} i\rho \\ -\sigma^2\lambda^\star \end{pmatrix} = \begin{pmatrix} \rho \\ (i\sigma^2)\lambda^\star \end{pmatrix}. \tag{4.184}$$

The point of the expression (4.184) is simply that we can express the conjugate as $i\gamma^2\psi^\star$. This will be useful later.

Further, it is clear that $C\bar{\psi}^T$ has the same form as ψ – it is a column spinor with a left-handed Weyl field on top and a right-handed Weyl field on the bottom.[15] So, we take as our conjugate field that we can use to impose a reality condition to be

$$\psi^{(c)} = C\bar{\psi}^T. \tag{4.185}$$

[15] You are encouraged to work out (4.183) in terms of the spinor indices discussed in Sect. 3.3.5.

Notice that going from ψ to $\psi^{(c)}$ involves merely switching the λ and ρ. Thus, our reality condition is

$$\psi^{(c)} = C\bar{\psi}^T = \psi, \tag{4.186}$$

which reads

$$\begin{pmatrix} \lambda \\ (i\sigma^2)\rho^\star \end{pmatrix} = \begin{pmatrix} \rho \\ (i\sigma^2)\lambda^\star \end{pmatrix}. \tag{4.187}$$

We call this the **Majorana Constraint**.

In general, we call fields satisfying this constraint **Majorana Fields**, which we denote ψ_M, and fields not necessarily satisfying this constraint **Dirac Fields**, which we denote ψ_D. In other words, we have

$$\begin{aligned} \psi_D &= \begin{pmatrix} \lambda \\ (i\sigma^2)\rho^\star \end{pmatrix}, \\ \psi_M &= \begin{pmatrix} \lambda \\ (i\sigma^2)\lambda^\star \end{pmatrix}. \end{aligned} \tag{4.188}$$

We generally talk about ψ_M as a "real" field and ψ_D as a "complex field".

Next, we can write out the Lagrangian for a Majorana field (as opposed to a Dirac field). We can start with (4.128) with the additional constraint

$$\lambda = \psi_L = (i\sigma^2)\psi_R. \tag{4.189}$$

We then get (copying directly from (4.128))

$$\mathcal{L}_M = i(i\sigma^2\lambda^\star)^\dagger\sigma^\mu\partial_\mu(i\sigma^2\lambda^\star) + i\lambda^\dagger\bar{\sigma}^\mu\partial_\mu\lambda - m\big((i\sigma^2\lambda^\star)^\dagger\lambda + \lambda^\dagger(i\sigma^2\lambda^\star)\big). \tag{4.190}$$

Looking at this first term,

$$(i\sigma^2\lambda^\star)^\dagger\sigma^\mu\partial_\mu(i\sigma^2\lambda^\star) = \lambda^T(\sigma^2\sigma^\mu\sigma^2\partial_\mu)\lambda^\star. \tag{4.191}$$

Then, looking at the term in parentheses,

$$\begin{aligned} \sigma^2\sigma^\mu\sigma^2\partial_\mu &= \sigma^2(\sigma^0\partial_0 + \sigma^1\partial_1 + \sigma^2\partial_2 + \sigma^3\partial_3)\sigma^2 \\ &= \sigma^0\partial_0 - \sigma^1\partial_1 + \sigma^2\partial_2 - \sigma^3\partial_3 \\ &= (\sigma^0\partial_0 - \sigma^1\partial_1 - \sigma^2\partial_2 - \sigma^3\partial_3)^\star \\ &= (\bar{\sigma}^\mu\partial_\mu)^\star. \end{aligned} \tag{4.192}$$

So,

$$i(i\sigma^2\lambda^\star)^\dagger\sigma^\mu\partial_\mu(i\sigma^2\lambda^\star) = i\lambda^T(\bar{\sigma}^\mu\partial_\mu)^\star\lambda^\star = i\lambda^\dagger\bar{\sigma}^\mu\partial_\mu\lambda, \tag{4.193}$$

where we have used the fact that the Lagrangian must be real to get the last equality.

Therefore, the first two terms in the Lagrangian are

$$\begin{aligned} i(i\sigma^2\lambda^\star)^\dagger\sigma^\mu\partial_\mu(i\sigma^2\lambda^\star) + i\lambda^\dagger\bar{\sigma}^\mu\partial_\mu\lambda &= i\lambda^\dagger\bar{\sigma}^\mu\partial_\mu\lambda + i\lambda^\dagger\bar{\sigma}^\mu\partial_\mu\lambda \\ &= 2i\lambda^\dagger\bar{\sigma}^\mu\partial_\mu\lambda. \end{aligned} \tag{4.194}$$

The second half of the Lagrangian, the mass terms, can be written as

$$\begin{aligned} m\big((i\sigma^2\lambda^\star)^\dagger\lambda + \lambda^\dagger(i\sigma^2\lambda^\star)\big) &= m\big(-\lambda^T(i\sigma^2)\lambda + \lambda^\dagger(i\sigma^2)\lambda^\star\big) \\ &= im\left(\lambda^\star\sigma^2\lambda^\star - \lambda^T\sigma^2\lambda\right). \end{aligned} \tag{4.195}$$

Writing out the full Majorana Lagrangian (with a new normalization), we have

$$\mathcal{L}_M = i\lambda^\dagger\bar{\sigma}^\mu\partial_\mu\lambda - \frac{i}{2}m\left(\lambda^\star\sigma^2\lambda^\star - \lambda^T\sigma^2\lambda\right). \tag{4.196}$$

We could then work out the equations of motion and find solutions from this Lagrangian to get the Majorana fields. However doing so is very similar to what we've already done, so we'll hold off for now.

A final question is to ask the physical meaning of the transformation that takes

$$\psi \longrightarrow C\bar{\psi}^T. \tag{4.197}$$

It turns out that it is called **Charge Conjugation** for reasons that will be clear later (we'll explain why in Sect. 4.5.4).

4.3.13 Summary of Spin-1/2 Fields

At this point we have all of the details we need regarding spin-1/2 fields. But before moving on to discuss quantization, we need one more topic – gauge theory, which we'll spend the next section on. However, before moving on to gauge theories, we will (as promised) take a few moments to review everything we've just discussed regarding spinors and spin-1/2 fields.

We started by reviewing spin. A "spinor" is an object that essentially specifies the axis of rotation a spin-1/2 particle is "spinning" around, in a similar way as a unit vector with an angular momentum coefficient specifies how a classical object is rotating in spacetime (i.e. $\boldsymbol{\omega} = \omega\hat{\boldsymbol{n}}$). Further, just as the description of a classical object's rotation transforms under $SO(3)$, spinors transform under $SU(2)$.

It is this $SU(2)$ structure that brings the Pauli matrices into the picture; they are the generators of the spinor rotations, and the physical states are expressed in terms of their eigenvectors. We then looked at a geometrical way of picturing spin via the stereographic projection. This gave us a way of understanding not only where spinors come from, but how left- and right-handed spinors are related.

Then we set out to find an equation that would govern spin-1/2 particles. We wanted this equation to be relativistic, which we took to mean that it should "imply" the Klein-Gordon equation (which is simply a field version of the usual relativistic relation $E^2 = m^2 + \mathbf{p}^2$). We found that the only way we could make our equation relativistic was to use a set of matrices γ^μ that obeyed a certain anti-commutation relation, namely the Clifford algebra. This was encouraging because we already knew that the Clifford algebra is related to Lorentz transformations. However, to our surprise we found that our spin-1/2 field (which obviously should only have two components) doesn't work; we were forced by the Clifford algebra to have no less than *four* components. However, we accepted this and pressed on with the Dirac equation using the γ^μ matrices. We found that, with the γ^μ matrices satisfying the Clifford algebra, the Dirac equation is the "square root" of the Klein-Gordon equation. We then spent some time discussing specific solutions of the Clifford algebra for the γ^μ matrices.

Next we set out to find an action for our spin-1/2 fields. We approached this through the knowledge that an action should be a scalar, and we therefore needed to find a way of making a scalar out of the spinor fields. The fact that the spinor representation of the Lorentz group isn't unitary caused significant problems for us, but we were able to find a combination of spinor fields that is a scalar, and another combination that is a vector (which we could contract with the differential operator to form another scalar). With these expressions it was straightforward to write down a Lagrangian and show that it did indeed recover the Dirac equation.

We then moved on to discuss why we needed four components in our field. The reason was that we were trying to make a relativistically acceptable theory, and that means that we must be invariant under *all* Lorentz transformations, even if they aren't proper transformations. This meant that we had to account for parity (switching handedness). As we had discussed when we looked at the geometrical nature of the spinors through the stereographic projection, it is precisely parity that swaps between left- and right-handed spinors. So, because our theory had to account for parity, we can't have *only* left- or *only* right-handed fields. We had to have both. Thus, our fields have four components: two for the left-handed field and two for the right-handed field. As we saw, the Lorentz transformations for left- vs. right-handed fields are exactly related to each other via parity.

Armed with that knowledge we looked at how these two spinors – the left- and right-handed spinors – looked in terms of the general four component field. We found that which parts of ψ are left-handed and which parts are right-handed depends on which representation of the γ^μ matrices we choose to use. A particularly nice representation is the chiral representation, in which the top two components are left-handed and the bottom two are right-handed. We worked out the specific forms of the Lorentz transformations in terms of the γ^μ matrices and found that they do indeed recover the forms we knew from the previous chapters on the Lorentz group.

We then discussed how we can separate out the left- and right-handed parts in representations other than the chiral representation. There are two projection operators that are built from the product of all of the γ^μ matrices, which can be used to tell us where both handed fields are "hiding" in ψ for a given representation.

We then spent a great deal of time looking at solutions to the Dirac equation in various γ^μ matrix representations and in various frames. We showed that the results are independent of our choice of γ^μ matrices (as they should be since physics shouldn't depend on our choice of basis), and therefore will usually use the chiral representation because it gives the answer a nice form.

We found that for a particle sitting still, the left- and right-handed parts are related (how they are related depends on whether you're looking at the positive or negative frequency solutions – they are either equal or differ by a minus sign). We then looked at solutions for moving particles, as well as massless particles or ultra-relativistic massive particles. This led to a discussion of helicity in that we found that massless particles or particles moving very close to the speed of light can only spin around the axis of their motion.

We then looked at two different interpretations of our left- and right-handed particles – Dirac's original (brilliant but incorrect) interpretation, and the QFT interpretation which allows them to simply be normal particles.

Then, finally, we discussed the spinor version of "real" vs. "complex". We found that we can enforce something analogous to a reality condition by imposing the "Majorana constraint". Fields satisfying this are called Majorana fields and fields not satisfying it are called Dirac fields. We'll talk more about the physical interpretation of Majorana fields later.

4.4 Spin-1 Fields

It will be helpful to discuss one final type of field before moving on: vector fields. This will be a field A_μ with four components that transform as a spacetime vector under Lorentz transformations. The electromagnetic potential from Sects. 1.6 and 1.7 (which will become the photon) is an example of this type of field. We'll approach such fields a bit less mathematically and a bit more intuitively.

4.4.1 Building a Lagrangian for Vector Fields

We know that we'll need Lorentz invariant (scalar) terms in order to build a Lagrangian out of A_μ. To second order there are only three such terms,

$$A_\mu A^\mu, \qquad (\partial_\mu A_\nu)(\partial^\mu A^\nu), \qquad (\partial_\mu A^\mu)(\partial_\nu A^\nu). \tag{4.198}$$

We can then build our Lagrangian out of these terms, but we don't know what the coefficient of each should be. So we'll leave them as variables for now. Our Lagrangian for A_μ is then

$$\mathcal{L}_{vec} = -\frac{1}{2}\big((\partial_\mu A_\nu)(\partial^\mu A^\nu) + \alpha(\partial_\mu A^\mu)(\partial_\nu A^\nu) + \beta A_\mu A^\mu\big), \tag{4.199}$$

where α and β are coefficients that we haven't determined yet (we can choose the overall normalization so we're taking the coefficient on the first term to be 1).

We can take a variation with respect to A_μ and get

$$\begin{aligned}\frac{\partial \mathcal{L}_{vec}}{\partial A_\mu} &= -\beta A^\mu,\\ \partial_\nu\left(\frac{\partial \mathcal{L}_{vec}}{\partial(\partial_\nu A_\mu)}\right) &= -\partial_\nu\partial^\nu A^\mu - \alpha\partial^\mu\partial^\nu A_\nu,\\ \Longrightarrow \partial_\nu\partial^\nu A^\mu + \alpha\partial^\mu\partial^\nu A_\nu - \beta A^\mu &= 0.\end{aligned} \tag{4.200}$$

Let's assume that the solution has the general form of a wave,

$$A_\mu = a_\mu e^{ip\cdot x}, \tag{4.201}$$

where a_μ is an arbitrary amplitude vector and $p\cdot x = p^\mu x_\mu$ where p^μ is the energy-momentum 4-vector for the field. Plugging this into the equation of motion (4.200) gives

$$\begin{aligned}\partial_\nu\partial^\nu(a^\mu e^{ip\cdot x}) + \alpha\partial^\mu\partial^\nu(a_\nu e^{ip\cdot x}) - \beta(a^\mu e^{ip\cdot x}) &= 0,\\ \Longrightarrow p_\nu p^\nu a^\mu + \alpha p^\mu p^\nu a_\nu + \beta a^\mu &= 0,\\ \Longrightarrow p_\mu(p^2 a^\mu + \alpha p^\mu p\cdot a + \beta a^\mu) &= 0,\\ \Longrightarrow \big((1+\alpha)p^2 + \beta\big)p\cdot a &= 0.\end{aligned} \tag{4.202}$$

There are various choices we can make for α and β, but there is one choice that gives us a particularly nice set of equations. Namely we want the field to have amplitude only perpendicular to its energy-momentum vector. This is achieved if we set $\alpha = -1$. Then the equation of motion is

$$p\cdot a = 0, \tag{4.203}$$

which eliminates the amplitude in the direction of p^μ (i.e. if the field is sitting still and only moving through time, then the amplitude will be purely spatial, etc.). Also, from the general form of the β term we can take β to be the mass of the field (cf. the mass terms in the spin-0 and spin-1/2 fields). So we take $\beta = m^2$.

Our Lagrangian is now

$$\mathcal{L}_{vec} = -\frac{1}{2}(\partial_\mu A_\nu)(\partial^\mu A^\nu) - \frac{1}{2}m^2 A_\mu A^\mu + \frac{1}{2}(\partial_\mu A^\mu)(\partial_\nu A^\nu). \tag{4.204}$$

The new equation of motion (from (4.200)) is

$$\begin{aligned}&\partial_\nu \partial^\nu A^\mu - \partial^\mu \partial^\nu A_\nu - m^2 A^\mu = 0,\\ \Longrightarrow\ &\partial_\nu(\partial^\nu A^\mu - \partial^\mu A^\nu) - m^2 A^\mu = 0.\end{aligned} \tag{4.205}$$

This is called the **Proca Equation**. This form suggests that we can simplify our notation if we define

$$F^{\mu\nu} = \partial^\mu A^\nu - \partial^\nu A^\mu. \tag{4.206}$$

The equation of motion then becomes

$$\partial_\mu F^{\mu\nu} - m^2 A^\nu = 0. \tag{4.207}$$

We can then rewrite the Lagrangian in terms of this new quantity as

$$\mathcal{L}_{vec} = -\frac{1}{4}F_{\mu\nu}F^{\mu\nu} - \frac{1}{2}m^2 A_\mu A^\mu. \tag{4.208}$$

(You are encouraged to work out the details to show that this is the correct Lagrangian.) Notice that this is the same as (1.127) except now our fields have mass and we aren't including a source term. We have in a sense recovered electromagnetism, but generalized to massive fields.

4.4.2 Vector Fields in the Massless Limit

We now want to consider what happens in the massless limit. Two things will happen. First of all, (4.202) is satisfied automatically, regardless of p^μ and a^μ. So, one of the three transverse amplitude degrees of freedom becomes longitudinal. Second, the theory develops a symmetry. Namely, under the **Gauge Transformation**

$$A_\mu \longrightarrow A_\mu + \partial_\mu \chi, \tag{4.209}$$

where χ is absolutely any function at all, the field strength $F_{\mu\nu}$ is unchanged and the Lagrangian changes only by a total derivative (cf. Sect. 1.8). Notice that it is the presence of a non-zero mass term that "obstructs" the gauge freedom; the gauge symmetry is only there when there is no mass. We will eventually find that such a relationship – of mass "spontaneously breaking" a symmetry – to be extremely important.

At this point we have happened upon gauge symmetry, which in Chap. 1 we said was enormously important. Now that we have reviewed the three major types of fields we'll be working with, it is time to outline exactly why gauge transformations are so important.

4.5 Gauge Theory

We've now discussed everything we need to determine the field equations for spin-0, spin-1/2, and vector fields. However, before proceeding to quantization, there is one extraordinarily powerful idea that we need to discuss: gauge theories. These next few sections literally form the basis for nearly everything in particle physics (actually, nearly everything in physics), and will be the primary subject of the next several books in this series. You should read the following sections very, very carefully.

4.5.1 Conserved Currents

In Chap. 1 we discussed how symmetries and conserved quantities are related. Let's consider a few examples of this using the Lagrangians we just defined. Consider a massless Klein-Gordon scalar particle, described by

$$\mathcal{L} = -\frac{1}{2}\partial^\mu \phi \partial_\mu \phi. \tag{4.210}$$

Following what we did starting with equation (1.30), consider the transformation $\phi \to \phi + \epsilon$, where ϵ is a constant. Because

$$\partial^\mu \phi \to \partial^\mu \phi + \partial^\mu \epsilon = \partial^\mu \phi, \tag{4.211}$$

the Lagrangian is invariant. So (using $\delta\phi = 1$), our conserved quantity is

$$j^\mu = \frac{\partial \mathcal{L}}{\partial(\partial_\mu \phi)}\delta\phi = -\partial^\mu \phi. \tag{4.212}$$

Similarly, consider the Klein-Gordon Lagrangian with complex scalar fields ϕ and $\phi^\dagger$, which we write as

$$\mathcal{L} = -\partial^\mu \phi^\dagger \partial_\mu \phi - m^2 \phi^\dagger \phi. \tag{4.213}$$

We can make the transformations

$$\begin{aligned} \phi &\longrightarrow e^{i\alpha}\phi, \\ \phi^\dagger &\longrightarrow \phi^\dagger e^{-i\alpha}, \end{aligned} \tag{4.214}$$

(where α is an arbitrary real constant). This type of transformation is called a $U(1)$ transformation, because $e^{i\alpha}$ is an element of the group of all 1×1 unitary matrices, as discussed in Sect. 3.2.1. The conserved quantity associated with this $U(1)$ symmetry is

$$j^{\mu} = \frac{\partial \mathcal{L}}{\partial(\partial_{\mu}\phi)}\delta\phi + \frac{\partial \mathcal{L}}{\partial(\partial_{\mu}\phi^{\dagger})}\delta\phi^{\dagger} = i(\phi\partial^{\mu}\phi^{\dagger} - \phi^{\dagger}\partial^{\mu}\phi). \tag{4.215}$$

Consider also the Dirac Lagrangian. Notice that it is invariant under the $U(1)$ transformation $\psi \to e^{i\alpha}\psi$, with current

$$j^{\mu} = \bar{\psi}\gamma^{\mu}\psi. \tag{4.216}$$

In both of the previous examples, notice that the $U(1)$ symmetry changes the field at *all* points in space at once, and all in the same way. In other words, it is a single overall constant phase $e^{i\alpha}$. We therefore call such a symmetry a **Global Symmetry**. The implications of this are likely not clear at this point. We merely wish to call your attention to the fact that $e^{i\alpha}$ has no spacetime dependence – it's just a phase.

4.5.2 The Dirac Equation with an Electromagnetic Field

Previously we found the Lagrangian for an electromagnetic field (1.126). Our goal now is to find a Lagrangian that describes the electromagnetic field and a spin-1/2 particle that couples to the electromagnetic field, and additionally the interaction between them. We start by writing down a Lagrangian without any interaction. This will simply be the sum of the two terms (1.126) and (4.112),

$$\mathcal{L} = \mathcal{L}_D + \mathcal{L}_{EM} = \bar{\psi}(i\gamma^{\mu}\partial_{\mu} - m)\psi - \frac{1}{4}F_{\mu\nu}F^{\mu\nu} - J^{\mu}A_{\mu}. \tag{4.217}$$

Because the Dirac part has no terms in common with the electromagnetic part, the equations of motion and the conserved quantities for both ψ and A^{μ} will be exactly the same, as if the other weren't present at all. In other words, both fields go about their way, ignoring the other; there is no interaction in this theory. Because this makes for a boring universe (and horrible phenomenology[16]), we need to find some way of coupling the two fields together to produce some sort of interaction.

---

[16]For readers new to particle physics, "phenomenology" is the study of relating the abstract and mathematical world of theory to things that can be experimentally measured in a lab. In other words, phenomenology involves pulling physically measurable values out of a theory and comparing them to what is actually measured. Obviously a theory that doesn't have any interacting particles, while perhaps providing an interesting mathematical theory, doesn't match up with what we see in nature and is therefore phenomenologically inadequate.

Interaction is added to a physical theory by adding another term to the Lagrangian called the **Interaction Term**. So, the final Lagrangian will have the form

$$\mathcal{L} = \mathcal{L}_D + \mathcal{L}_{EM} + \mathcal{L}_{int}. \tag{4.218}$$

Now, for reasons that will become clear in the next section (and even more clear when we quantize), we do this by coupling the electromagnetic field A^μ to the current resulting from the $U(1)$ symmetry in $\mathcal{L}_D$, which we discussed in Sect. 4.5.1, and wrote out in equation (4.216). In other words, our interaction term will be proportional to $A^\mu j_\mu$.

Adding a constant of proportionality q (which we will see has the physical interpretation of a coupling constant, weighting the probability of an interaction to take place, or equivalently the physical interpretation of electric charge), our Lagrangian is now

$$\begin{aligned}\mathcal{L} &= \bar{\psi}(i\gamma^\mu\partial_\mu - m)\psi - \frac{1}{4}F_{\mu\nu}F^{\mu\nu} - J^\mu A_\mu - qj^\mu A_\mu \\ &= \bar{\psi}(i\gamma^\mu\partial_\mu - m)\psi - \frac{1}{4}F_{\mu\nu}F^{\mu\nu} - (J^\mu + q\bar{\psi}\gamma^\mu\psi)A_\mu.\end{aligned} \tag{4.219}$$

Notice that $\mathcal{L}$ is still invariant under the global $U(1)$ symmetry, and the $U(1)$ current is still $j^\mu = \bar{\psi}\gamma^\mu\psi$.

Also, notice that the Lagrangians in (4.217) and (4.219) are the same except for a shift in the current term, $J^\mu \to J^\mu + qj^\mu$. Recall that physically, $J^\mu = (\rho, \mathbf{J})$ represents the charge and current creating the field. The fact that J^μ has shifted in (4.219) simply means that the spin-1/2 particle in this theory contributes to the field, which is exactly what we would expect it to do.

If we set $q = e$, the electric charge, this Lagrangian becomes upon quantization the Lagrangian of Quantum Electrodynamics (QED), which to date makes the most accurate experimental predictions of any scientific theory in history. In the next section we will re-derive this Lagrangian in a more fundamental way.

4.5.3 Gauging the Symmetry

In terms of physics, this section is probably the most important of this book. Please read and reread this section until you understand every step. One comment before beginning: you can think of this section as another way of getting the result (4.219) from the previous section. You should forget everything we've said about electromagnetism in this book while reading this section. We are going to recover the entire theory from more fundamental principles.

Consider once again the Dirac Lagrangian (4.80). As we said in Sect. 4.5.1, it is invariant under the *global* $U(1)$ transformation $\psi \to e^{i\alpha}\psi$. It is global in that it acts on the field the exact same way at every point in spacetime. The idea behind this section is that we are going to make this symmetry **Local**, so that α depends on spacetime

$$\alpha = \alpha(x^\mu), \tag{4.220}$$

and then try to force the Lagrangian to maintain its invariance under the *local* $U(1)$ transformation. Making a global symmetry local is referred to as **Gauging** the symmetry.

We start by making the local $U(1)$ transformation:

$$\mathcal{L} = \bar{\psi}(i\gamma^\mu \partial_\mu - m)\psi \rightarrow \bar{\psi}e^{-i\alpha(x)}(i\gamma^\mu \partial_\mu - m)e^{i\alpha(x)}\psi, \tag{4.221}$$

and because the differential operators will now act on $\alpha(x)$ as well as ψ, we get extra terms:

$$\begin{aligned}\mathcal{L} \rightarrow \bar{\psi}e^{-\alpha(x)}(i\gamma^\mu \partial_\mu - m)e^{i\alpha(x)}\psi &= \bar{\psi}(i\gamma^\mu \partial_\mu - m)\psi - \bar{\psi}\gamma^\mu \psi \partial_\mu \alpha(x) \\ &= \bar{\psi}(i\gamma^\mu \partial_\mu - m - \gamma^\mu \partial_\mu \alpha(x))\psi. \end{aligned} \tag{4.222}$$

If we want to demand that $\mathcal{L}$ still be invariant under this local $U(1)$ transformation, we must find a way of canceling the $\bar{\psi}\gamma^\mu \psi \partial_\mu \alpha(x)$ term. We do this in the following way. Define some new field A_μ which under the $U(1)$ transformation $e^{i\alpha(x)}$ transforms according to

$$A_\mu \rightarrow A_\mu - \frac{1}{q}\partial_\mu \alpha(x). \tag{4.223}$$

Note that we are *defining* A_μ to transform this way; there is nothing about it that makes this transformation natural.[17] We call A_μ the **Gauge Field** for reasons that will be clear soon, and q is a constant we have included for later convenience.

We introduce A_μ by replacing the standard derivative ∂_μ with the **Covariant Derivative**

$$D_\mu = \partial_\mu + iqA_\mu. \tag{4.224}$$

If you have studied general relativity or differential geometry at any point, you are familiar with covariant derivatives. If you haven't seen general relativity or differential geometry, don't worry about it for now. Just accept that a covariant derivative is the normal partial derivative with a term proportional to a gauge field added to it. The deeper meaning will be given later.

As a comment regarding vocabulary, to say that a particle "carries charge" mathematically means that it has the corresponding term in its covariant derivative. Thus,

---

[17] Actually, there is something natural about this definition, but we won't get there until the next book.

if a particle's covariant derivative is equal to the normal differential operator ∂^μ, then the particle has no charge, and it will not interact with any gauge fields. However, if it carries charge, it will have a term corresponding to that charge in its covariant derivative. This will become clearer as we proceed.

Thus, our Lagrangian is now

$$\begin{aligned} \mathcal{L} &= \bar{\psi}(i\gamma^\mu D_\mu - m)\psi \\ &= \bar{\psi}(i\gamma^\mu[\partial_\mu + iqA_\mu] - m)\psi \\ &= \bar{\psi}(i\gamma^\mu\partial_\mu - m - q\gamma^\mu A_\mu)\psi. \end{aligned} \tag{4.225}$$

Under the local $U(1)$ we have

$$\begin{aligned} \mathcal{L} \to \bar{\psi}e^{-i\alpha(x)} & \left(i\gamma^\mu\partial_\mu - m - q\gamma^\mu\left[A_\mu - \frac{1}{q}\partial_\mu\alpha(x)\right]\right)e^{i\alpha(x)}\psi \\ &= \bar{\psi}(i\gamma^\mu\partial_\mu - m - \gamma^\mu\partial_\mu\alpha(x) - q\gamma^\mu A_\mu + \gamma^\mu\partial_\mu\alpha(x))\psi \\ &= \bar{\psi}(i\gamma^\mu\partial_\mu - m - q\gamma^\mu A_\mu)\psi = \bar{\psi}(i\gamma^\mu D_\mu - m)\psi \\ &= \mathcal{L}. \end{aligned} \tag{4.226}$$

So, the addition of the field A_μ has indeed restored the $U(1)$ symmetry. Notice that now it is not only invariant under the local $U(1)$, but also still under the global $U(1)$ with which we started, and has the same conserved $U(1)$ current, $j^\mu = \bar{\psi}\gamma^\mu\psi$. This allows us to rewrite the Lagrangian as

$$\mathcal{L} = \bar{\psi}(i\gamma^\mu D_\mu - m)\psi = \bar{\psi}(i\gamma^\mu\partial_\mu - m)\psi - qj^\mu A_\mu \tag{4.227}$$

However, we have a problem. If we want to know what the dynamics of A_μ will be, we naturally take the variation of the Lagrangian with respect to A_μ. Because there are no derivatives of A_μ, the Euler-Lagrange equation is merely

$$\frac{\partial\mathcal{L}}{\partial A_\mu} = -q\bar{\psi}\gamma^\mu\psi = -qj^\mu = 0. \tag{4.228}$$

Thus, the equation of motion for A_μ says that the current vanishes, or that $j^\mu = 0$, and so the Lagrangian is reduced back to (4.112), which was not invariant under the local $U(1)$.

We can state this problem in another way. All physical fields have some sort of dynamics. If they don't, then they are merely a constant background field that never changes and does nothing. As it is written, equation (4.227) has a field A_μ, but A_μ has no kinetic term, and therefore no dynamics.

To fix this problem we must include some sort of dynamics, or kinetic terms, for A_μ. The way to do this turns out to involve a considerable amount of geometry which would be out of place in this book. We will cover the necessary ideas later in this series and derive the following expressions. For now we merely give the results and ask you for patience until we have the machinery to derive them.

For an arbitrary field A_μ, the appropriate gauge-invariant kinetic term is

$$\mathcal{L}_{Kin,A} = -\frac{1}{4} F_{\mu\nu} F^{\mu\nu} \tag{4.229}$$

where

$$F^{\mu\nu} = \frac{i}{q}[D^\mu, D^\nu] \tag{4.230}$$

and q is the constant of proportionality introduced in the transformation of A_μ in equation (4.223). D^μ is the covariant derivative defined in (4.224).

Writing out (4.230) (and using an arbitrary test function $f(x)$),

$$\begin{aligned}
F^{\mu\nu} f(x) &= \frac{i}{q}[D^\mu, D^\nu] f(x) \\
&= \frac{i}{q}\big[(\partial^\mu + iqA^\mu)(\partial^\nu + iqA^\nu) - (\partial^\nu + iqA^\nu)(\partial^\mu + iqA^\mu\big] f(x) \\
&= \frac{i}{q}\big[\partial^\mu \partial^\nu f(x) + iq\partial^\mu(A^\nu f(x)) + iqA^\mu \partial^\nu f(x) - q^2 A^\mu A^\nu f(x) \\
&\qquad - \partial^\nu \partial^\mu f(x) + iq\partial^\nu(A^\mu f(x)) + iqA^\nu \partial^\mu f(x) - q^2 A^\nu A^\mu f(x)\big] \\
&= \frac{i}{q}\big[iqf(x)\partial^\mu A^\nu + iqA^\nu \partial^\mu f(x) + iqA^\mu \partial^\nu f(x) - q^2 A^\mu A^\nu f(x) \\
&\qquad - iqf(x)\partial^\nu A^\mu - iqA^\mu \partial^\nu f(x) - iqA^\nu \partial^\mu f(x) + q^2 A^\nu A^\mu f(x)\big] \\
&= \big[\partial^\mu A^\nu - \partial^\nu A^\mu + iq[A^\mu, A^\nu]\big] f(x).
\end{aligned} \tag{4.231}$$

For each value of μ, A^μ is a scalar function, so the commutator term vanishes, leaving (dropping the test function $f(x)$)

$$F^{\mu\nu} = \frac{i}{q}[D^\mu, D^\nu] = \partial^\mu A^\nu - \partial^\nu A^\mu. \tag{4.232}$$

Therefore, writing out the entire Lagrangian we have

$$\mathcal{L} = \bar{\psi}(i\gamma^\mu D_\mu - m)\psi - \frac{1}{4} F_{\mu\nu} F^{\mu\nu}. \tag{4.233}$$

Finally, because A^μ is obviously a physical field, we can naturally assume that there is some source term causing it, or a source term, which we simply call J^μ. This makes our final Lagrangian

$$\mathcal{L} = \bar{\psi}(i\gamma^\mu D_\mu - m)\psi - \frac{1}{4} F_{\mu\nu} F^{\mu\nu} - J^\mu A_\mu. \tag{4.234}$$

Comparing this to (4.219), we see that they are exactly the same. So what have we done? We started with nothing but a Lagrangian for a spin-1/2 particle, which had a global $U(1)$ symmetry. Then, all we did was promote the $U(1)$ symmetry to a local symmetry (we gauged the symmetry), and then imposed what was necessary to get a consistent (invariant) theory. The gauge field A_μ was forced upon us, and the form of the kinetic term for A_μ is demanded automatically by geometric considerations we did not delve into (but will in the next book).

In other words, we started with nothing but a non-interacting particle, and by specifying *nothing* but $U(1)$ we have created a theory with not only that same particle, but also electromagnetism. The A_μ field, which upon quantization will be the photon, is a direct consequence of the $U(1)$. This is what we meant at the end of Sect. 3.2.11 when we said that electromagnetism is described by $U(1)$. We will talk more about the weak and strong forces later, as well as the groups that give rise to them. Theories of this type, where we generate forces by specifying a Lie group, are called **Gauge Theories**, or **Yang-Mills Theories**.

Finally, notice that (4.223) has exactly the same form as (1.130). This is why we call A_μ a gauge field. The gauge symmetry in electromagnetism is a sort of remnant of the much deeper and more fundamental $U(1)$ structure of the theory.

4.5.4 A Final Comment: Charge Conjugation

Before moving on, we'll make a brief comment about something we referenced earlier. Namely, at the end of Sect. 4.3.12 we mentioned that the operation

$$\psi \longrightarrow C\bar{\psi}^T \tag{4.235}$$

is the charge conjugation operation. We're now ready to see why.

In the Dirac Lagrangian (4.112) the fundamental field was the 4 component spinor which, in the chiral representation of the γ^μ matrices had the form

$$\psi = \begin{pmatrix} \psi_L \\ \psi_R \end{pmatrix}. \tag{4.236}$$

We also had the field

$$\bar{\psi} = \psi^\dagger \gamma^0 = \begin{pmatrix} \psi_R^\dagger & \psi_L^\dagger \end{pmatrix}. \tag{4.237}$$

Let's now take a closer look at the nature of the Dirac equation (4.80) with charge q in an electric field A_μ (cf. Sect. 4.5.3, where we are ignoring the source J^μ and the kinetic term for A_μ):

$$(i\gamma^\mu D_\mu - m)\psi = \big(i\gamma^\mu(\partial_\mu + iqA_\mu) - m\big)\psi = 0. \tag{4.238}$$

Next, take the complex conjugate of this equation:

$$\big(-i(\gamma^{\mu})^{\star}(D_{\mu})^{\star}-m\big)\psi^{\star}=\big(-i(\gamma^{\mu})^{\star}(\partial_{\mu}-iqA_{\mu})-m\big)\psi^{\star}=0. \quad (4.239)$$

Looking at the gamma matrices in the chiral representation (cf. (4.89)), we have

$$\begin{aligned}(\gamma^0)^{\star}&=\gamma^0,\\(\gamma^1)^{\star}&=\gamma^1,\\(\gamma^2)^{\star}&=-\gamma^2,\\(\gamma^3)^{\star}&=\gamma^3.\end{aligned} \quad (4.240)$$

Only γ^2 is affected by this. Multiply (4.239) by γ^2 and then use the anti-commutation relations (4.78):

$$\begin{aligned}\big[-i\big(\gamma^2\gamma^0D_0^{\star}+\gamma^2\gamma^1D_1^{\star}-\gamma^2\gamma^2D_2^{\star}+\gamma^2\gamma^3D_3^{\star}\big)-m\big]\psi^{\star}&=0,\\\big[-i\big(-\gamma^0\gamma^2D_0^{\star}-\gamma^1\gamma^2D_1^{\star}-\gamma^2\gamma^2D_2^{\star}-\gamma^2\gamma^3D_3^{\star}\big)-m\big]\psi^{\star}&=0,\\\big[i\big(\gamma^0D_0^{\star}+\gamma^1D_1^{\star}+\gamma^2D_2^{\star}+\gamma^3D_3^{\star}\big)-m\big](\gamma^2\psi^{\star})&=0,\\\big(i\gamma^{\mu}D_{\mu}^{\star}-m\big)(\gamma^2\psi^{\star})&=0,\\(i\gamma^{\mu}(\partial_{\mu}-iqA_{\mu})-m)(\gamma^2\psi^{\star})&=0,\\(i\gamma^{\mu}(\partial_{\mu}-iqA_{\mu})-m)(-i\gamma^2\psi^{\star})&=0,\\(i\gamma^{\mu}(\partial_{\mu}-iqA_{\mu})-m)\psi^{(c)}&=0,\end{aligned} \quad (4.241)$$

where we have used (4.184) to get the last line. We see that if ψ satisfies the Dirac equation for a particle with charge q, then $\psi^{(c)}$ satisfies the Dirac equation for a particle with charge $-q$. Charge conjugation does indeed swap the charge of the field.

4.6 Quantization

4.6.1 *Review of What Quantization Means*

Quantization is done by taking certain dynamical quantities and making use of the **Heisenberg Uncertainty Principle**. Normally we take position x and momentum p and, according to Heisenberg, the measurement of the particle's position will effect its momentum and vice-versa.

To make this more precise, we promote x and p from merely being variables to being Hermitian operators $\hat{x}$ and $\hat{p}$ (which can be represented by matrices) acting on some vector space. Calling a vector in this space $|\psi\rangle$, physically measurable quantities (like position or momentum) become the eigenvalues of the operators $\hat{x}$ and $\hat{p}$,

$$\hat{x}|\psi\rangle = x|\psi\rangle,$$
$$\hat{p}|\psi\rangle = p|\psi\rangle. \tag{4.242}$$

Heisenberg Uncertainty says that measuring x will affect the value of p, and vice-versa. Specifically, the accuracy to which x is measured sets a limit to the accuracy to which p can simultaneously be measured, and vice versa. The lower the uncertainty in the measurement of x, the greater the uncertainty in the measurement of p, and vice-versa. It is the act of measuring which enacts this effect. It is not an engineering problem in the sense that there is no better measurement technique which would undo this. It is a fundamental fact of quantum mechanics (and therefore of our universe) that measurement of one variable affects another.

If we measure x (using $\hat{x}$) and then p (using $\hat{p}$), we will in general get different values for both than if we measured p and then x. More mathematically, $\hat{x}\hat{p} \neq \hat{p}\hat{x}$. Put another way,

$$[\hat{x}, \hat{p}] = \hat{x}\hat{p} - \hat{p}\hat{x} \neq 0. \tag{4.243}$$

For reasons learned in an introductory quantum course, the actual relation is

$$[\hat{x}, \hat{p}] = i\hbar, \tag{4.244}$$

where $\hbar$ is Planck's Constant. We call (4.244) the **Canonical Commutation Relation**, and it is this structure which allows us to determine the physical structure of the theory.

More generally, we choose some set of operators that all commute with each other, and then label a physical state by its eigenvectors. For example $\hat{x}$, $\hat{y}$ and $\hat{z}$ all commute with each other, so we may label a physical state by its eigenvectors $|\psi_r\rangle = |x, y, z\rangle$. Or, because $\hat{p}_x$, $\hat{p}_y$, and $\hat{p}_z$ all commute, we may call the state $|\psi_p\rangle = |p_x, p_y, p_z\rangle$. We may also include some other values like spin and angular momentum, to have (for example) $|\psi\rangle = |x, y, z, s_z, L_z, \ldots\rangle$. As a note, this is absolutely identical to what we were doing in Sect. 3.2.12. We are just emphasizing the physics rather than the mathematics for now.

As discussed in section 4.1, when we make the jump to QFT, the fields are no longer the states but the operators. We are therefore going to impose commutation relations on the fields, not on the coordinates. This means that it is not possible to know both the field and its canonical momentum to infinite precision at the same spacetime point. The more precise we know one, the less precise we can know the other.

Furthermore, whereas before the states were eigenvectors of the coordinate operators, we now will expand the fields in terms of the eigenvectors of the Hamiltonian.

4.6.2 Canonical Quantization of Scalar Fields

We'll start with real scalar fields. We begin with the Klein-Gordon Lagrangian in equation (4.11), but we make the slight modification of adding an arbitrary constant Ω,

$$\mathcal{L}_{KG} = -\frac{1}{2}\partial^{\mu}\phi\partial_{\mu}\phi - \frac{1}{2}m^2\phi^2 + \Omega. \tag{4.245}$$

Note that Ω has absolutely no affect whatsoever on the physics because any derivatives of it will vanish, so it won't show up in the equations of motion.

Quantization then comes about by defining the field momentum and Hamiltonian (using (1.114) and (1.115)) to get

$$\Pi = \frac{\partial\mathcal{L}}{\partial\dot{\phi}(x)} = \dot{\phi}, \tag{4.246}$$

$$\mathcal{H} = \Pi\dot{\phi} - \mathcal{L} = \frac{1}{2}\Pi^2 + \frac{1}{2}(\nabla\phi)^2 + \frac{1}{2}m^2\phi^2 - \Omega. \tag{4.247}$$

Now, using the canonical commutation relations (4.244) as guides, we impose

$$\begin{aligned}
[\phi(t,\mathbf{x}), \phi(t',\mathbf{x}')] &= 0,\\
[\Pi(t,\mathbf{x}), \Pi(t',\mathbf{x}')] &= 0,\\
[\phi(t,\mathbf{x}), \Pi(t',\mathbf{x}')] &= i\delta(t-t')\delta(\mathbf{x}-\mathbf{x}'),
\end{aligned} \tag{4.248}$$

(where we have set $\hbar = 1$). The point is that ϕ commutes with itself at any spacetime points (just like any two spatial coordinates commute), Π commutes with itself at any spacetime points (just like any two momentum operators commute), but the field and its momentum don't commute at the same spacetime points (just like position and momentum don't commute).

To get a better feel for what these commutation relations do, let's expand out the full solution to the Klein-Gordon equation (4.23). The *full* solution will not only be the sum over the positive and negative frequency parts, but also the sum over all energy-momentum values p_{μ} (because (4.23) is a solution for any value of p_{μ}). Hence, the general solution is

$$\phi(t,\mathbf{x}) = \int \frac{d^3\mathbf{p}}{f(\mathbf{p})}\left[a(\mathbf{p})e^{-iEt+i\mathbf{p}\cdot\mathbf{x}} + b(\mathbf{p})e^{iEt-i\mathbf{p}\cdot\mathbf{x}}\right], \tag{4.249}$$

where $f(\mathbf{p})$ is a redundant function which we have included for later convenience (i.e. we didn't need to include it; we are because we know it will be useful later). For now, both $a(\mathbf{p})$ and $b(\mathbf{p})$ are the arbitrary Fourier coefficients used to expand $\phi(t, \mathbf{x})$ in terms of individual solutions.

Now we want to make sure the ϕ is real. This requires

$$\phi^{\dagger} = \phi, \tag{4.250}$$

which for scalar fields is simply

$$\phi^{\star} = \phi. \tag{4.251}$$

Imposing this on the field gives (suppressing the measure part of the integral to be notationally concise)

$$\int \left[a(\mathbf{p})e^{-iEt+i\mathbf{p}\cdot\mathbf{x}} + b(\mathbf{p})e^{iEt-i\mathbf{p}\cdot\mathbf{x}}\right] = \int \left[a^{\star}(\mathbf{p})e^{iEt-i\mathbf{p}\cdot\mathbf{x}} + b^{\star}(\mathbf{p})e^{-iEt+i\mathbf{p}\cdot\mathbf{x}}\right] \tag{4.252}$$

Thus, the reality of ϕ means that we must require

$$b(\mathbf{p}) = a^{\star}(\mathbf{p}). \tag{4.253}$$

So our (real) field is (using the notation $p \cdot x = p_{\mu}x^{\mu} = -Et + \mathbf{p}\cdot\mathbf{x}$ once again)

$$\phi(x) = \int \frac{d^3\mathbf{p}}{f(\mathbf{p})}\left[a(\mathbf{p})e^{ip\cdot x} + a^{\dagger}(\mathbf{p})e^{-ip\cdot x}\right], \tag{4.254}$$

where we are using the dagger $\dagger$ rather than $\star$ once again.

Now we need to deal with a problem. Notice that the integration measure, $d^3\mathbf{p}$, is not invariant under Lorentz transformations (because it integrates over the spatial part but not over the time part). We therefore choose $f(\mathbf{p})$ to restore Lorentz invariance (that's why we included it in the first place).

We know that the measure d^4p would be invariant, as would δ functions and Θ (step) functions. So consider the invariant combination

$$d^4p\,\delta(p^2 + m^2)\Theta(p^0). \tag{4.255}$$

The δ function is requiring

$$p^2 + m^2 = p_{\mu}p^{\mu} + m^2 = -E^2 + \mathbf{p}^2 + m^2 = -m^2 + m^2 = 0, \tag{4.256}$$

which is obviously satisfied as long as the field satisfies (1.98), which it always should. The Θ function is then simply preserving causality. This is a relativistically invariant integration measure that is physically acceptable.

How can we choose $f(\mathbf{p})$ to make the measure in $\phi(x)$ invariant? Recall the general δ function identity,

$$\int_{-\infty}^{\infty} dx\delta(g(x)) = \sum_i \frac{1}{\left|\frac{dg(x)}{dx}\Big|_{x=x_i}\right|}, \tag{4.257}$$

where the x_i's are the zeros of the function $g(x)$. We can do the $p^0 = E$ integral over the measure (4.255), and using the fact that the zeros of $p^2 + m^2 = -E^2 + \mathbf{p}^2 + m^2$ in terms of $p^0 = E$ are

$$(p^0)^2 = E^2 = \mathbf{p}^2 + m^2, \tag{4.258}$$

we get

$$\int d^3\mathbf{p} dp^0 \delta(p^2 + m^2)\Theta(p^0) = \int \frac{d^3\mathbf{p}}{2E}. \tag{4.259}$$

Adding a factor of $(2\pi)^3$ for later convenience, we take our Lorentz invariant measure to be

$$\frac{d^3\mathbf{p}}{(2\pi)^3 2E}. \tag{4.260}$$

The end result is,

$$\phi(x) = \int \widetilde{dp}\big[a(\mathbf{p})e^{ip\cdot x} + a^\dagger(\mathbf{p})e^{-ip\cdot x}\big], \tag{4.261}$$

where we have defined

$$\widetilde{dp} = \frac{d^3\mathbf{p}}{(2\pi)^3 2E}. \tag{4.262}$$

We started with the general solution of the Klein-Gordon equation and, after imposing the reality condition, determined that the second coefficient ($b(\mathbf{p})$) is the conjugate of the first. Then we found an integration measure that is Lorentz invariant. Now, finally, we want to see what commutation relations are imposed on a and $a^\dagger$ by the canonical commutation relations (4.248). You can work out the details yourself to find that a and $a^\dagger$ will obey

$$\begin{aligned} [a(\mathbf{p}), a(\mathbf{p}')] &= 0, \\ [a^\dagger(\mathbf{p}), a^\dagger(\mathbf{p}')] &= 0, \\ [a(\mathbf{p}), a^\dagger(\mathbf{p}')] &= (2\pi)^3(2E)\delta^3(\mathbf{p} - \mathbf{p}'), \end{aligned} \tag{4.263}$$

(showing this is fairly tedious, but we encourage you to work it out). We are using $\dagger$ instead of $\star$ to emphasize that, in the quantum theory, we are talking about Hermitian operators. In this case (with scalar fields) it makes no difference.

We can now write the Hamiltonian H in terms of (4.261):

$$
\begin{aligned}
H &= \int d^3x\mathcal{H} = \int d^3x\left(\frac{1}{2}\Pi^2 + \frac{1}{2}(\nabla\phi)^2 + \frac{1}{2}m^2\phi^2 - \Omega\right) \\
&= \frac{1}{2}\int \widetilde{dp}\widetilde{dp'}d^3x\big[(-iEa(\mathbf{p})e^{ip\cdot x} + iEa^\dagger(\mathbf{p})e^{-ip\cdot x})(-iE'a(\mathbf{p}')e^{i\mathbf{p}'\cdot x} \\
&\qquad + iE'a^\dagger(\mathbf{p}')e^{-ip'\cdot x}) + (i\mathbf{p}a(\mathbf{p})e^{ip\cdot x} - i\mathbf{p}a^\dagger(\mathbf{p})e^{-ip\cdot x}) \\
&\qquad \times\ (i\mathbf{p}'a(\mathbf{p}')e^{ip'\cdot x} - i\mathbf{p}'a^\dagger(\mathbf{p}')e^{-ip'\cdot x})m^2(a(\mathbf{p})e^{ip\cdot x} \\
&\qquad + a^\dagger(\mathbf{p})e^{-ip\cdot x})(a(\mathbf{p}')e^{ip'\cdot x} + a^\dagger(\mathbf{p}')e^{-ip'\cdot x})\big] - \int d^3x\Omega \\
&= \frac{1}{2}\int \widetilde{dp}\widetilde{dp'}d^3x\big[\big(-EE'a(\mathbf{p})a(\mathbf{p}')e^{i(p+p')\cdot x} + EE'a(\mathbf{p})a^\dagger(\mathbf{p}')e^{i(p-p')\cdot x} \\
&\quad + EE'a^\dagger(\mathbf{p})a(\mathbf{p}')e^{-i(p-p')\cdot} - EE'a^\dagger(\mathbf{p})a^\dagger(\mathbf{p}')e^{-i(p+p')\cdot x}\big) \\
&\quad + \big(-\mathbf{p}\cdot\mathbf{p}'a(\mathbf{p})a(\mathbf{p}')e^{i(p+p')\cdot x} + \mathbf{p}\cdot\mathbf{p}'a(\mathbf{p})a^\dagger(\mathbf{p}')e^{i(p-p')\cdot x} \\
&\qquad + \mathbf{p}\cdot\mathbf{p}'a^\dagger(\mathbf{p})a(\mathbf{p}')e^{-i(p-p')\cdot x} - \mathbf{p}\cdot\mathbf{p}'a^\dagger(\mathbf{p})a^\dagger(\mathbf{p}')e^{-i(p+p')\cdot x}\big) \\
&\quad + m^2\big(a(\mathbf{p})a(\mathbf{p}')e^{i(p+p')\cdot x} + a(\mathbf{p})a^\dagger(\mathbf{p}')e^{i(p-p')\cdot x} \\
&\qquad + a^\dagger(\mathbf{p})a(\mathbf{p}')e^{-i(p-p')\cdot x} + a^\dagger(\mathbf{p})a^\dagger(\mathbf{p}')e^{-i(p+p')\cdot x}\big)\big] \\
&\quad - V\Omega,
\end{aligned} \tag{4.264}
$$

where V is the volume of the space resulting from the $\int d^3x$ integral. Then, from the fact that (in general)

$$
\int d^3x e^{i\mathbf{x}\cdot\mathbf{y}} = (2\pi)^3\delta^3(\mathbf{y}), \tag{4.265}
$$

we have

$$
\begin{aligned}
H &= \frac{1}{2}(2\pi)^3\int \widetilde{dp}\widetilde{dp'}\big[\delta^3(\mathbf{p}-\mathbf{p}')(EE' + \mathbf{p}\cdot\mathbf{p}' + m^2)(a^\dagger(\mathbf{p})a(\mathbf{p}')e^{-i(E-E')t} \\
&\quad + a(\mathbf{p})a^\dagger(\mathbf{p}')e^{-i(E-E')t}) + \delta^3(\mathbf{p}+\mathbf{p}')(-EE' - \mathbf{p}\cdot\mathbf{p}' + m^2) \\
&\quad \times (a(\mathbf{p})a(\mathbf{p}')e^{-i(E+E')t} + a^\dagger(\mathbf{p})a^\dagger(\mathbf{p}')e^{i(E+E')t})\big] - V\Omega \\
&= \frac{1}{2}\int \widetilde{dp}\frac{1}{2E}\big[(E^2 + \mathbf{p}^2 + m^2)(a^\dagger(\mathbf{p})a(t) + a(\mathbf{p})a^\dagger(\mathbf{p})) + (-E^2 + \mathbf{p}^2 + m^2) \\
&\qquad \times (a(\mathbf{p})a(-\mathbf{p})e^{-2iEt} + a^\dagger(\mathbf{p})a^\dagger(-\mathbf{p})e^{2iEt})\big] - V\Omega \\
&= \frac{1}{2}\int \widetilde{dp}\frac{1}{2E}\big[2E^2(a^\dagger(\mathbf{p})a(\mathbf{p}) + a(\mathbf{p})a^\dagger(\mathbf{p}))\big] - V\Omega.
\end{aligned} \tag{4.266}
$$

Using (4.263), we can rewrite this as

$$\begin{aligned}
H &= \frac{1}{2}\int \widetilde{dp}\; E(a^\dagger(\mathbf{p})a(\mathbf{p}) + a(\mathbf{p})a^\dagger(\mathbf{p})) - V\Omega \\
&= \frac{1}{2}\int \widetilde{dp}\; E(a^\dagger(\mathbf{p})a(\mathbf{p}) + (2\pi)^3 2E\delta^3(\mathbf{p}-\mathbf{p}) + a^\dagger(\mathbf{p})a(\mathbf{p})) - V\Omega \\
&= \int \widetilde{dp}\; Ea^\dagger(\mathbf{p})a(\mathbf{p}) + \int \widetilde{dp}\; E(2\pi)^3\delta^3(0) - V\Omega \\
&= \int \widetilde{dp} Ea^\dagger(\mathbf{p})a(\mathbf{p}) + \int \frac{d^3\mathbf{p}}{(2\pi)^3 2E} E(2\pi)^3\delta^3(0) - V\Omega \\
&= \int \widetilde{dp} Ea^\dagger(\mathbf{p})a(\mathbf{p}) + \frac{1}{2}\delta^3(0)\int d^3\mathbf{p} - V\Omega .
\end{aligned} \tag{4.267}$$

Notice that both the second and third terms are infinite (assuming the volume V of the space we are in is infinite). This may be troubling, but remember that Ω is an arbitrary constant we can set to be anything we want. So, let's define

$$\Omega = \frac{1}{2V}\delta^3(0)\int d^3\mathbf{p}, \tag{4.268}$$

leaving

$$H = \int \widetilde{dp}\; Ea^\dagger(\mathbf{p})a(\mathbf{p}). \tag{4.269}$$

This use of Ω may bother you. That's okay, because it should bother you. It means something is wrong. However, it's not the serious kind of wrong. In fact, it's not a problem at all.[18] We'll try to explain why.

Remember that (in non-gravitational theories) physics can only depend on *changes* in energy, and therefore the infinity we subtracted off does not affect the value we will measure experimentally. What we have done here, by subtracting off the infinite part in a way that doesn't change the physics, is a very primitive example of **Renormalization**. Often, for various reasons, measurable quantities in QFT are plagued by different types of infinities. However, it is possible to subtract off those infinities in a well-defined way, leaving a finite part. It turns out that this finite part is the correct value seen in nature. The reasons for this are very deep, and we will not discuss them (or general renormalization theory) in much depth in this book. For correlating theoretical results with experiment, being able to renormalize results

[18]Actually, it is a problem, but one inherent to all of quantum field theory. It doesn't really affect our ability to make accurate predictions about what's going to happen in a lab, but it does point to a deeper insufficiency in quantum field theory. In fact, this is a simple example of one of the QFT problems that string theory is proposed to solve. More on that (much) later.

correctly is vital. However, our goal in this book is to study mostly non-perturbative topics in particle physics, and therefore we will leave renormalization (which is a perturbative topic) for a later book.

We now have our field expansion (4.261) and commutation relations (4.263). Notice that (4.263) have the exact form of a simple harmonic oscillator, which you learned about in introductory quantum mechanics. Therefore, because these operators have the same structure as the harmonic oscillator, the corresponding states will also have the same structure. By doing nothing but imposing relativity, we have found that scalar fields, which are Hermitian operators, act as raising and lowering (or synonymously creation and annihilation) operators on the vacuum (just like the simple harmonic oscillator).

Comparing (4.263) with the standard harmonic oscillator operators, you can convince yourself that $a^\dagger(\mathbf{p})$ *creates* a ϕ particle with momentum $\mathbf{p}$ and energy E, whereas $a(\mathbf{p})$ annihilates a ϕ particle with momentum $\mathbf{p}$ and energy E. A physical state will be

$$|\mathbf{p}\rangle = a^\dagger(\mathbf{p})|0\rangle, \tag{4.270}$$

with expectation value given by

$$\begin{aligned}
\langle\mathbf{p}'|\mathbf{p}\rangle &= \langle 0|a(\mathbf{p}')a^\dagger(\mathbf{p})|0\rangle \\
&= \langle 0|\big[(2\pi)^3 2E\delta^3(\mathbf{p}-\mathbf{p}') + a^\dagger(\mathbf{p})a(\mathbf{p}')\big]|0\rangle \\
&= (2\pi)^3 2E\delta^3(\mathbf{p}-\mathbf{p}')\langle 0|0\rangle \\
&= (2\pi)^3 2E\delta^3(\mathbf{p}-\mathbf{p}'),
\end{aligned} \tag{4.271}$$

where we are assuming $\langle 0|0\rangle = 1$ and $|0\rangle$ denotes the vacuum.[19] Furthermore, we have

$$a(\mathbf{p})|0\rangle = 0. \tag{4.272}$$

The entire spectrum of states can be studied by acting on $|0\rangle$ with creation operators, and probability amplitudes for one state to be found in another, $\langle\mathbf{p}_f|\mathbf{p}_i\rangle$, are straightforward to calculate (and positive semi-definite). Naturally this theory does not discuss any interactions between particles, and therefore we will have to do a great deal of modification before we are done. But by this simple exercise of merely imposing the standard commutation relations (4.248) between the field and its momentum, we have gained complete knowledge of the quantum mechanical states of the theory.

[19]The probability of finding nothing when you look at nothing is 1, of course. This is simply assuming that something can't come from nothing.

Before moving on we'll briefly consider the complex scalar field. In this case we don't need to require that ϕ satisfy $\phi = \phi^\dagger = \phi^\star$ (because we're assuming that it is a *complex* field). As we said in Sect. 4.2.2, the complex field has two degrees of freedom (whereas the real field only has one, which is why only a (and its conjugate) appeared in the expansion for the real field). Thus, we take the expansion for the complex field ϕ to have two different coefficients. It is similar to the real case, but with two different types of operators – an a operator and a b operator.

$$\phi(x) = \int \widetilde{dp}\big[a(\mathbf{p})e^{ip\cdot x} + b^\dagger(\mathbf{p})e^{-ip\cdot x}\big],$$

$$\phi^\dagger(x) = \int \widetilde{dp}\big[a^\dagger(\mathbf{p})e^{-ip\cdot x} + b(\mathbf{p})e^{ip\cdot x}\big]. \tag{4.273}$$

When written in this form, the canonical commutation relations (4.248) will make both a and b (and their conjugates) separately obey the same commutation relations as in (4.263). So $a^\dagger(\mathbf{p})$ and $b^\dagger(\mathbf{p})$ create a and b particles with momentum $\mathbf{p}$ and $a(\mathbf{p})$ and $b(\mathbf{p})$ destroy a and b particles with momentum $\mathbf{p}$. The two types of particles are related. They form a particle/anti-particle pair with matching masses, but opposite charges.

4.6.3 The Spin-Statistics Theorem

Notice that the states coming from (4.270) will include the two particle state

$$|\mathbf{p};\mathbf{p}'\rangle = a^\dagger(\mathbf{p})a^\dagger(\mathbf{p}')|0\rangle. \tag{4.274}$$

The commutation relations (4.263) tell us that

$$a^\dagger(\mathbf{p})a^\dagger(\mathbf{p}') = a^\dagger(\mathbf{p}')a^\dagger(\mathbf{p}). \tag{4.275}$$

So, this theory also allows the state

$$|\mathbf{p}';\mathbf{p}\rangle = a^\dagger(\mathbf{p}')a^\dagger(\mathbf{p})|0\rangle. \tag{4.276}$$

Recall from a chemistry or modern physics course that particles with half-integer spin obey the Pauli Exclusion Principle, whereas particles of integer spin do not. Our Klein-Gordon scalar fields ϕ are spinless ($j = 0$), and therefore we would expect that they do not obey Pauli exclusion. The fact that our commutation relations have allowed both states (4.274) and (4.276) is therefore expected. This is an indication that we quantized correctly.

Notice that this statistical result (that the scalar fields do not obey Pauli exclusion) is entirely a result of the commutation relations. Therefore, if we attempt to quantize a spin-$1/2$ field in the same way, they will obviously not obey Pauli exclusion either. We must therefore quantize spin-$1/2$ differently.

It turns out that the correct way to quantize spin-$1/2$ fields is to use, instead of commutation relations like we used for scalar fields, *anticommutation relations*. If the operators of our spin-$1/2$ fields obey

$$\left\{a_1^\dagger, a_2^\dagger\right\} = a_1^\dagger a_2^\dagger + a_2^\dagger a_1^\dagger = 0 \Rightarrow a_1^\dagger a_2^\dagger = -a_2^\dagger a_1^\dagger. \tag{4.277}$$

Then if we try to act twice with the same operator, we have

$$a_1^\dagger a_1^\dagger |0\rangle = -a_1^\dagger a_1^\dagger |0\rangle \Rightarrow a_1^\dagger a_1^\dagger |0\rangle = 0. \tag{4.278}$$

In other words, if we quantize with anticommutation relations, it is not possible for two particles to occupy the same state simultaneously.

This relationship between the spin of a particle and the statistics it obeys (which demands that integer spin particles be quantized by commutation relations and half-integer spin particles to be quantized with anticommutation relations) is called the **Spin-Statistics Theorem**. Because particles not obeying Pauli exclusion are said to have **Bose-Einstein** statistics, and particles that do obey Pauli exclusion are said to have **Fermi-Dirac** statistics, we call particles with integer spin **Bosons**, and particles with half-integer spin **Fermions**.

4.6.4 Canonical Quantization of Fermions

Recall that the difference between the Dirac and Majorana fields is that Majorana fields are "real" whereas Dirac fields are "complex" (cf. Sect. 4.3.12). So we expect the general form of the Dirac field to be similar to (4.273), and the Majorana field to be similar to (4.261).

We'll start with the Dirac field. One final comment is that we must keep in mind that a spin $1/2$ particle will be a superposition of *two* parts: a spin up part and a spin down part. We write the Fermion field as a sum over these two parts. The general solution will be

$$\psi_D(x) = \sum_{s=1}^{2} \int \widetilde{dp}[b_s(\mathbf{p})v_s(\mathbf{p})e^{ip\cdot x} + d_s^\dagger(\mathbf{p})u_s(\mathbf{p})e^{-ip\cdot x}], \tag{4.279}$$

where $s = 1, 2$ are the two spin states, b_s and $d_s^\dagger$ are (respectively) the lowering operator for the particle and the raising operator for the antiparticle. The charge conjugate of ψ_D will have the raising operator for the particle and the lowering operator for the antiparticle. The v_s and u_s are constant 4-component spinors we discussed in Sect. 4.3.9.

We quantize, as we said in Sect. 4.6.3, using *anti*-commutation relations. Writing only the non-zero relation,

$$\{\psi_\alpha(t,\mathbf{x}),\bar{\psi}_\beta(t,\mathbf{x})\} = \delta^3(\mathbf{x}-\mathbf{x}')(\gamma^0)_{\alpha\beta}. \tag{4.280}$$

These imply that the only non-zero commutation relations in terms of the operators are

$$\begin{aligned}\left\{b_s(\mathbf{p}),b_{s'}^\dagger(\mathbf{p}')\right\} &= (2\pi)^3\delta^3(\mathbf{p}-\mathbf{p}')2\omega\delta_{ss'},\\ \{d_s^\dagger(\mathbf{p}),d_{s'}(\mathbf{p}')\} &= (2\pi)^3\delta^3(\mathbf{p}-\mathbf{p}')2\omega\delta_{ss'}.\end{aligned} \tag{4.281}$$

Once again, these form the algebra of a simple harmonic oscillator, and we can therefore find the entire spectrum of states by acting on $|0\rangle$ with $b_s^\dagger$ and $d_s^\dagger$. Then, following a series of calculations nearly identical to the ones in Sect. 4.6.2, we arrive at the Hamiltonian

$$H = \sum_{s=1}^{2}\int \widetilde{dp}\; E[b_s^\dagger(\mathbf{p})b_s(\mathbf{p}) + d_s^\dagger(\mathbf{p})d_s(\mathbf{p})] - \lambda, \tag{4.282}$$

where λ is an infinite constant we can merely subtract off and therefore ignore.

Comparing (4.269) and (4.282), we see that they both have essentially the same form; E (energy) to the left of the creation operator, which is to the left of the annihilation operator. To understand the meaning of this, we will see how it generates energy eigenvalues. We will use equation (4.269) for simplicity. Consider acting with the Hamiltonian operator on some arbitrary state $|\mathbf{p}\rangle$ with momentum $\mathbf{p}$. Using (4.270) and putting the momentum subscript on E so it is clear what the momentum associated with E is (i.e. E_p satisfies $E^2 = \mathbf{p}^2 + m^2$ while E_k satisfies $E^2 = \mathbf{k}^2 + m^2$, where k is just another expression for momentum),

$$\begin{aligned}H|\mathbf{p}\rangle &= \int \widetilde{dk}\,E_k a^\dagger(\mathbf{k})a(\mathbf{k})|\mathbf{p}\rangle = \int \widetilde{dk}\,E_k a^\dagger(\mathbf{k})a(\mathbf{k})\sqrt{2E_p}a^\dagger(\mathbf{p})|0\rangle\\ &= \int \widetilde{dk}\,E_k\sqrt{2E_p}a^\dagger(\mathbf{k})\big((2\pi)^3 2E_p\delta^3(\mathbf{k}-\mathbf{p}) + a^\dagger(\mathbf{p})a(\mathbf{k})\big)|0\rangle\\ &= \int \widetilde{dk}\,E_k\sqrt{2E_p}a^\dagger(\mathbf{k})(2\pi)^3 2E_p\delta^3(\mathbf{k}-\mathbf{p})|0\rangle\\ &= \int \frac{d^3\mathbf{k}}{(2\pi)^3 2E_k}E_k\sqrt{2E_p}a^\dagger(\mathbf{k})(2\pi)^3 2E_p\delta^3(\mathbf{k}-\mathbf{p})|0\rangle\\ &= \int d^3\mathbf{k}\sqrt{2E_p}a^\dagger(\mathbf{k})E_p\delta^3(\mathbf{k}-\mathbf{p})|0\rangle\\ &= E_p\sqrt{2E_p}a^\dagger|0\rangle\\ &= E_p|\mathbf{p}\rangle.\end{aligned} \tag{4.283}$$

Hence, $H|\mathbf{p}\rangle = E_p|\mathbf{p}\rangle$, where $E_p^2 = \mathbf{p}^2 + m^2$. That is, the Hamiltonian operator gives the appropriate energy eigenvalue on our physical quantum states.

If we denote the particle/anti-particle states (b and d) of a Dirac Hamilton as $|\mathbf{p}_b, s_b\rangle$ and $|\mathbf{p}_d, s_d\rangle$, (specifying the momentum and spin of each state), then

$$H|\mathbf{p}_b, s_b\rangle = E_{p_b}|\mathbf{p}_b, s_b\rangle,$$
$$H|\mathbf{p}_d, s_d\rangle = E_{p_d}|\mathbf{p}_d, s_d\rangle. \tag{4.284}$$

For Majorana fields things are simpler. We only have one type of particle, so

$$\psi_M(x) = \sum_{s=1}^{2} \int \widetilde{dp}[b_s(\mathbf{p})u_s(\mathbf{p})e^{ip\cdot x} + b_s^\dagger(\mathbf{p})v_s(\mathbf{p})e^{-ip\cdot x}]. \tag{4.285}$$

Quantization with anticommutation relations will give

$$H = \sum_{s=1}^{2} \int \widetilde{dp}\; E\; b_s^\dagger(\mathbf{p})b_s(\mathbf{p}). \tag{4.286}$$

4.6.5 Symmetries in Quantum Mechanics

We'll now take a brief pause from the path we're on to discuss an important point regarding symmetries, conserved quantities, and how they show up in quantum mechanics. In Sect. 1.5 we discussed classical fields. For now we'll be interested only in the case of multiple fields. We saw there that given some Lagrangian $\mathcal{L}$ and an infinitesimal transformation

$$\phi_a \to \phi_a + \epsilon\delta\phi_a, \tag{4.287}$$

(for $a = 1, \ldots, n$, assuming there are n fields) that leaves the Lagrangian unchanged, we have the conserved current

$$j^\mu = \frac{\partial \mathcal{L}}{\partial(\partial_\mu \phi_a)}\delta\phi_a, \tag{4.288}$$

where j^0 is the "charge" density and j^i is the current density.[20] The total conserved charge is then the integral over all space of the charge density,

$$Q = \int_{all\ space} d^3x\; j^0. \tag{4.289}$$

[20]The words "charge" and "current" in this context do not necessarily refer to electric charge and current, but rather to whatever physical quantity is conserved under the relevant symmetry transformation.

Then, finally, we saw that the momentum associated with the field is

$$\Pi^a = \frac{\partial \mathcal{L}}{\partial \dot{\phi}_a}. \tag{4.290}$$

Now let's say that our Lagrangian is invariant under some Lie group transformation U,

$$\phi_a \longrightarrow U_a^b \phi_b. \tag{4.291}$$

If we call the generators of this Lie algebra T (suppressing an index to label each generator because we won't need it for what we're doing here – just know that in general there will be multiple generators), then (taking α to be the set of parameters specifying the "angle" of the transformation – there will of course be multiple α's, one for each generator, but again we're suppressing this index)

$$U = e^{i\alpha T}. \tag{4.292}$$

If we take an infinitesimal transformation of this Lie group (which is the relevant limit for a symmetry transformation, i.e. (4.287)), we have (only needing to keep up to first order),

$$\phi_a \longrightarrow U_a^b \phi_b \approx \left(\delta_a^b + i\epsilon T_a^b\right)\phi_b = \phi_a + \epsilon\left(i T_a^b \phi_b\right), \tag{4.293}$$

(where the indices on T denote the elements of the matrix T). Comparing this to (4.287) we see that

$$\delta\phi_a = i T_a^b \phi_b. \tag{4.294}$$

Let's now look again at the conserved current. Specifically, we'll look at the time component. Using the expression for the momentum and the expression for $\delta\phi_a$ that we just found, we have

$$j^0 = \frac{\partial \mathcal{L}}{\partial(\partial_0 \phi_a)}\delta\phi_a = \frac{\partial \mathcal{L}}{\partial \dot{\phi}_a}\delta\phi_a = \Pi^a \delta\phi_a = i\Pi^a T_a^b \phi_b. \tag{4.295}$$

Therefore the total charge Q is (writing the measure d^3x to indicate all space),

$$Q = \int d^3x\, j^0 = i\int d^3x\, \Pi^a T_a^b \phi_b. \tag{4.296}$$

Next we look again at the canonical commutation relations for scalars in (4.248). Focusing on the non-zero relation (and extending to multiple fields)

$$[\phi_a(t,\mathbf{x}), \Pi^b(t',\mathbf{x}')] = i\delta_a^b \delta(t-t')\delta(\mathbf{x}-\mathbf{x}'). \tag{4.297}$$

We can use this to find the commutation relation of the field ϕ_a with the conserved charge Q. It is (setting $t' = t$ up front for simplicity)

$$\begin{aligned}
[Q, \phi_a(t, \mathbf{x})] &= \left[\int d^3y \; i\Pi^b(t, \mathbf{y}) T_b^c \phi_c(t, \mathbf{y}), \phi_a(t, \mathbf{x}) \right] \\
&= i \int d^3y \; \left[\Pi^b(t, \mathbf{y}), \phi_a(t, \mathbf{x})\right] T_b^c \phi_c(t, \mathbf{y}) \\
&= -i \int d^3y \; \left(i\delta_a^b \delta(\mathbf{x} - \mathbf{y})\right) T_b^c \phi_c(t, \mathbf{y}) \\
&= T_a^c \phi_c(t, \mathbf{x}).
\end{aligned} \tag{4.298}$$

This indicates that if we make a Lie group symmetry transformation to a Lagrangian, the total charge that results from that transformation will generate the original transformation on the field via commutation. We won't be using this fact directly in this book, but it is an important aspect of symmetry in quantum field theory and we will be using it in later books.

4.6.6 *Insufficiencies of Canonical Quantization*

While the Canonical quantization procedure we have carried out in the past several sections has given us a tremendous amount of information (the entire spectrum of states for bosons, Dirac fermions, and Majorana Fermions), it is still lacking quite a bit. As we said at the beginning of Sect. 4.1, we ultimately want a relativistic quantum mechanical theory of interactions. Canonical quantization has provided a relativistic quantum mechanical theory, but we aren't close to being able to incorporate interactions into our theory. While it is possible to incorporate interactions with canonical quantization, it is very difficult. In order to simplify we will need a new way of quantizing.

4.6.7 *Path Integrals and Path Integral Quantization*

Perhaps the most fundamental experiment in quantum mechanics is the **Double Slit** experiment. In brief, what this experiment tells us is that, when a single electron moves through a screen with two slits, and no observation is made regarding which slit it goes through, it actually goes through *both* slits, and until a measurement is made (for example, when it hits the observation screen behind the double slit), it exists in a superposition of *both* paths. As a result, the particle exhibits a wave nature, and the pattern that emerges on the observation screen is an interference pattern – the same as if a classical wave was passing through the double slit. All paths in the superposition of the single electron are interfering with each other, both destructively and constructively. Once the electron is observed on the observation screen, it collapses probabilistically into one of its possible states (a particular location on the observation screen).

If, on the other hand, you set up some mechanism to observe *which* of the two slits the electron travels through, then the observation has been made *before* the observation screen, and you no longer have the superposition, and therefore you no longer see any indication of an interference pattern. The electrons are behaving, in a sense, classically from the double slit to the observation screen in this case.

The meaning of this is that a particle that has not been observed will actually take every possible path at once. Once an observation has been made, there is some probability associated with each path. Some paths are very likely, and others are less likely (some are nearly impossible). But until observation, it actually exists in a superposition of all possible states/paths.

So, to quantize, we will create a mathematical expression for a "sum over all possible paths". This expression is called a **Path Integral**, and will prove to be a much more useful way to quantize a physical system.

We begin this construction by considering merely the amplitude for a particle at position q_1 at time t_1 to propagate to q_2 at time t_2. This amplitude will be given by

$$\langle q_2, t_2 | q_1, t_1 \rangle = \langle q_2 | e^{iH(t_2 - t_1)} | q_1 \rangle. \tag{4.299}$$

To evaluate this, we begin by dividing the time interval $T = t_2 - t_1$ into $N+1$ equal intervals of length $\delta t = \frac{T}{N+1}$ each. Then we can insert N complete sets of position eigenstates,

$$\langle q_2, t_2 | q_1, t_1 \rangle = \int_{-\infty}^{\infty} \prod_{i=1}^{N} dQ_i \langle q_2 | e^{-iH\delta t} | Q_N \rangle \langle Q_N | e^{-iH\delta t} | Q_{N-1} \rangle \cdots \langle Q_1 | e^{-iH\delta t} | q_1 \rangle. \tag{4.300}$$

Let's look at a single one of these amplitudes. We know that in nearly all physical theories, we can break the Hamiltonian up as

$$H = \frac{P^2}{2m} + V(Q). \tag{4.301}$$

Using the completeness of momentum eigenstates (this is just inserting the identity operator repeatedly),

$$\begin{aligned}
\langle Q_{i+1} | e^{-iH\delta t} | Q_i \rangle &= \langle Q_{i+1} | e^{-i\left(\frac{P^2}{2m} + V(Q)\right)\delta t} | Q_i \rangle \\
&= \langle Q_{i+1} | e^{-i\delta t \frac{P^2}{2m}} e^{-i\delta t V(Q)} | Q_i \rangle \\
&= \int dP' \langle Q_{i+1} | e^{-i\delta t \frac{P^2}{2m}} | P' \rangle \langle P' | e^{-i\delta t V(Q)} | Q_i \rangle
\end{aligned}$$

$$= \int dP' e^{-i\delta t \frac{P'^2}{2m}} e^{-i\delta t V(Q_i)} \langle Q_{i+1}|P'\rangle\langle P'|Q_i\rangle$$

$$= \int dP' e^{-i\delta t \frac{P'^2}{2m}} e^{-i\delta t V(Q_i)} \frac{e^{iP'Q_{i+1}}}{\sqrt{2\pi}} \frac{e^{-iP'Q_i}}{\sqrt{2\pi}}$$

$$= \int \frac{dP'}{2\pi} e^{iH\delta t} e^{iP'(Q_{i+1}-Q_i)}$$

$$= \int \frac{dP'}{2\pi} e^{i\left[P'(Q_{i+1}-Q_i)-H\delta t\right]}$$

$$= \int \frac{dP'}{2\pi} e^{i\delta t\left[P'\left(\frac{Q_{i+1}-Q_i}{\delta t}\right)-H\right]}. \tag{4.302}$$

Taking the limit as $\delta t \to 0$ gives

$$\frac{Q_{i+1}-Q_i}{\delta t} \to \dot{Q}_i. \tag{4.303}$$

Therefore,

$$\int \frac{dP'}{2\pi} e^{i\delta t\left[P'\left(\frac{Q_{i+1}-Q_i}{\delta t}\right)-H\right]} = \int \frac{dP'}{2\pi} e^{i dt_{i+1}[P'\dot{Q}_i - H]}. \tag{4.304}$$

where the subscript on dt merely indicates where the infinitesimal time interval "ends". Next we can plug this into (4.300) and taking the limit as $\delta t \to 0$,

$$\langle q_2, t_2|q_1, t_1\rangle = \int_{-\infty}^{\infty} \prod_{i=1}^{N} dQ_i \,\langle q_2|e^{-iH\delta t}|Q_N\rangle\langle Q_N|e^{-iH\delta t}|Q_{N-1}\rangle \cdots \langle Q_1|e^{-iH\delta t}|q_1\rangle$$

$$= \lim_{N\to\infty} \int_{-\infty}^{\infty} \prod_{i=1}^{N} dQ_i$$

$$\times \int \frac{dP'_i}{2\pi} e^{idt_2[P'_N\dot{Q}_N - H]} e^{idt_N[P'_{N-1}\dot{Q}_{N-1}-H]} \cdots e^{idt_1[P'_1\dot{Q}_1-H]}$$

$$= \int_{-\infty}^{\infty} \mathcal{D}p\mathcal{D}q \; e^{i\int_{t_1}^{t_2} dt(p\dot{q}-H)}, \tag{4.305}$$

where

$$\mathcal{D}p = \prod_{i=1}^{\infty} dp_i \tag{4.306}$$

and

$$\mathcal{D}q = \prod_{i=1}^{\infty} dq_i. \tag{4.307}$$

If p shows up quadratically (as it always does, i.e. $\frac{p^2}{2m}$), then we can merely do the Gaussian integral over p, resulting in an overall constant which we can just absorb back into the measure when we normalize. Then, recognizing that the integrand in the exponent is $p\dot{q} - H = L$ (cf. Eq. (1.45)), we have

$$\langle q_2, t_2 | q_1, t_1 \rangle = \int \mathcal{D}q \; e^{i \int_{t_1}^{t_2} dt L} = \int \mathcal{D}q e^{iS}. \tag{4.308}$$

Formally, the measure of (4.308) has an infinite number of differentials, and therefore evaluating it would require doing an infinite number of integrals. This is to be expected, since the point of the path integral is a sum over every possible path, of which there are an infinite number. Because we obviously can't do an infinite number of integrals, we will have to find a clever way of evaluating (4.308). Nevertheless, before doing so, we discuss what the path integral means.

4.6.8 *Interpretation of the Path Integral*

Equation (4.308) says that, given an initial and final configuration (q_1, t_1) and (q_2, t_2), absolutely *any* path between them is possible. This is the content of the $\mathcal{D}q$ part: it is the sum over all paths. Then, for each of those paths, the integral assigns a *statistical weight* of e^{iS} to it, where the action S is calculated using *that* path (recall our comments in Sect. 1.1 about S being a functional, not a function).

Consider an arbitrary path q_0, which receives statistical weight $e^{iS[q_0]}$. Now, also consider a path q' very close to q_0, only varying by a small amount: $q' = q_0 + \epsilon\delta q_0$. This will have statistical weight

$$e^{iS[q_0+\epsilon\delta q_0]} = e^{iS[q_0]+i\epsilon\delta q_0 \frac{\delta S[q_0]}{\delta q}}, \tag{4.309}$$

where $\frac{\delta S}{\delta q}$ is the Euler Lagrange derivative (1.17)

$$\frac{\delta S}{\delta q} = \frac{d}{dt}\frac{\partial S}{\partial \dot{q}} - \frac{\partial S}{\partial q}. \tag{4.310}$$

To make our intended result more obvious, we do a Wick rotation,[21] taking $t \to it$, so $dt \to idt$, and

$$S = \int dt\mathcal{L} \to i \int dt\mathcal{L} = iS, \tag{4.311}$$

---

[21]Don't worry if you've never heard of a Wick rotation – it's just taking $t \to it$. It won't be important after this section.

and therefore

$$e^{iS} \to e^{-S}. \tag{4.312}$$

Now, the path $q' = q_0 + \epsilon\delta q_0$ gets weight

$$e^{-S[q_0]}e^{-\epsilon\delta q_0 \frac{\delta S[q_0]}{\delta q}}. \tag{4.313}$$

If $\frac{\delta S}{\delta q}$ is very large, then the weight becomes exponentially small. In other words, the larger the variation of the action is, the less probable that path is. *that* means that the *most* probable path is the one for the *smallest* value of $\frac{\delta S}{\delta t}$, or the path at which

$$\frac{\delta S}{\delta q} = 0. \tag{4.314}$$

As we discussed in 1.1, this is the path of **Least Action**. Thus, we have recovered classical mechanics as the first-order approximation of quantum mechanics. The classical result is simply the *most likely* of all the quantum possibilities.

Now imagine a single quantum particle. The probability of finding it in an "unusual" state – one far from the least action (or classically expected) state – may be large enough to find it somewhere (classically) unexpected fairly frequently. However, the probably of finding *two* particles far from the classical state will be a bit smaller. The chances are smaller still for three particles, and so on. Hence, in a macroscopic object, with a huge number of particles coupled together (i.e. $\approx 10^{23}$ particles for a mole), the probability of observing that macroscopic object doing anything other than what is most likely (with the least action) is essentially zero.

The meaning of the path integral is that all imaginable paths are possible for the particle to travel in moving from one configuration to another. However, not all paths are equally probable. The likelihood of a given path is given by the action exponentiated, and therefore the most probable paths are the ones which minimize the action. This is the reason that, macroscopically, the world appears classical. The likelihood of every particle in, say, a baseball, simultaneously taking a path noticeably far from the path of least action is negligibly small.

We will find that path integral quantization provides an extremely powerful tool with which to create our relativistic quantum theory of interactions.

4.6.9 Expectation Values

Now that we have a way of finding $\langle q_2, t_2|q_1, t_1\rangle$, the natural question to ask next is how do we find expectation values like $\langle q_2, t_2|Q(t')|q_1, t_1\rangle$ or $\langle q_2, t_2|P(t')|q_1, t_1\rangle$. By doing a similar derivation as in the last section, it is straightforward to show that

$$\langle q_2, t_2|Q(t')|q_1, t_1\rangle = \cdots = \int \mathcal{D}q\; Q(t')e^{iS}. \tag{4.315}$$

We will find that evaluating integrals of this form is simplified greatly through making use of **Functional Derivatives**. For some function $f(x)$, the functional derivative is defined by

$$\frac{\delta}{\delta f(y)} f(x) = \delta(x - y). \tag{4.316}$$

Next, we modify our path integral by adding an **Auxiliary External Source** function, so that

$$\mathcal{L} \to \mathcal{L} + f(t)Q(t) + h(t)P(t). \tag{4.317}$$

From this, we now have

$$\langle q_2, t_2 | q_1, t_1 \rangle_{f,h} = \int \mathcal{D}q \; e^{\int dt(\mathcal{L}+fQ+hP)}, \tag{4.318}$$

which allows us to write out expectation values in the simple form

$$\begin{aligned} \langle q_2, t_2 | Q(t') | q_1, t_1 \rangle &= \frac{1}{i} \frac{\delta}{\delta f(t')} \langle q_2, t_2 | q_1, t_1 \rangle_{f,h} \Big|_{f,h=0} \\ &= \int \mathcal{D}q Q(t') e^{iS + i \int dt(fQ+hP)} \Big|_{f,h=0} \\ &= \int \mathcal{D}q Q(t') e^{iS}, \end{aligned} \tag{4.319}$$

or

$$\begin{aligned} \langle q_2, t_2 | P(t') | q_1, t_1 \rangle &= \frac{1}{i} \frac{\delta}{\delta h(t')} \langle q_2, t_2 | q_1, t_1 \rangle_{f,h} \Big|_{f,h=0} \\ &= \int \mathcal{D}q P(t') e^{iS + i \int dt(fQ+hP)} \Big|_{f,h=0} \\ &= \int \mathcal{D}q P(t') e^{iS}. \end{aligned} \tag{4.320}$$

Once we have $\langle q_2, t_2 | q_1, t_1 \rangle$, we can find any expectation value we want simply by taking successive functional derivatives.

4.6.10 Path Integrals with Fields

Because we can build whatever state we want by acting on the vacuum, the important quantity for us to work with will be the **Vacuum to Vacuum** expectation value, or VEV, $\langle 0|0 \rangle$, and the various expectation values we can build through functional derivatives ($\langle 0|\phi\phi|0\rangle$, $\langle 0|\psi\phi\phi|0\rangle$, etc.).

For simplicity let's consider a scalar boson ϕ. The Lagrangian is given in equation (4.11). Using this, we can write the path integral

$$\langle 0|0\rangle = \int \mathcal{D}\phi e^{i\int d^4x\left[-\frac{1}{2}\partial^\mu\phi\partial_\mu\phi-\frac{1}{2}m^2\phi^2\right]} = \int \mathcal{D}\phi e^{i\int d^4x\mathcal{L}_0}. \tag{4.321}$$

We will eventually want to find expectation values, so we introduce the auxiliary field J, creating

$$\langle 0|0\rangle_J = \int \mathcal{D}\phi e^{i\int d^4x(\mathcal{L}_0+J\phi)}. \tag{4.322}$$

For example, the expectation value of ϕ is

$$\langle 0|\phi|0\rangle = \frac{1}{i}\frac{\delta}{\delta J}\langle 0|0\rangle_J\big|_{J=0}. \tag{4.323}$$

Of course, we still have a path integral with an infinite number of integrals to evaluate, but we are finally able to discuss how we can do the evaluation. We define $Z_0(J) = \langle 0|0\rangle_J$. Then, making use of the Fourier transform of ϕ,

$$\widetilde{\phi}(k) = \int d^4x\, e^{-ikx}\phi(x), \qquad \textit{and} \qquad \phi(x) = \int \frac{d^4k}{(2\pi)^4}e^{ikx}\widetilde{\phi}(k), \tag{4.324}$$

we begin with the $\mathcal{L}_0$ part:

$$\begin{aligned}
S_0 &= \int d^4x\mathcal{L}_0 = \int d^4x\Big(-\frac{1}{2}\partial^\mu\phi\partial_\mu\phi - \frac{1}{2}m^2\phi^2\Big)\\
&= \int d^4x\Big[-\frac{1}{2}\partial^\mu\Big(\int\frac{d^4k}{(2\pi)^4}e^{ik\cdot x}\widetilde{\phi}(k)\Big)\partial_\mu\Big(\int\frac{d^4k'}{(2\pi)^4}e^{ik'\cdot x}\widetilde{\phi}(k')\Big)\\
&\qquad -\frac{1}{2}m^2\Big(\int\frac{d^4k}{(2\pi)^4}e^{ik\cdot x}\widetilde{\phi}(k)\Big)\Big(\int\frac{d^4k'}{(2\pi)^4}e^{ik'\cdot x}\widetilde{\phi}(k')\Big)\Big]\\
&= \int d^4x\Big[\frac{1}{2}\int\frac{d^4kd^4k'}{(2\pi)^8}e^{ik\cdot x}e^{ik'\cdot x}\widetilde{\phi}(k)\widetilde{\phi}(k')(k^\mu k'_\mu - m^2)\Big]\\
&= \frac{1}{2}\int\frac{d^4kd^4k'}{(2\pi)^8}\widetilde{\phi}(k)\widetilde{\phi}(k')(k^\mu k'_\mu - m^2)\int d^4x e^{i(k+k')\cdot x}\\
&= \frac{1}{2}\int\frac{d^4kd^4k'}{(2\pi)^8}\widetilde{\phi}(k)\widetilde{\phi}(k')(k^\mu k'_\mu - m^2)(2\pi)^4\delta^4(k+k')\\
&= -\frac{1}{2}\int\frac{d^4k}{(2\pi)^4}\widetilde{\phi}(k)(k^2+m^2)\widetilde{\phi}(-k).
\end{aligned} \tag{4.325}$$

Next, we transform the auxiliary field part,

$$\begin{aligned}\int d^4x J(x)\phi(x) &= \int d^4x \Big(\int \frac{d^4k}{(2\pi)^4} e^{ik\cdot x}\widetilde{J}(k)\Big)\Big(\int \frac{d^4k'}{(2\pi)^4} e^{ik'\cdot x}\widetilde{\phi}(k')\Big) \\ &= \int \frac{d^4k d^4k'}{(2\pi)^8}\widetilde{J}(k)\widetilde{\phi}(k')\int d^4x e^{i(k+k')\cdot x} \\ &= \int \frac{d^4k d^4k'}{(2\pi)^8}\widetilde{J}(k)\widetilde{\phi}(k')(2\pi)^4\delta^4(k+k') \\ &= \int \frac{d^4k}{(2\pi)^4}\widetilde{J}(k)\widetilde{\phi}(-k). \end{aligned} \tag{4.326}$$

Because the integral is over all k^μ, we can rewrite this as

$$\int \frac{d^4k}{(2\pi)^4}\widetilde{J}(k)\widetilde{\phi}(-k) = \frac{1}{2}\int \frac{d^4k}{(2\pi)^4}\big(\widetilde{J}(k)\widetilde{\phi}(-k) + \widetilde{J}(-k)\widetilde{\phi}(k)\big) \tag{4.327}$$

(we did this to get the factor of $1/2$ out front in order to have the same coefficient as the $\mathcal{L}_0$ part from above). Therefore, we have

$$S = \frac{1}{2}\int \frac{d^4k}{(2\pi)^4}\left[-\widetilde{\phi}(k)(k^2+m^2)\widetilde{\phi}(-k) + \widetilde{J}(k)\widetilde{\phi}(-k) + \widetilde{J}(-k)\widetilde{\phi}(k)\right]. \tag{4.328}$$

Now we make a change of variables to complete the square on $\tilde{\phi}$,

$$\widetilde{\chi}(k) = \widetilde{\phi}(k) - \frac{\widetilde{J}(k)}{k^2+m^2}. \tag{4.329}$$

(Note that this leaves the measure of the path integral unchanged: $\mathcal{D}\phi \to \mathcal{D}\chi$.) Plugging this in, we have,

$$\begin{aligned} S = \frac{1}{2}\int \frac{d^4k}{(2\pi)^4}\Big[&- \Big(\widetilde{\chi}(k) + \frac{\widetilde{J}(k)}{k^2+m^2}\Big)(k^2+m^2)\Big(\widetilde{\chi}(-k) + \frac{\widetilde{J}(-k)}{k^2+m^2}\Big) \\ &+ \widetilde{J}(k)\Big(\widetilde{\chi}(-k) + \frac{\widetilde{J}(-k)}{k^2+m^2}\Big) + \widetilde{J}(-k)\Big(\widetilde{\chi}(k) + \frac{\widetilde{J}(k)}{k^2+m^2}\Big) \\ = \frac{1}{2}\int \frac{d^4k}{(2\pi)^4}\Big[&- \widetilde{\chi}(k)(k^2+m^2)\widetilde{\chi}(-k) + \frac{\widetilde{J}(k)\widetilde{J}(-k)}{k^2+m^2}\Big]. \end{aligned} \tag{4.330}$$

(The point of all of this is that, in this form, we have all of the ϕ, or equivalently χ, dependence in the first term, with no ϕ or χ dependence on the second term.)

Finally, our path integral (4.322) is

$$\langle 0|0\rangle_J = \int \mathcal{D}\chi e^{\frac{i}{2}\int \frac{d^4k}{(2\pi)^4}\left[-\widetilde{\chi}(k)(k^2+m^2)\widetilde{\chi}(-k)+\frac{\widetilde{J}(k)\widetilde{J}(-k)}{k^2+m^2}\right]}. \tag{4.331}$$

Now, using some clever physical reasoning, we can see how to evaluate the infinite number of integrals in this expression. Notice that if we set $J = 0$, we have a free theory in which no interactions take place. This means that if we start with nothing (the vacuum), the probability of having nothing later is 100%. Or,

$$\langle 0|0\rangle_J\Big|_{J=0} = 1 = \int \mathcal{D}\chi e^{\frac{i}{2}\int \frac{d^4k}{(2\pi)^4}\left[-\widetilde{\chi}(k)(k^2+m^2)\widetilde{\chi}(-k)\right]}. \tag{4.332}$$

If that part is 1, then we have

$$\langle 0|0\rangle_J = \int \mathcal{D}\chi e^{\frac{i}{2}\int \frac{d^4k}{(2\pi)^4}\frac{\widetilde{J}(k)\widetilde{J}(-k)}{k^2+m^2}}. \tag{4.333}$$

Remarkably, the integrand *has no χ dependence*! Therefore, the infinite number of integrals over all possible paths becomes nothing more than a constant we can absorb into the normalization, leaving

$$\langle 0|0\rangle_J = e^{\frac{i}{2}\int \frac{d^4k}{(2\pi)^4}\frac{\widetilde{J}(k)\widetilde{J}(-k)}{k^2+m^2}}.$$

We can Fourier transform back to coordinate space to get

$$Z_0(J) = \langle 0|0\rangle_J = e^{\frac{i}{2}\int d^4x d^4x' J(x)\Delta(x-x')J(x')}, \tag{4.334}$$

where

$$\Delta(x-x') = \int \frac{d^4k}{(2\pi)^4}\frac{e^{ik\cdot(x-x')}}{k^2+m^2} \tag{4.335}$$

is called the **Feynman Propagator** for the scalar field in position space. We can then find expectation values by operating on this with $\frac{1}{i}\frac{\delta}{\delta J}$ as described in Sect. 4.6.9. We'll work out examples of this in the next sections.

We can repeat everything we have just done for fermions, and while it is a great deal more complicated (and tedious), it is in essence the same calculation. We begin by adding the auxiliary functions $\bar{\eta}\psi + \bar{\psi}\eta$, to get expectation values of $\bar{\psi}$ and ψ by using $\frac{1}{i}\frac{\delta}{\delta\eta}$ and $\frac{1}{i}\frac{\delta}{\delta\bar{\eta}}$, respectively. We then Fourier transform every term in the exponent and find that we can separate out the $\bar{\psi}$ and ψ dependence, allowing us to set the term which does depend on ψ and $\bar{\psi}$ equal to 1. Fourier transforming back then gives

$$Z_0(\eta,\bar{\eta}) = e^{i\int d^4x d^4x' \bar{\eta}(x)S(x-x')\eta(x')}, \tag{4.336}$$

where

$$S(x - x') = \int \frac{d^4k}{(2\pi)^4} \frac{(-\gamma^\mu k_\mu + m)e^{ik\cdot(x-x')}}{k^2 + m^2}. \tag{4.337}$$

is the Feynman propagator for fermion fields.

Recall that we are calling the auxiliary fields J, η, and $\bar{\eta}$ **Source Fields**. Comparing the form of the Lagrangian in equation (4.322) to (1.126) reveals why. J, η, and $\bar{\eta}$ behave mathematically as sources, giving rise to the field they are coupled to, in the same way that the electromagnetic source J^μ gives rise to the electromagnetic field A^μ. The meaning behind equations (4.334) (and (4.336)) is that J (or η and $\bar{\eta}$) act as sources for the fields, creating a ϕ (or ψ and $\bar{\psi}$) at spacetime point x, and absorbing it at point x'. The terms $\Delta(x - x')$ and $S(x - x')$ then represent the expression giving the probability amplitude for that particular event to occur. In other words, the propagator is the statistical weight of a particle going from x to x'.

4.6.11 Interacting Scalar Fields and Feynman Diagrams

We can now consider how to incorporate interactions into our formalism, allowing us to finally have our relativistic quantum theory of interactions. Beginning with the free scalar Lagrangian (4.11), we can add an interaction term $\mathcal{L}_1$. At this point, we only have one type of particle, ϕ, so we can only have ϕ's interacting with other ϕ's. Terms proportional to ϕ or ϕ^2 are either constant or linear in the equations of motion, and therefore aren't valid candidates for interaction terms. So, the simplest expression we can have is

$$\mathcal{L}_1 = \frac{1}{3!} g\phi^3, \tag{4.338}$$

where $\frac{1}{3!}$ is a conventional normalization, and g is a **Coupling Constant**. So our total Lagrangian is

$$\mathcal{L} = \mathcal{L}_0 + \mathcal{L}_1 = -\frac{1}{2}\partial^\mu\phi\partial_\mu\phi - \frac{1}{2}m^2\phi^2 + \frac{1}{6}g\phi^3, \tag{4.339}$$

and the path integral is

$$\begin{aligned} Z(J) &= \langle 0|0\rangle_J \\ &= \int \mathcal{D}\phi e^{i\int d^4x[\mathcal{L}_0 + \mathcal{L}_1 + J\phi]} \\ &= \int \mathcal{D}\phi e^{i\int d^4x \mathcal{L}_1} e^{i\int d^4x[\mathcal{L}_0 + J\phi]} \\ &= \int \mathcal{D}\phi e^{i\int d^4x \mathcal{L}_1} Z_0(J) = \int \mathcal{D}\phi e^{i\int d^4x \mathcal{L}_1} \langle 0|0\rangle_J. \end{aligned} \tag{4.340}$$

Recall that we can bring out a factor of ϕ from $\langle 0|0\rangle_J$ using the functional derivative $\frac{1}{i}\frac{\delta}{\delta J}$. So, we can make the replacement

$$\mathcal{L}_1(\phi) \to \mathcal{L}_1\left(\frac{1}{i}\frac{\delta}{\delta J}\right) \Rightarrow \frac{1}{6}g\phi^3 \to \frac{g}{6}\left(\frac{1}{i}\frac{\delta}{\delta J}\right)^3. \tag{4.341}$$

Notice also that once this is done, there is no longer any ϕ dependence in $Z(J)$. With the free theory, we were able to remove the ϕ dependence, leading to (4.334). Here we were able to remove it from the interaction term as well. Thus, once again, the infinite number of integrals in (4.308) will merely give a constant which we can absorb into the normalization. This leaves the result

$$\begin{aligned} Z(J) &= e^{\frac{i}{6}g\int d^4x\left(\frac{1}{i}\frac{\delta}{\delta J(x)}\right)^3} Z_0(J) \\ &= e^{-\frac{1}{6}g\int d^4x\left(\frac{\delta}{\delta J(x)}\right)^3} e^{\frac{i}{2}\int d^4x d^4x' J(x)\Delta(x-x')J(x')}. \end{aligned} \tag{4.342}$$

We now Taylor expand each of the two exponentials,

$$\begin{aligned} Z(J) = &\sum_{V=0}^{\infty}\frac{1}{V!}\left[-\frac{g}{6}\int d^4x\left(\frac{\delta}{\delta J(x)}\right)^3\right]^V \\ &\times\sum_{P=0}^{\infty}\frac{1}{P!}\left[\frac{i}{2}\int d^4y d^4z J(y)\Delta(y-z)J(z)\right]^P. \end{aligned} \tag{4.343}$$

Now, recall that a functional derivative $\frac{1}{i}\frac{\delta}{\delta J}$, will remove a J term. Furthermore, after taking the functional derivatives, we will set $J = 0$ to get the physical result. Therefore, for a term to survive, the $2P$ sources must all be exactly removed by the $3V$ functional derivatives.

Hence, using (4.343), we can expand in orders of g (the coupling constant), keeping only the terms which survive, and after removing the sources, evaluate the integrals over the propagators Δ. The value of the integral will then be the physical amplitude for a particular event.

In practice, a slightly different formalism is used to organize and keep track of each term in this expansion – we draw diagrams that correspond to the integrals. Note that there will be P propagators of Δ. We can represent each of these terms diagrammatically, by making each source a solid dot, each propagator a line, and let the g terms be vertices joining the lines together. There will be a total of V vertices, each joining three lines (matching the fact that we are looking at ϕ^3 theory; there would be four lines at each vertex for ϕ^4 theory, etc.).

For example, for $V = 0$ and $P = 1$,

$$Z(J) = \frac{i}{2}\int d^4y d^4z J(y)\Delta(y-z)J(z). \tag{4.344}$$

We have two sources, one located at z and the other located at y, so we draw two dots, corresponding to those locations. Then, the propagator $\Delta(y-z)$ connects them together, so we draw a line between the two dots. The diagram should look like this:

Of course, once we set $J = 0$, this will vanish because it contains two sources.

As another example, consider $V = 0$ and $P = 2$. Now,

$$Z(J)=\frac{1}{2!}\left(\frac{i}{2}\right)^2 \int d^4y d^4z d^4y' d^4z' \big(J(y)\Delta(y-z)J(z)\big)\big(J(y')\Delta(y'-z')J(z')\big). \tag{4.345}$$

This corresponds to four sources, located at y, z, y' and z', with propagator lines connecting y to z, and connecting y' to z'. But there are no lines connecting an unprimed source to a primed source, so this results in two disconnected diagrams:

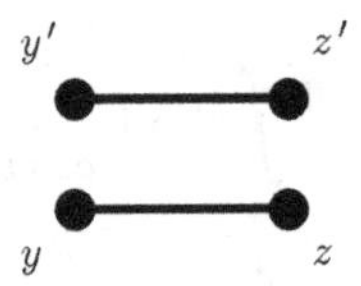

This too will vanish when we set $J = 0$.

As another example, consider $V = 1$ and $P = 2$,

$$\begin{aligned} Z(J) &= -\frac{g}{6}\int d^4x \left(\frac{\delta}{\delta J(x)}\right)^3 \\ &\quad \times \frac{1}{2!}\left(\frac{i}{2}\right)^2 \int d^4y d^4z d^4y' d^4z' \big(J(y)\Delta(y-z)J(z)\big)\big(J(y')\Delta(y'-z')J(z')\big) \\ &= \frac{g}{48}\int d^4x d^4y d^4z d^4y' d^4z' \delta(y-x)\Delta(y-z)\delta(z-x)\delta(y'-x)\Delta(y'-z')J(z') \\ &= \frac{g}{48}\int d^4x d^4z' \Delta(x-x)\Delta(x-z')J(z'). \end{aligned} \tag{4.346}$$

This will correspond to

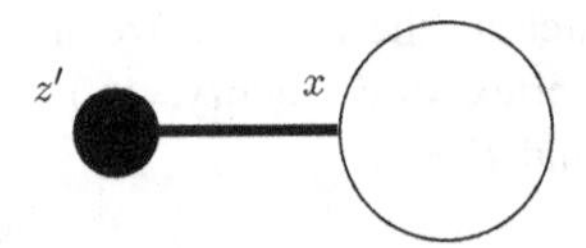

where the source J is located at the dot, and the vertex joining the line to the loop is at x.

You can work out the following, and see that there are multiple diagrams possible for $V = 3,\ P = 5$,

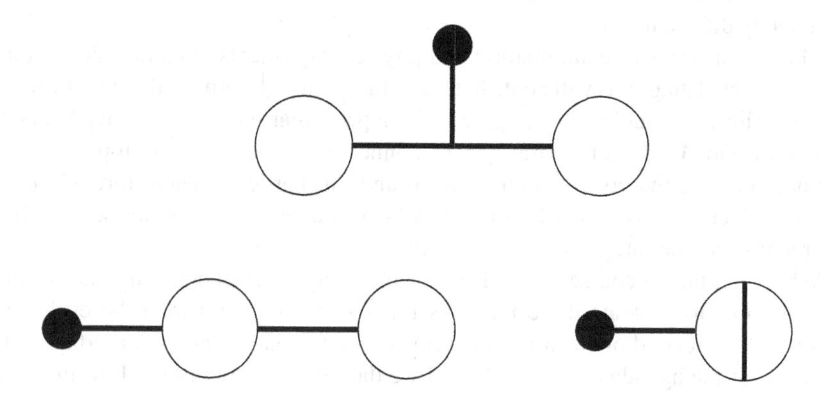

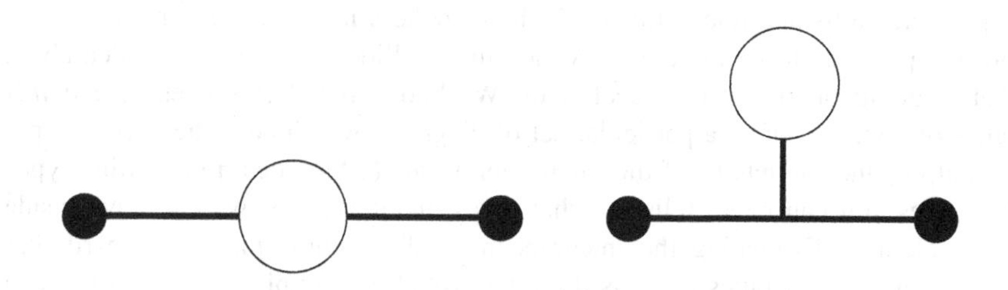

while for $V = 2$, $P = 4$,

and for $V = 1$, $P = 3$,

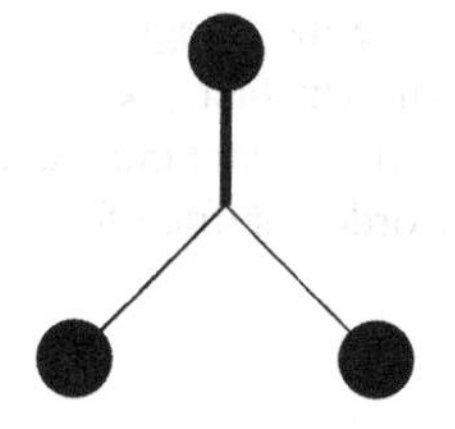

and so on.

An important observation to make is that in each case, every vertex has exactly *three* lines coming into/out of it. This is because our interaction term (4.338) is proportional to ϕ^3 – that is, the interactions in this theory involve three ϕ particles.

Had we chosen the interaction term to instead be proportional to ϕ^4, every vertex would have had exactly *four* lines coming into/out of it. Obviously this would be a completely different set of diagrams, which makes sense – that would be a completely different theory.

Through a series of combinatoric and physical arguments, it can be shown that only connected diagrams will contribute, and the $\frac{1}{P!}$ and $\frac{1}{V!}$ terms will always cancel exactly. Thus, to calculate the amplitude for a particular interaction to happen (say N ϕ's in and M ϕ's out), draw every connected diagram that is topologically distinct, and has the correct number of in and out particles. Then, through a set of rules which you will learn formally in a QFT course (or later in this series), you can reconstruct the integrals which we started with in (4.343).

When you take a course on QFT, you will spend a tremendous amount of time learning how to evaluate these integrals for low order (they cannot be evaluated beyond about second order in most cases). While this is extremely important, it is not vital for the agenda of this book, and we therefore do not discuss how they are evaluated.

The idea is that each diagram represents one of the possible paths the particle can take, along with the possible interactions it can be a part of. Because this is a quantum mechanical theory, we know the particle is actually in a superposition of all possible paths and interactions. We don't make a measurement or observation until the particles leave the area in which they collide, so we have no idea about what is going on inside the accelerator. We know that if *this* goes in and *this* comes out, we can draw a particular set of diagrams which have the correct input and output, and the nature of the interaction terms (which determines what types of vertices you can have) tells us what types of interactions we can have inside the accelerator. Evaluating the integrals then tells us how much that particular event/diagram contributes towards the total probability amplitude. If you want to know how likely a certain incoming/outgoing set of particles is, write down all the diagrams, evaluate the corresponding integrals, and add them up.

As we pointed out above, the classical behavior (which is more probable) is closer to the first-order approximation of the quantum behavior. Therefore, even though in general we can't evaluate the integrals past about second order, the first few orders tell us to a reasonable (in fact, exceptional in most cases) degree of accuracy what the amplitude is.[22] If we want more accuracy, we can seek to evaluate higher orders, but usually lower orders suffice for experiments at energy levels we can currently attain.

[22]Once again, the word "amplitude" simply means "probability".

4.6.12 Interacting Fermion Fields

The analysis we performed above for scalar fields ϕ is almost identical for fermions, and we therefore won't repeat it. The main difference is that the interaction terms will have a field ψ interacting with $\bar{\psi}$, and so the vertices will be slightly different. We won't bother with those details for now.

Finally, we can have a Lagrangian with *both* scalars and fermions. Then, naturally, you could have interaction terms where the scalars interact with fermions. While there are countless interaction terms of this type, the one that will be the most interesting to us is the **Yukawa** term,

$$\mathcal{L}_{Yuk} = g\phi\bar{\psi}\psi. \tag{4.347}$$

If we represent ϕ by a dotted line, ψ by a line with an arrow in the forward time direction, and $\bar{\psi}$ with an arrow going backwards in time, this interaction term will show up in a Feynman diagram as

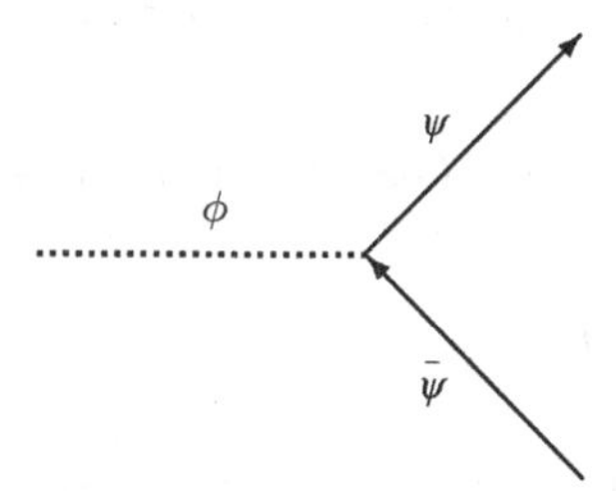

Once each diagram is drawn, there are well defined rules to write down an integral corresponding to each diagram.

4.6.13 A Brief Glance at Renormalization

One of the difficulties encountered with evaluating these integrals is that you almost always find that they yield infinite amplitudes. Since an amplitude (which is a probability) should be between 0 and 1, this is obviously unacceptable. The process of finding the infinite parts and separating them from the finite parts of the amplitude is a very well defined mathematical construct called **Renormalization**. The basic idea is that any infinite term consists of a pure infinity and a finite part. For example the following infinite sum is

$$\sum_{n=1}^{\infty} n = \lim_{x \to 0} \frac{1}{x^2} - \frac{1}{12}. \tag{4.348}$$

The derivation of this is not at all important for the purposes of this book, but people frequently ask to see it, so here is one fairly simple way of deriving it.[23] We'll make frequent use of the identity

$$(1+x)^{-1} = 1 - x + x^2 - x^3 + x^4 + \cdots \tag{4.349}$$

We have

$$\begin{aligned}
\sum_{n=1}^{\infty} n &= \sum_{n=0}^{\infty} n \\
&= \lim_{\epsilon\to 0} \sum_{n=0}^{\infty} n e^{-\epsilon n} \\
&= -\lim_{\epsilon\to 0} \frac{d}{d\epsilon} \sum_{n=0}^{\infty} e^{-\epsilon n} \\
&= -\lim_{\epsilon\to 0} \frac{d}{d\epsilon} (1 + e^{-\epsilon} + e^{-2\epsilon} + e^{-3\epsilon} + e^{-4\epsilon} + \cdots) \\
&= -\lim_{\epsilon\to 0} \frac{d}{d\epsilon} (1 - (-e^{-\epsilon}) + (-e^{-\epsilon})^2 - (-e^{-\epsilon})^3 + (-e^{-\epsilon})^4 - \cdots) \\
&= -\lim_{\epsilon\to 0} \frac{d}{d\epsilon} (1 - e^{-\epsilon})^{-1} \\
&= -\lim_{\epsilon\to 0} \frac{d}{d\epsilon} \left(1 - \left[1 + (-\epsilon) + \frac{1}{2}(-\epsilon)^2 + \frac{1}{3!}(-\epsilon)^3 + \cdots\right]\right)^{-1} \\
&= -\lim_{\epsilon\to 0} \frac{d}{d\epsilon} \left(\epsilon - \frac{\epsilon^2}{2} + \frac{\epsilon^3}{3!} - \frac{\epsilon^4}{4!} + \cdots\right)^{-1} \\
&= -\lim_{\epsilon\to 0} \frac{d}{d\epsilon} \frac{1}{\epsilon} \left(1 - \frac{\epsilon}{2} + \frac{\epsilon^2}{6} - \frac{\epsilon^3}{24} + \cdots\right)^{-1} \\
&= -\lim_{\epsilon\to 0} \frac{d}{d\epsilon} \frac{1}{\epsilon} \left(1 - \left[-\frac{\epsilon}{2} + \frac{\epsilon^2}{6} - \frac{\epsilon^3}{24} + \cdots\right] + \left[-\frac{\epsilon}{2} + \frac{\epsilon^2}{6} - \frac{\epsilon^3}{24} + \cdots\right]^2 + \cdots\right) \\
&= -\lim_{\epsilon\to 0} \frac{d}{d\epsilon} \frac{1}{\epsilon} \left(1 + \frac{\epsilon}{2} + \frac{\epsilon^2}{12} + \cdots\right) \\
&= -\lim_{\epsilon\to 0} \frac{d}{d\epsilon} \left(\frac{1}{\epsilon} + \frac{1}{2} + \frac{\epsilon}{12} + \cdots\right)
\end{aligned}$$

[23] There are other, perhaps better, ways of deriving this result. We're using this one because we believe it is one of the clearest.

$$= -\lim_{\epsilon \to 0}\left(-\frac{1}{\epsilon^2} + \frac{1}{12}\right)$$
$$= \lim_{\epsilon \to 0} \frac{1}{\epsilon^2} - \frac{1}{12}. \tag{4.350}$$

The point is that there is a part which is a pure infinity (the first term on the right-hand side), and a term which is finite. While this may seem strange and extremely unfamiliar (and a bit like hand waving), it is actually a very rigorous and very well understood mathematical idea.

Much of what particle physicists attempt to do is find theories (and types of theories) that can be renormalized and theories that cannot. For example, the action which leads to general relativity leads to a quantum theory which cannot be renormalized. Renormalization is a fascinating and deep topic, and will be covered in great, great depth in a later volume in this series.

4.7 Final Ingredients

The purpose of the previous section was merely to introduce the idea of **Feynman Diagrams** as a tool to calculate amplitudes for physical processes. In doing so, we have met the goal set out in Sect. 4.1, a relativistic quantum mechanical theory of interactions. We achieve such a theory by finding a Lagrangian of a classical theory (both with and without interaction terms), and using equation (4.343) (and the analogous equation for fermions) to write down integrals which, when evaluated, give a contribution to a total amplitude. It is important to remember that we will eventually set all sources J to zero, and a functional derivative (as contained in the interaction term $\mathcal{L}_1$) will set any term without J's to zero. So, the only non-zero terms will be the ones where all of the J's are exactly removed by the functional derivatives, and those non-zero terms are evaluated by simply doing the integral.

That is, in a (very loose) sense, all there is to quantum field theory. The rest of the details of QFT are largely tied up in the extremely difficult and nuanced aspects of actually setting up and evaluating those integrals. We won't delve into those details of **Perturbative Quantum Field Theory**, where amplitudes are studied order by order. The goal of this book is merely to provide an overview of how, once given a Lagrangian, that Lagrangian can be turned into a physically measurable quantity.

With this done, we now set out to find the Lagrangian for the Standard Model of Particle Physics, the theory which seems to explain our universe (at low enough energies and apart from gravity). Once this Lagrangian has been explained, we trust you have a general concept of what to do with it from the previous sections. We'll also come back to the "difficult and nuanced" details of perturbative topics later in this series.

Before we are able to explain the Standard Model Lagrangian, there are a few final concepts we need. They will be the subject of this section. Namely, we will be studying the ideas of **Spontaneous Symmetry Breaking** and **Gauge Theories**. In Sect. 4.5.3, we discussed the simple $U(1)$ gauge theory, where we made a global $U(1)$ symmetry of the free Dirac Lagrangian. We then discussed a local $U(1)$ symmetry, or a gauged symmetry, and showed that consistency demanded the introduction of a gauge field A^{μ}, and consequently a kinetic term and a source term. Thus we recovered the entire electromagnetic force from nothing but a $U(1)$. Later in this section, we'll generalize this to an arbitrary Lie group. Because $U(1)$ is an Abelian group, we refer to the gauge theory of Sect. 4.5.3 as an Abelian gauge theory. For a more general, non-Abelian group, we refer to the resulting theory as a **Non-Abelian Gauge Theory**. Such theories introduce a great deal of complexity, and we therefore consider them in detail in this section before moving on to the Standard Model.

First, however, we begin with the idea of spontaneous symmetry breaking.

4.7.1 Spontaneous Symmetry Breaking

Consider a complex scalar boson ϕ and $\phi^{\dagger}$. The Lagrangian will be

$$\mathcal{L} = -\frac{1}{2}\partial^{\mu}\phi^{\dagger}\partial_{\mu}\phi - \frac{1}{2}m^{2}\phi^{\dagger}\phi. \tag{4.351}$$

Naturally we can write this as

$$\mathcal{L} = -\frac{1}{2}\partial^{\mu}\phi^{\dagger}\partial_{\mu}\phi - V(\phi^{\dagger},\phi), \tag{4.352}$$

where

$$V(\phi^{\dagger},\phi) = \frac{1}{2}m^{2}\phi^{\dagger}\phi. \tag{4.353}$$

This Lagrangian has the $U(1)$ symmetry we discussed in Sect. 4.5.3.

Also, notice that we can graph $V(\phi^{\dagger},\phi)$, plotting V vs. $|\phi|$,

We see a "bowl" with $V_{minimum}$ at $|\phi|^2 = 0$. The vacuum of any theory ends up being at the lowest potential point, and therefore the vacuum of this theory is at $\phi = 0$, as we would expect. The vacuum corresponds to the complete absence of any particle.

Let's now change the potential. Consider

$$V(\phi^{\dagger},\phi) = \frac{1}{2}\lambda m^{2}(\phi^{\dagger}\phi - \Phi^{2})^{2}, \tag{4.354}$$

where λ and Φ are real constants. Notice that the Lagrangian will still have the global $U(1)$ symmetry from before. Now when we graph V vs. $|\phi|$, we get

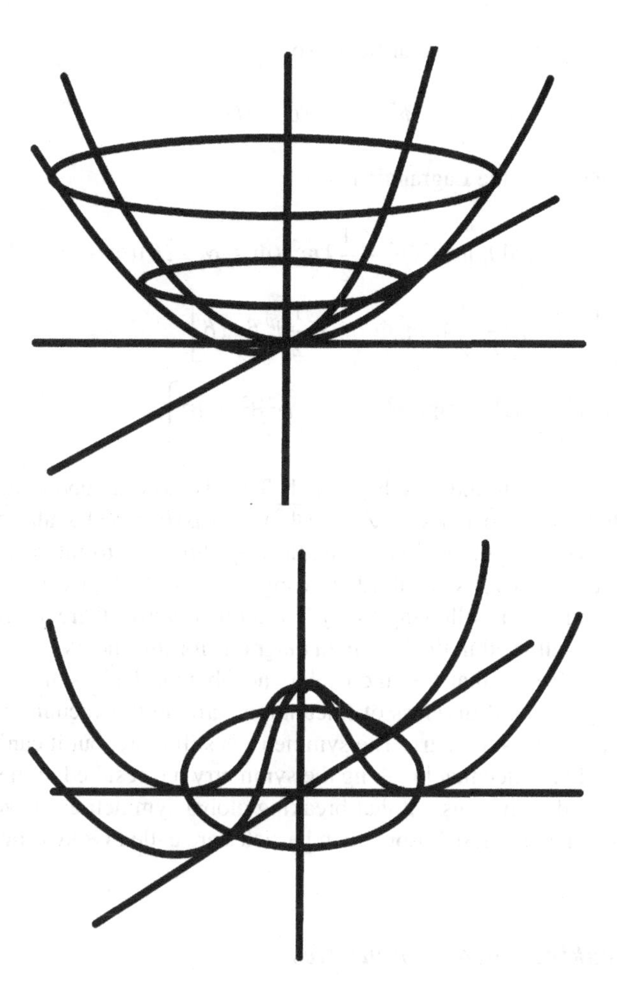

where now the vacuum $V_{minimum}$ is represented by the circle at $|\phi| = \Phi$. In other words, there are an infinite number of vacuums in this theory – one for every point on the circle at $|\phi| = \Phi$. And because the circle drawn in the figure above represents a rotation through field space, this degenerate vacuum is parameterized by $e^{i\alpha}$, the global $U(1)$. There will be a vacuum for every value of α, located at $|\phi| = \Phi$.

In order to make sense of this theory, we must *choose* a vacuum by hand. Because the theory is completely invariant under the value of $e^{i\alpha}$, we can choose any α and *define* that as our true vacuum. We will choose α to make our vacuum at $\phi = \Phi$, that is, where ϕ is real and equal to Φ. We have thus **Gauge Fixed** the symmetry in the Lagrangian, and the $U(1)$ symmetry is no longer manifest.

Now we need to rewrite this theory in terms of our new vacuum. This is simply a change of basis. We expand around the constant vacuum value Φ with the new field

$$\phi = \Phi + \alpha + i\beta, \tag{4.355}$$

where α and β are new real scalar fields, so

$$\phi^\dagger = \Phi + \alpha - i\beta. \tag{4.356}$$

We can now write out the Lagrangian as

$$\begin{aligned}\mathcal{L} &= -\frac{1}{2}\partial^\mu[\alpha - i\beta]\partial_\mu[\alpha + i\beta] - \frac{1}{2}\lambda m^2[(\Phi + \alpha - i\beta)(\Phi + \alpha + i\beta) - \Phi^2]^2 \\ &= \left[-\frac{1}{2}\partial^\mu\alpha\partial_\mu\alpha - \frac{1}{2}4\lambda m^2\Phi^2\alpha^2 - \frac{1}{2}\partial^\mu\beta\partial_\mu\beta\right] \\ &\quad -\frac{1}{2}\lambda m^2\left[4\Phi\alpha^3 + 4\Phi\alpha\beta^2 + \alpha^4 + \alpha^2\beta^2 + \beta^4\right].\end{aligned} \tag{4.357}$$

Something fairly profound has happened. This is now a theory of a *massive* real scalar field α (with mass = $\sqrt{4\lambda m^2\Phi^2}$), a *massless* real scalar field β, and five different types of interactions (one allowing three α's to interact, the second allowing one α and two β's, the third allowing four α's, the fourth allowing two α's and two β's, and the last allowing four β's.) In other words, there are five different types of vertices allowed in the Feynman diagrams for this theory.

Furthermore, notice that this theory has no obvious $U(1)$ symmetry. For this reason, writing the field in terms of fluctuations around the vacuum we choose is called "breaking" the symmetry. The symmetry is still there, but it can't be seen in this form. Finally, notice that breaking the symmetry has resulted in the addition of the massless field β. It turns out that breaking global symmetries as we have done *always* results in a massless boson. Such particles are called **Goldstone Bosons**.

4.7.2 Breaking Local Symmetries

In the previous section, we broke a global $U(1)$ symmetry. In this section, we will break a local $U(1)$ and see what happens. We begin with the Lagrangian for a complex scalar field with a gauged $U(1)$ (cf. (4.234)):

$$\mathcal{L} = -\frac{1}{2}[(\partial^\mu - iqA^\mu)\phi^\dagger][(\partial_\mu + iqA_\mu)\phi] - \frac{1}{4}F_{\mu\nu}F^{\mu\nu} - V(\phi^\dagger, \phi), \tag{4.358}$$

where we have taken the external source $J^\mu = 0$. Let's once again assume $V(\phi^\dagger, \phi)$ has the form of equation (4.354), so the vacuum has the $U(1)$ degeneracy at $|\phi| = \Phi$.

Because our $U(1)$ is now local, we can choose the gauge $\alpha(x)$ so that not only is the vacuum real, but also so that ϕ is always real. We therefore expand

$$\phi = \Phi + h, \tag{4.359}$$

where h is a real scalar field representing fluctuations around the vacuum we chose. Now,

$$\begin{aligned}\mathcal{L} &= -\frac{1}{2}[(\partial^{\mu} - iqA^{\mu})(\Phi + h)][(\partial_{\mu} + iqA_{\mu})(\Phi + h)] - \frac{1}{4}F_{\mu\nu}F^{\mu\nu} \\ &\quad -\frac{1}{2}\lambda m^2[(\Phi + h)(\Phi + h) - \Phi^2]^2 \\ &= \cdots \\ &= -\frac{1}{2}\partial^{\mu}h\partial_{\mu}h - \frac{1}{2}4\lambda m^2\Phi^2 h^2 - \frac{1}{4}F^{\mu\nu}F_{\mu\nu} - \frac{1}{2}q^2\Phi^2 A^2 + \mathcal{L}_{interactions},\end{aligned} \tag{4.360}$$

where the allowed interaction terms include a vertex connecting an h and two A^{μ}'s, four h's, and three h's.

Before breaking, we had a complex scalar field ϕ and a *massless* vector field A^{μ} with two polarization states (because it is a photon). Now, we have a single real scalar h with mass $= \sqrt{4\lambda m^2\Phi^2}$ and a field A^{μ} with mass $= q\Phi$. In other words, our force-carrying particle A^{μ} has gained mass! We started with a theory with no mass, and by merely breaking the symmetry, we have introduced mass into our theory. This mechanism for introducing mass into a theory, called the **Higgs Mechanism**, was first discovered by Peter Higgs, and the resulting field h is called the **Higgs Boson**. Whereas the consequence of global symmetry breaking is a massless boson called a Goldstone boson, the consequence of a local symmetry breaking is that the gauge field, which came about as a result of the symmetry being local, acquires mass.

4.7.3 Non-Abelian Gauge Theory

We are now ready to generalize the gauging of a symmetry we did in Sect. 4.5.3 to an arbitrary Lie group. Consider a Lagrangian $\mathcal{L}$ with N scalar (or spinor) fields ϕ_i ($i = 1, \ldots, N$) that is invariant under a continuous $SO(N)$ or $SU(N)$ symmetry,

$$\phi_i \to U_{ij}\phi_j, \tag{4.361}$$

where U_{ij} is an $N \times N$ matrix of $SO(N)$ or $SU(N)$.

In Sect. 4.5.3, we saw that if the group is $U(1)$, gauging it demands the introduction of the gauge field A^{μ} to preserve the symmetry, which shows up in the covariant derivative $D_{\mu} = \partial_{\mu} - ieA_{\mu}$. To say a field carried some sort of charge means that it has the corresponding term in its covariant derivative. We then added a kinetic term for A^{μ} as well as an external source J^{μ}. Then, higher-order interaction terms can be included in whatever way is appropriate for the theory.

To generalize this, let's say for the sake of concreteness that our Lie group is $SU(N)$. An arbitrary element of $SU(N)$ is

$$U(x) = e^{ig\theta^a(x)T^a}, \tag{4.362}$$

where g is a constant we have added for later convenience, θ^a are the $N^2 - 1$ parameters of the group (cf. Sect. 3.2.15), and the T^a are the generator matrices for the group. Notice that we have gauged the symmetry (in that $\theta(x)$ is a function of spacetime).

By definition, we know that the generators T^a will obey the commutation relations

$$[T^a, T^b] = if_{abc}T^c, \tag{4.363}$$

(cf. equation (3.99)), where f_{abc} are the structure constants of the group (cf. equation (3.99)).

When gauging the $U(1)$ in Sect. 4.5.3, the transformation of the gauge field was given by equation (4.223). For the more general transformation $\phi_i \to U_{ij}\phi_j$, the gauge field transforms according to[24]

$$A^\mu \to U(x)A^\mu U^\dagger(x) + \frac{i}{g}U(x)\partial^\mu U^\dagger(x), \tag{4.364}$$

(where we have removed the indicial notation and it is understood that matrix multiplication is being discussed). If $U(x)$ is an element of $U(1)$ (so it is $e^{ig\theta(x)}$), then this transformation reduces to

$$A^\mu \to e^{ig\theta(x)}A^\mu e^{-ig\theta(x)} + \frac{i}{g}e^{ig\theta(x)}(-ig\partial^\mu\theta(x))e^{-ig\theta(x)} = A^\mu + \partial^\mu\theta(x), \tag{4.365}$$

which is exactly what we had in (1.130). For general $SU(N)$, however, the U's are elements of a non-Abelian group, and the A^μ's are matrices of the same size, making this a more complicated expression.

Generalizing, we find that an element of the $SU(N)$ has the form of (changing notation slightly)

$$U(x) = e^{-ig\Gamma^a(x)T^a} \tag{4.366}$$

with $N^2 - 1$ real parameters Γ^a. We then build the covariant derivative in the exact same way as in equation (4.224), by adding a term proportional to the gauge field

$$D_\mu = \mathbb{I}^{N\times N}\partial_\mu - igA_\mu. \tag{4.367}$$

---

[24]Don't worry about where this came from. It's another fairly geometrical idea that we'll derive in the next book.

(Remember that each component of A_μ is an $N \times N$ matrix. They were scalars for $U(1)$ because $U(1)$ is a 1×1 matrix). Thus, acting on the fields, the covariant derivative is

$$(D_\mu \phi)_j = \partial_\mu \phi_j(x) - ig[A_\mu(x)]_{jk}\phi_k(x), \tag{4.368}$$

where k is to be summed on the last term. It will be understood from now on that the normal partial derivative term (the first term) has an $N \times N$ identity matrix multiplied by it.

Then, just as in (4.230), we have the field strength

$$F_{\mu\nu}(x) = \frac{i}{g}[D_\mu, D_\nu] = \partial_\mu A_\nu - \partial_\nu A_\mu - ig[A_\mu, A_\nu], \tag{4.369}$$

where the commutator term doesn't vanish for arbitrary Lie group as it did for Abelian $U(1)$.

Recall from equation (1.132) that for $U(1)$, $F_{\mu\nu}$ is invariant under the gauge transformation (1.130) on its own, because the commutator term vanishes. In general, however, the commutator term does not vanish, and we must therefore be careful in writing down the correct kinetic term. It turns out that the correct choice is[25]

$$\mathcal{L}_{Kin} = -\frac{1}{2}Tr(F_{\mu\nu}F^{\mu\nu}). \tag{4.370}$$

It may not be obvious, but this form is actually a consequence of (3.150). There is algebraic machinery working under the surface of this that, while extremely interesting, is unfortunately beyond the scope of these introductory notes. We will discuss all of these ideas in much greater depth later in this series.

Starting with a non-interacting Lagrangian that is invariant under the global $SU(N)$, we can gauge the $SU(N)$ to create a theory with a gauge field (or synonymously a "force carrying" field) A^μ, which is an $N \times N$ matrix. Hence, every Lie group gives rise to a particular gauge field (which is a force carrying particle, like the photon), and therefore a particular force.

For this reason, we discuss forces in terms of Lie groups, or synonymously **Gauge Groups**. Each group defines a force. As we said at the very end of Sect. 3.2.11, $U(1)$ represents the electromagnetic force (as we have seen in Sect. 4.5.3), while $SU(2)$ describes the weak force, and $SU(3)$ describes the strong color force.

[25] Again, this comes from deeper geometrical reasoning. We'll get there later.

4.7.4 Representations of Gauge Groups

As we discussed in Sect. 3.2, given a set of structure constants f_{abc} which define the Lie algebra of some Lie group, we can form a representation of that group that we denote R. So, R will be a set of $D(R)\times D(R)$ matrices, where D is the dimension of the representation R. We then call the generators of the group (in the representation R) T_R^a, and they naturally obey

$$[T_R^a, T_R^b] = if_{abc}T_R^c. \tag{4.371}$$

One representation which exists for any of the groups we have considered is the representation of $SO(N)$ or $SU(N)$ consisting of $N \times N$ matrices. We denote this the **Fundamental Representation** (also called **Defining Representation** in some books). The fundamental representations of $SO(2)$, $SO(3)$, $SU(2)$, and $SU(3)$ are the 2×2, 3×3, 2×2, and 3×3 matrix representations, respectively. We will denote the fundamental representation for a given group by writing the number in bold. Thus, the fundamental representation of $SU(2)$ will be denoted **2**, and the generators for $SU(2)$ in the fundamental representation will be denoted $T_{\mathbf{2}}^a$. Obviously, the fundamental representation of $SU(3)$ will be **3** with generators $T_{\mathbf{3}}^a$.

Furthermore, let's say we have some arbitrary representation generated by T_R^a obeying

$$[T_R^a, T_R^b] = if_{abc}T_R^c. \tag{4.372}$$

We can take the complex conjugate of the commutation relations to get

$$[T_R^{\star a}, T_R^{\star b}] = -if_{abc}T_R^{\star c}. \tag{4.373}$$

If we define the *new* set of generators

$$T_R^{\prime a} = -T_R^{\star a}, \tag{4.374}$$

then the $T_R^{\prime a}$ will obey the correct commutation relations, and will therefore form a representation of the group as well. If it turns out that

$$T_R^{\prime a} = -(T_R^a)^\star = T_R^a, \tag{4.375}$$

or if there is some unitary similarity transformation

$$T_R^a \to U^{-1}T_R^a U \tag{4.376}$$

such that

$$T_R^{\prime a} = -(T_R^a)^\star = T_R^a, \tag{4.377}$$

then this is actually an equivalent representation (cf. equation (3.62) ff). In this case we call the representation **Real** because the complex conjugate of the representation is equal to the original representation. However, if no such transformation exists, then the complex conjugate is actually a *new* representation, called the **Complex Conjugate** representation to R, or the **Anti-R** representation, which we denote $\bar{R}$.

For example, there is the fundamental representation of $SU(3)$, denoted **3**, generated by $T^a_{\mathbf{3}}$, and then there is the anti-fundamental representation $\bar{\mathbf{3}}$, generated by $T^a_{\bar{\mathbf{3}}}$. We can find the generators $T^a_{\bar{\mathbf{3}}}$ of $\bar{\mathbf{3}}$ by simply taking the negative of the complex conjugate of (3.208) as in (4.374) – they are $\frac{1}{2}\lambda^a$ for:

$$\lambda^1 = \begin{pmatrix} 0 & -1 & 0 \\ -1 & 0 & 0 \\ 0 & 0 & 0 \end{pmatrix}, \; \lambda^2 = \begin{pmatrix} 0 & -i & 0 \\ i & 0 & 0 \\ 0 & 0 & 0 \end{pmatrix}, \; \lambda^3 = \begin{pmatrix} -1 & 0 & 0 \\ 0 & 1 & 0 \\ 0 & 0 & 0 \end{pmatrix}, \; \lambda^4 = \begin{pmatrix} 0 & 0 & -1 \\ 0 & 0 & 0 \\ -1 & 0 & 0 \end{pmatrix},$$

$$\lambda^5 = \begin{pmatrix} 0 & 0 & -i \\ 0 & 0 & 0 \\ i & 0 & 0 \end{pmatrix}, \; \lambda^6 = \begin{pmatrix} 0 & 0 & 0 \\ 0 & 0 & -1 \\ 0 & -1 & 0 \end{pmatrix}, \; \lambda^7 = \begin{pmatrix} 0 & 0 & 0 \\ 0 & 0 & -i \\ 0 & i & 0 \end{pmatrix},$$

$$\lambda^8 = \frac{1}{\sqrt{3}} \begin{pmatrix} -1 & 0 & 0 \\ 0 & -1 & 0 \\ 0 & 0 & 2 \end{pmatrix}. \tag{4.378}$$

The weight vectors will be (from λ^3 and λ^8)

$$\mathbf{t}_1 = \begin{pmatrix} -\frac{1}{2} \\ -\frac{1}{2\sqrt{3}} \end{pmatrix}, \qquad \mathbf{t}_2 = \begin{pmatrix} \frac{1}{2} \\ -\frac{1}{2\sqrt{3}} \end{pmatrix}, \qquad \mathbf{t}_3 = \begin{pmatrix} 0 \\ \frac{1}{\sqrt{3}} \end{pmatrix}. \tag{4.379}$$

We can graph these as we did before

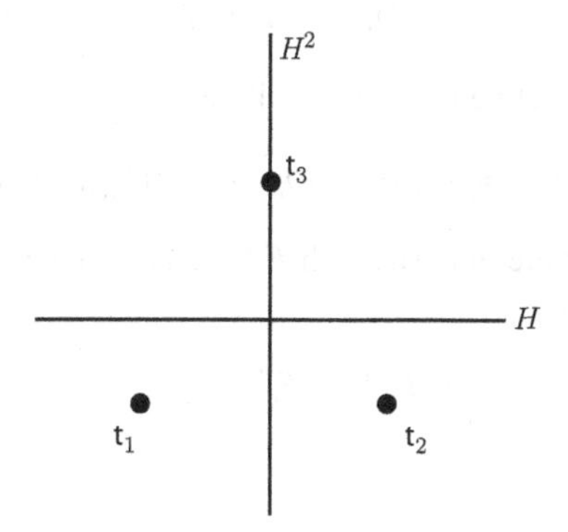

You can work out the details of the non-Cartan generators and the relevant diagram yourself.

The representations of a group which will be important to us are the fundamental, anti-fundamental, and adjoint.

4.7.5 Symmetry Breaking Revisited

As we said in Sect. 4.7.3, given a field transforming in a particular representation R, the gauge fields A^{μ} will be $D(R) \times D(R)$ matrices.

Once we know what representation we are working in, and therefore know the generators T_R^a, it turns out that it is always possible to write the gauge fields in terms of the generators. Recall in Sects. 3.2.2 and 3.2.12, we encouraged you to think of the generators as basis vectors which span the parameter space for the group. Because the gauge fields live in the $N \times N$ space as well, we can write them in terms of the generators. That is, instead of the gauge fields being $N \times N$ matrices on their own, we will use the $N \times N$ matrix generators as basis vectors, and then the gauge fields can be written as scalar coefficients of each generator:

$$A^{\mu} = A_a^{\mu} T_R^a, \tag{4.380}$$

where a is understood to be summed, and each A_a^{μ} is now a scalar function rather than a $D(R) \times D(R)$ matrix (the advantage of this is that we can continue to think of the gauge fields as scalars with an extra index, rather than as matrices).

As a note, we haven't done anything particularly profound here. We are merely writing each component of the $D(R) \times D(R)$ matrix A^{μ} in terms of the $D(R) \times D(R)$ generators, allowing us to work with a *scalar* field A_a^{μ} rather than the matrix field A^{μ}. We now actually view each A_a^{μ} as a separate field. So, if a group has N generators, we say there are N gauge fields associated with it, each one having 4 spacetime components μ. In matrix components, this will be

$$(A^{\mu})_{ij} = (A_a^{\mu} T_R^a)_{ij}. \tag{4.381}$$

Then, the covariant derivative in (4.368) will be

$$(D_{\mu}\phi)_j = \partial_{\mu}\phi_j(x) - ig[A_{\mu a}(x) T_R^a]_{jk}\phi_k(x). \tag{4.382}$$

We may assume that the field strength $F^{\mu\nu}$ can also be expressed in terms of the generators, so that we have

$$F^{\mu\nu} = F_a^{\mu\nu} T^a, \tag{4.383}$$

or

$$(F^{\mu\nu})_{ij} = (F_a^{\mu\nu} T^a)_{ij}. \tag{4.384}$$

Now, using (3.150) (and taking $\kappa = 1/2$), we can write (4.370) in terms of the new basis:

$$\begin{aligned}
\mathcal{L}_{Kin} = -\frac{1}{2}Tr(F_{\mu\nu}F^{\mu\nu}) &= -\frac{1}{2}Tr(F_a^{\mu\nu}T^a F_{\mu\nu b}T^b) \\
&= -\frac{1}{2}F_a^{\mu\nu}F_{\mu\nu b}Tr(T^a T^b) \\
&= -\frac{1}{2}F_a^{\mu\nu}F_{\mu\nu b}\kappa\delta^{ab} \\
&= -\frac{1}{2}F_a^{\mu\nu}F_{\mu\nu}^a\kappa \\
&= -\frac{1}{4}F_a^{\mu\nu}F_{\mu\nu}^a.
\end{aligned} \tag{4.385}$$

(We have raised the index a on the second field strength term in the last two lines simply to explicitly imply the summation over it. The fact that it is raised doesn't change its value in this case; it is merely notational.)

Furthermore, we can use (3.150) to invert (4.383):

$$\begin{aligned}
F^{\mu\nu} = F_a^{\mu\nu}T^a &\Rightarrow F^{\mu\nu}T^b = F_a^{\mu\nu}T^a T^b \\
&\Rightarrow Tr(F^{\mu\nu}T^b) = F_a^{\mu\nu}Tr(T^a T^b) \\
&\Rightarrow Tr(F^{\mu\nu}T^b) = F_a^{\mu\nu}\kappa\delta^{ab} \\
&\Rightarrow Tr(F^{\mu\nu}T^b) = \frac{1}{2}F_b^{\mu\nu} \\
&\Rightarrow F_a^{\mu\nu} = 2Tr(F^{\mu\nu}T^a).
\end{aligned} \tag{4.386}$$

In Sects. 4.7.1 and 4.7.2, we broke the $U(1)$ symmetry, which only had one generator. However, if we break larger groups we may only break part of it. For example, we will see that $SU(3)$ has an $SU(2)$ subgroup. It is actually possible to break all but the $SU(2)$ part of the $SU(3)$. Five of the $SU(3)$ generators are broken (those not corresponding to the $SU(2)$ subgroup/subalgebra), and the other three (generating $SU(2)$) are unbroken. Because we are now writing our gauge fields using the generators as a basis, this means that five of the gauge fields are broken, while three of the gauge fields are not.

Finally, recall from Sect. 4.7.2 that breaking a local symmetry results in a gauge field gaining mass. We seek now to elucidate the relationship between breaking a symmetry and a field gaining mass. We can summarize the relationship as follows: *Gauge fields corresponding to broken generators acquire mass, while those corresponding to unbroken generators do not. The unbroken generators form a new gauge group that is smaller (being of lower rank and/or lower dimension) than the original group that was broken.*

In 4.7.2, we saw that breaking a symmetry gave the gauge field mass. Now, we see that the converse holds as well; giving a gauge field mass will break the symmetry. To make this clearer, we begin with a very simple example, then move on to a more complicated example.

4.7.6 Simple Examples of Symmetry Breaking

Consider a theory with three real massless scalar fields ϕ_i ($i = 1, 2, 3$) and with Lagrangian

$$\mathcal{L} = -\frac{1}{2}\partial^\mu \phi_i \partial_\mu \phi^i, \tag{4.387}$$

which is invariant under the global $SO(3)$ rotation

$$\phi_i \to R^{ij}\phi_j, \tag{4.388}$$

where R^{ij} is an element of $SO(3)$, because the Lagrangian is merely a dot product in field space, and we know that dot products are invariant under $SO(3)$.

Now, let's say that one of the fields, say ϕ_1, gains mass. The new Lagrangian will then be

$$\mathcal{L} = -\frac{1}{2}\partial^\mu \phi_i \partial_\mu \phi^i - \frac{1}{2}m^2\phi_1^2. \tag{4.389}$$

This Lagrangian is no longer invariant under the full $SO(3)$ group, which mixes any two of the three fields. Rather, it is only invariant under rotations in field space that mix ϕ_2 and ϕ_3 or $SO(2)$. In other words, giving one field mass broke $SO(3)$ to the smaller $SO(2)$.

As another simple example, we start with five massless complex scalar fields ϕ_i, with Lagrangian

$$\mathcal{L} = -\frac{1}{2}\partial^\mu \phi_i^\dagger \partial_\mu \phi^i. \tag{4.390}$$

This will be invariant under any $SU(5)$ transformation. Then let's say we give two of the fields, ϕ_1 and ϕ_2 (equal) mass. The new Lagrangian will be

$$\mathcal{L} = -\frac{1}{2}\partial^\mu \phi_i^\dagger \partial_\mu \phi^i - \frac{1}{2}m\left(\phi_1^\dagger\phi_1 + \phi_2^\dagger\phi_2\right). \tag{4.391}$$

We now no longer have the full $SU(5)$ symmetry, but we do have the special unitary transformations mixing ϕ_3, ϕ_4, and ϕ_5. This is an $SU(3)$ subgroup. Also, we can do a special unitary transformation mixing ϕ_1 and ϕ_2. This is an $SU(2)$ subgroup. So, we have broken

$$SU(5) \to SU(3) \times SU(2). \tag{4.392}$$

Before considering a more complicated example of this, we further discuss the connection between symmetry breaking and fields gaining mass. When we introduced spontaneous symmetry breaking in Sect. 4.7.1, recall that we shifted the potential minimum from $V_{minimum}$ at $\phi = 0$ to $V_{minimum}$ at $|\phi| = \Phi$. But we were discussing this in very classical language. We can interpret all of this in a more "quantum" way in terms of VEV's. As we said, the vacuum of a theory is defined as the minimum potential field configuration. For the $V_{minimum}$ at $\phi = 0$ potential, the VEV of the field ϕ was at 0, or

$$\langle 0|\phi|0\rangle = 0. \tag{4.393}$$

However, for the $V_{minimum}$ at $|\phi| = \Phi$ potential, we have

$$\langle 0|\phi|0\rangle = \Phi. \tag{4.394}$$

Thus, in quantum mechanical language, symmetry breaking can occur when a field, or some components of a field, take on a non-zero VEV.

This seems to be what is happening in nature. At higher energies, there is some "Master Theory" with some gauge group defining the physics, and all of the fields involved have 0 VEV's. At lower energies, for whatever reason (which is not well understood at the time of this writing), some of the fields take on non-zero VEV's, which break the symmetry into smaller groups, giving mass to certain fields through the Higgs Mechanism discussed in Sect. 4.7.2. We call the theory with the unbroken gauge symmetry at higher energies the more fundamental theory (analogous to equation (4.352)), and the Lagrangian which results from breaking the symmetry (analogous to (4.357)) the **Low Energy Effective Theory**.

This is how mass is introduced into the Standard Model. It turns out that if a theory is renormalizable one can prove that any lower energy effective theory that results from breaking the original theory's symmetry is also renormalizable, even if it doesn't appear to be. Because the actions that appear to describe the universe at the energy level we live at (and the levels attainable by current experiment) are not renormalizable when they have quadratic mass terms, we work with a larger theory which has no massive particles but can be renormalized, and use the Higgs Mechanism to give various particles mass through third-order terms. Hence, whereas the physics we see at low energies may not appear renormalizable, if we can find a renormalizable theory which breaks down to our physics, we are safe.

Now, we consider a slightly more complicated (and realistic) example of symmetry breaking.

4.7.7 A More Complicated Example of Symmetry Breaking

Consider the gauge group $SU(N)$, acting on N complex scalar fields ϕ_i ($i = 1, \ldots, N$) in the fundamental representation $\mathbf{N}$. Recall that in Sect. 4.7.2, in order to get equation (4.359), we made use of the $U(1)$ symmetry to make the vacuum, or the VEV, real. We can now do something similar: we make use of the $SU(N)$ to not only make the VEV real, but also to rotate it to a single component of the field, ϕ_N. In other words, we do an $SU(N)$ rotation so that

$$\begin{aligned} \langle 0|\phi_i|0\rangle &= 0, \quad \text{for} \quad i = 1, \dots, N-1, \\ \langle 0|\phi_N|0\rangle &= \Phi. \end{aligned} \tag{4.395}$$

We then expand ϕ_N around this new vacuum:

$$\begin{aligned} \phi_i &= \phi_i, \quad \text{for} \quad i = 1, \dots, N-1, \\ \phi_N &= \Phi + \chi. \end{aligned} \tag{4.396}$$

This means that in the vacuum configuration, the fields will have the form

$$\begin{pmatrix} \phi_1 \\ \phi_2 \\ \vdots \\ \phi_N \end{pmatrix}_{vacuum} = \begin{pmatrix} 0 \\ 0 \\ \vdots \\ \Phi \end{pmatrix}. \tag{4.397}$$

How will the action of $SU(N)$ be affected by this VEV? If we consider a general element of $SU(N)$ acting on this, then

$$\begin{pmatrix} U_{11} & U_{12} & \cdots & U_{1N} \\ U_{21} & U_{22} & \cdots & U_{2N} \\ \vdots & \vdots & \ddots & \cdots \\ U_{N1} & U_{N2} & \vdots & U_{NN} \end{pmatrix} \begin{pmatrix} 0 \\ 0 \\ \vdots \\ \Phi \end{pmatrix} = \begin{pmatrix} U_{1N}\Phi \\ U_{2N}\Phi \\ \vdots \\ U_{NN}\Phi \end{pmatrix}. \tag{4.398}$$

Thus, only elements of $SU(N)$ with non-zero elements in the last column will be affected by this VEV. But the other $N-1$ elements' rows and columns are unaffected. This means that we have an $SU(N-1)$ symmetry left. Or in other words, we have broken

$$SU(N) \rightarrow SU(N-1) \tag{4.399}$$

with this VEV.

Let's consider a specific example of this. Consider $SU(3)$. The generators are written out in (3.208). Notice that exactly three of them have all zeros in the last column; λ^1, λ^2, and λ^3. We expect these three to give an

$$SU(3-1) = SU(2) \tag{4.400}$$

subgroup. Looking at the upper left 2×2 boxes in those three generators, we can see that they are the Pauli matrices, the generators of $SU(2)$. If we give a non-zero VEV to the fields transforming under $SU(3)$, we see that they do indeed break the $SU(3)$ to $SU(2)$. The other five generators of $SU(3)$ will be affected by the VEV, and consequently the corresponding fields will acquire mass.

4.8 The Standard Model

4.8.1 Helpful Background

We are finally ready to study the **Standard Model of Particle Physics**, which since about 1974 has successfully explained our observations at particle colliders. The theory presented here is both a climax and an anticlimax. It is a drama of unraveling the invisible hidden symmetries chosen for the design of our universe. The main parts of the Standard Model were introduced with almost no fanfare by Salam and Weinberg independently around 1968. Weinberg's paper for example received only two citations in 1969 and 1970. The lack of early citations should indicate that the theory's beauty was not its winning quality. Although the correctness of the Standard Model was not plainly apparent, the Standard Model managed to predate the experimental evidence by years. The theory is considered ugly by most, but it is successful far beyond anyone's dreams. The Standard Model is our most confirmed blueprint to the universe's master design, yet its failures make it a tantalizing suggestion that the story is not yet over.

We have done quite a bit of math in this chapter. We now return our focus to the application of the math to describing nature for this section. In the previous sections, we have learned the basic principles and ingredients needed for the Standard Model (except renormalizability). What may not be as obvious is why they are key ingredients in the first place.

To place the Standard Model in context let us return to 1965. Tomonaga, Feynmann, and Schwinger have just won the Nobel prize for Quantum Electrodynamics. They calculated the magnetic moment of the electron and other observables using quantum field theory and renormalization to separate out the infinities of the theory from a finite contribution. Their calculations and the corresponding experiments verified the vital role of renormalization in fundamental physics. The anomalous magnetic moment of the electron, which is calculated using the renormalization toolbox, today agrees with experiment to more than 13 significant digits! Renormalized gauge theories agree with experiment to a miraculous precision. Renormalization works! Renormalization cannot be ignored!

Unfortunately in 1965 the models explaining radioactive decay and the strong interaction were not renormalizable. The leading theory was called the chiral $V - A$ universal model of weak decays featuring four-fermion interactions in the combination of vector minus axial currents. The $V - A$ model could not be mathematically broken down into a finite and infinite component. The rates of radioactive decay processes were only carried to first order using the Born-approximation. Because these are rare processes, the Born approximation is a good assumption. These rare processes could be estimated without resorting to higher-order terms where the infinities appear. However, everyone knew the infinities were there at the next order of the calculation; the QFT framework they were using was unstable and ready to crack at the first challenge.

Although gauge theory and renormalization explained the interaction of electrons with photons, gauge theory was not able to address the strong and weak forces. These forces were known to be short-range forces. To make a force have a short range in QFT, the mediating boson needed a mass. The Yukawa theory of scalar fields included such a term as an early model for the strong force with short range. The force law then fell off as $\exp(-rm)/r^2$ with both the classic inverse square law multiplied by an exponential dampening with distance parameterized by the mass m. To give a gauge boson A_μ a short range, the Lagrangian would need a mass term like $m_A^2 A_\mu A^\mu$. This term violates gauge symmetry because when $A'_\mu = A_\mu + \partial_\mu \chi$ we see that $A_\mu A^\mu \neq A'_\mu A'^\mu$. Naively, one would think that gauge symmetry blocks all gauge bosons from having mass; and therefore, all gauge theories (Abelian and the non-Abelian ones) would obey force laws that scale as $1/r^2$. This would mean that all gauge theories would represent long-range forces similar to gravity and electromagnetism (each of which is mediated with a massless boson). In 1954 when Yang was first giving a presentation on non-Abelian gauge theories, Pauli interrupted the talk. Pauli wanted to know what the mass of the non-Abelian gauge boson was. Pauli was so insistent that Yang eventually sat down. Pauli realized that a mass term violated gauge symmetry; the mass terms were needed for short-range forces; non-Abelian gauge theories seemed like they should have long-range forces; and therefore, they probably do not explain strong or weak forces. In short, people no less then Pauli felt gauge symmetry's properties made them unlikely candidates for the a short-range force needed to explain the strong and weak forces.

There are two solutions to this quandary: (1) the Higgs mechanism discussed in Sect. 4.7.2 which gives gauge bosons mass without violating gauge symmetry and (2) a spontaneously created mass gap phenomena associated with non-Abelian gauge theories which is yet not fully understood and seems related to the confinement of individual quarks. The Standard Model chooses (1) the Higgs mechanism. The strong force seems to use (2) the and confinement mass-gap phenomena.[26]

Citations to Weinberg's "Theory of Leptons" started after 1972 for two reasons. Neither of these was because people noticed the beauty of the Standard Model. The first reason is that 't Hooft and Veltman had shown that spontaneously broken

[26]The short-range nature of the $SU(3)$ strong force is a much bigger mystery. As of 2010, aspiring physicists can win one million dollars from the Clay Mathematics Institute Millennium Prize by explaining why the unbroken $SU(3)$ non-Abelian gauge theory requires particles to be $SU(3)$ color singlets and why there is a mass gap for gluonic excitations that makes the force effectively short range.

gauge theories were renormalizable. In other words, models that used the Higgs mechanism and non-Abelian gauge theories together provided for a renormalizable theory with a short range force. The second reason is that the observation of a rare event at CERN highlighted a new phenomena present in the Standard Model. A neutrino beam was being aimed at a bubble chamber. A track was recorded showing an electron that must have started at rest but was then apparently hit by the neutrino leaving a distinct track with a curvature that marked it as an electron. We will return to this event several times as we develop some understanding of the Standard Model's structure.

4.8.2 The Outline

We will cover the main points of the Standard Model, but there is tremendous detail we are skipping over. We will not be working out every step in as much detail because nearly all calculations done in this section have a similar tedious calculation worked out previously.

Before diving into this in detail, look over the general structure of the Standard Model on page 50. This gives the matter content and the symmetry transformation properties. The Standard Model has a symmetry group of $SU(3)_C \times SU(2)_L \times U(1)_Y$. The C on $SU(3)_C$ stands for color. The L on $SU(2)_L$ means it only acts on the left-handed states. The Y on $U(1)_Y$ stands for hypercharge and is to distinguish $U(1)_Y$ from the $U(1)$ for electromagnetism.

The gauge bosons content is therefore 8 'photon-like' fields from the adjoint of $SU(3)_C$, 3 from the adjoint of $SU(2)_L$ and 1 from the Abelian $U(1)_Y$. For each matter ingredient (not a gauge boson) we specify the representations and the charge under these three symmetry groups as a triplet of numbers. A 1 for $SU(3)_C$

A Yang-Mills (Gauge) Theory with Gauge Group

$$SU(3)_C \times SU(2)_L \times U(1)_Y$$

with left-handed Weyl fields fields in three copies of the representation

$$(1, \mathbf{2}, -1/2) \oplus (1, 1, 1) \oplus (\mathbf{3}, \mathbf{2}, 1/6) \oplus (\bar{\mathbf{3}}, 1, -2/3) \oplus (\bar{\mathbf{3}}, 1, 1/3)$$

where the first entry of each triplet is the $SU(3)_C$ representation,
the next entry is the $SU(2)_L$ representation,
and the last entry specifies the value of the $U(1)$ hypercharge Y,

and a single copy of a complex scalar field in the representation

$$(1, \mathbf{2}, -1/2)$$

or $SU(2)_L$ is a singlet (does not transform). The value in the third entry is the $U(1)_Y$ hypercharge. For example $(1, \mathbf{2}, -1/2)$ means that the field is a singlet under $SU(3)_C$ a doublet under $SU(2)_L$ and has a charge $-1/2$ under $U(1)_Y$.

We will begin by describing the origin of the short range force found in the gauge and Higgs sectors. Next we give the kinetic terms for the gauge bosons and the Higgs. Then we will explain the matter content for the leptons and how the Higgs field gives mass to fermions and then to the quarks. Finally we will discuss how the 3 generations are accounted for and the mixing that occurs between them.

4.8.3 A Short-Range Force: The Gauge and Higgs Sector

We begin our exposition with the **Electroweak** part of the Standard Model gauge group, the $SU(2)_L \times U(1)_Y$. We will use the Higgs mechanism to make a renormalizable gauge theory with a short range that also includes a long-range $U(1)_{EM}$ for electromagnetism.

The story of creating a renormalizable short-range force for the weak force of the Standard Model begins with the Higgs complex scalar doublet field ϕ in the $(\mathbf{2}, -1/2)$ representation of $SU(2)_L \times U(1)_Y$. Using the Higgs mechanism, ϕ will be given a potential that spontaneously breaks the symmetry of the vacuum. The resulting vacuum expectation value will give mass to the gauge bosons with a special pattern, with the resulting broken theory having the symmetry $U(1)_{EM}$ associated with QED. The pattern of masses of the massive gauge bosons will make the weak force a short-range force and provide a falsifiable prediction cementing the role of the Standard Model in nature.

The first step is to write down the covariant derivative as in (4.382). We denote the generators of the $\mathbf{2}$ representation of $SU(2)$ as

$$T_{\mathbf{2}}^a = \frac{1}{2}\sigma^a \tag{4.401}$$

(the Pauli matrices) and the gauge fields[27] as W_μ^a. The generator of $U(1)_Y$ is

$$Y = C\begin{pmatrix} 1 & 0 \\ 0 & 1 \end{pmatrix}, \tag{4.402}$$

where C is the hypercharge ($-1/2$ in this case), and the $U(1)_Y$ gauge field is B_μ. Hence, the covariant derivative is[28]

$$(D_\mu\phi)_i = \partial_\mu\phi_i - i\,[g_2 W_\mu^a T_{\mathbf{2}}^a + g_1 B_\mu Y]_{ij}\phi_j, \tag{4.403}$$

[27]We are using W^μ instead of A^μ to match with the most popular convention. The reasoning will be clear soon.

[28]Don't be intimidated by this expression. Take a deep breath and go back and reread Sects. 4.5 and 4.7 carefully – everything is there.

where g_1 and g_2 are coupling constants for the $U(1)_Y$ part and the $SU(2)_L$ part, respectively. If the reason we wrote it down this way isn't clear, compare this expression to equation (4.382), and remember that we are saying the field carries *two* charges; one for $SU(2)$ and one for $U(1)$. Therefore, it has two terms in its covariant derivative. And, as usual, μ is a spacetime index.

Knowing that the generators of $SU(2)$ are the Pauli matrices, we can expand the second part of the covariant derivative in matrix form,

$$\begin{aligned} g_2 W_\mu^a T_2^a + g_1 B_\mu Y &= \frac{g_2}{2}(W_\mu^1 \sigma^1 + W_\mu^2 \sigma^2 + W_\mu^3 \sigma^3) - \frac{g_1}{2} B_\mu \mathbb{I}^{2\times 2} \\ &= \frac{1}{2}\begin{pmatrix} g_2 W_\mu^3 - g_1 B_\mu & g_2(W_\mu^1 - iW_\mu^2) \\ g_2(W_\mu^1 + iW_\mu^2) & -g_2 W_\mu^3 - g_1 B_\mu \end{pmatrix}. \end{aligned} \tag{4.404}$$

Hence, the full covariant derivative is

$$(D_\mu \phi)_i \doteq \begin{pmatrix} D_\mu \phi_1 \\ D_\mu \phi_2 \end{pmatrix} = \begin{pmatrix} \partial_\mu \phi_1 + \frac{i}{2}(g_2 W_\mu^3 - g_1 B_\mu)\phi_1 + \frac{ig_2}{2}(W_\mu^1 - iW_\mu^2)\phi_2 \\ \partial_\mu \phi_2 + \frac{ig_2}{2}(W_\mu^1 + iW_\mu^2)\phi_1 - \frac{i}{2}(g_2 W_\mu^3 + g_1 B_\mu)\phi_2 \end{pmatrix}. \tag{4.405}$$

We know that the Lagrangian will have the kinetic term and some potential:

$$\mathcal{L}_\phi = -\frac{1}{2} D_\mu \phi_i^\dagger D^\mu \phi_i - V(\phi^\dagger, \phi). \tag{4.406}$$

Let's assume that the potential has a similar form as equation (4.354) (we add the factors of one-half here for the sake of convention; they don't amount to anything other than a rescaling of λ and v (where we are replacing Φ with v),

$$V(\phi^\dagger, \phi) = \frac{1}{4}\lambda\left(\phi^\dagger\phi - \frac{1}{2}v^2\right)^2. \tag{4.407}$$

With $\lambda > 0$ the minimum field configuration is not at $\phi = 0$, but at

$$|\phi| = \frac{v}{\sqrt{2}}. \tag{4.408}$$

So, following what we did in Sect. 4.7.7, we make a global $SU(2)$ transformation to put the entire VEV on the *first* component of ϕ, and then make a global $U(1)$ transformation to make the field real. Therefore,

$$\langle 0|\phi|0\rangle = \langle\phi\rangle = \frac{1}{\sqrt{2}}\begin{pmatrix} v \\ 0 \end{pmatrix}. \tag{4.409}$$

and we expand ϕ around this new vacuum:

$$\phi(x) = \frac{1}{\sqrt{2}}\begin{pmatrix} v + h(x) \\ 0 \end{pmatrix}. \tag{4.410}$$

Remember that we have chosen our $SU(2)$ gauge to keep the second component 0 and our $U(1)$ to keep the first component real. Thus, $h(x)$ is a real scalar field. The fluctuations of the other components turn out to be the longitudinal components of the massive gauge fields in the Higgs mechanism. The gauge choice that keeps the Higgs in the first term is called the **Unitarity Gauge**. Although it is easiest to see the physical content, the unitary gauge is the most difficult to renormalize and perform advanced calculations with.

Plugging this into the covariant derivative (4.405) will give the exact same expression as before, but with ϕ_1 replaced by $\frac{1}{\sqrt{2}}h(x)$ and ϕ_2 replaced by 0, plus an extra term for v. This is the vacuum expectation value (VEV) for the Lagrangian. When we plug the VEV into the kinetic term in the Lagrangian (4.406), we get

$$\mathcal{L}_{\langle\phi\rangle} = -\frac{1}{8}\begin{pmatrix} v & 0 \end{pmatrix}\begin{pmatrix} g_2 W_\mu^3 - g_1 B_\mu & g_2(W_\mu^1 - iW_\mu^2) \\ g_2(W_\mu^1 + iW_\mu^2) & -g_2 W_\mu^3 - g_1 B_\mu \end{pmatrix}^2 \begin{pmatrix} v \\ 0 \end{pmatrix}. \tag{4.411}$$

Next we want to find the masses of the gauge bosons W_μ^a and B^μ. There are four gauge fields, so we should find four masses. We can rewrite this matrix as:

$$\mathcal{L}_{\langle\phi\rangle} = -\frac{1}{8}v^2 V_\mu^T \begin{pmatrix} g_2^2 & 0 & 0 & 0 \\ 0 & g_2^2 & 0 & 0 \\ 0 & 0 & g_2^2 & -g_1 g_2 \\ 0 & 0 & -g_1 g_2 & g_1^2 \end{pmatrix} V^\mu \tag{4.412}$$

where $V_\mu^T = (W_\mu^1, W_\mu^2, W_\mu^3, B_\mu)$. From the mass matrix, we can read that W_μ^1 and W_μ^2 each have mass and are already diagonalized. The submatrix formed by the W_μ^3 and B_μ components has a determinant of 0 and therefore must have a zero eigenvalue. This means there is a massless gauge boson left with a corresponding symmetry that keeps it massless under renormalization effects. The eigenvalues are 0, $-\frac{1}{8}v^2 g_2^2$, $-\frac{1}{8}v^2 g_2^2$, and $-\frac{1}{8}v^2(g_1^2 + g_2^2)$. The normalized eigenvector for the massless state A_μ is $V_A^T = (0, 0, g_1, g_2)/\sqrt{g_1^2 + g_2^2}$. The eigenvector for the massive vector boson state Z_μ is $V_Z^T = (0, 0, g_2, -g_1)/\sqrt{g_1^2 + g_2^2}$. Notice that $Z_\mu = g_2 W_\mu^3 - g_1 B_\mu$ is linear combination in equation (4.411).

The format of the eigenvectors suggests a parameterization based on a right triangle with g_1 on one leg and g_2 on the other leg. Traditionally, this right triangle is used to describe the mixed states. The angle opposite the leg with length g_1 is called the **Weak Mixing Angle**

$$\theta_w = \tan^{-1}\left(\frac{g_1}{g_2}\right), \tag{4.413}$$

and the shorthand notation

$$s_w = \sin\theta_w = \frac{g_1}{\sqrt{g_1^2 + g_2^2}}, \qquad \text{and} \qquad c_w = \cos\theta_w = \frac{g_2}{\sqrt{g_1^2 + g_2^2}}. \tag{4.414}$$

We now define four new gauge fields as linear combinations of the four we have been using:

$$W_\mu^+ = \frac{1}{\sqrt{2}}(W_\mu^1 - iW_\mu^2), \tag{4.415}$$

$$W_\mu^- = \frac{1}{\sqrt{2}}(W_\mu^1 + iW_\mu^2), \tag{4.416}$$

$$Z_\mu = c_w W_\mu^3 - s_w B_\mu, \tag{4.417}$$

$$A_\mu = s_w W_\mu^3 + c_w B_\mu. \tag{4.418}$$

The $W^\pm$ are chosen so that each term in the final Lagrangian will have an explicit $U(1)_{EM}$ gauge symmetry associated with charge conservation. We can express the new fields as simple Euler rotations of the old fields:

$$\begin{pmatrix} Z_\mu \\ A_\mu \end{pmatrix} = \begin{pmatrix} W_\mu^3 \cos\theta_w - B_\mu \sin\theta_w \\ W_\mu^3 \sin\theta_w + B_\mu \cos\theta_w \end{pmatrix} \Rightarrow \begin{pmatrix} Z_\mu \\ A_\mu \end{pmatrix} = R(\theta_w) \begin{pmatrix} W_\mu^3 \\ B_\mu \end{pmatrix}. \tag{4.419}$$

These can easily be inverted to give the old fields in terms of the new fields,

$$W_\mu^1 = \frac{1}{\sqrt{2}}(W_\mu^+ + W_\mu^-), \tag{4.420}$$

$$W_\mu^2 = \frac{i}{\sqrt{2}}(W_\mu^+ - W_\mu^-), \tag{4.421}$$

$$W_\mu^3 = c_w Z_\mu + s_w A_\mu, \tag{4.422}$$

$$B_\mu = -s_w Z_\mu + c_w A_\mu. \tag{4.423}$$

The Z_μ is a massive linear combination of the W_μ^3 and B_μ, while A_μ is a massless linear combination of the two. We can do the same type of analysis for the $W_\mu^\pm$, where they are both massive linear combinations of W_μ^1 and W_μ^2. The Z_μ and A_μ are both made up of a mixture of the $SU(2)$ and $U(1)$ gauge groups, whereas the $W_\mu^\pm$ come solely from the $SU(2)$ part.

We make a few observations about these fields before moving on to work out the normalized mass of the fields. First of all, they are merely linear combinations of the gauge fields introduced in equation (4.403). Second, notice that the two fields $W_\mu^\pm$ are both linear combinations of fields corresponding to non-Cartan generators of $SU(2)$, whereas Z_μ and A_μ are both linear combinations of fields corresponding to Cartan generators of $SU(2)$ and $U(1)$. According to our discussion in Sect. 3.2.16, we expect that Z_μ and A_μ will interact but not change the charge, and that $W_\mu^\pm$ will interact and change the charge. Incidentally, notice that $W_\mu^\pm$ has the exact form of the raising and lowering operators defined in (3.122).

Next we study the mass terms for the gauge fields. With these fields defined, we can now rewrite (4.411) as

$$-\frac{1}{8}g_2^2\,(v\ 0)\begin{pmatrix}\frac{1}{c_w}Z_\mu & \sqrt{2}W_\mu^+\\ \sqrt{2}W_\mu^- & \star\end{pmatrix}^2\begin{pmatrix}v\\0\end{pmatrix} = -M_w^2W^{+\mu}W_\mu^- - \frac{1}{2}M_Z^2Z^\mu Z_\mu, \quad (4.424)$$

(the $\star$ is there because that matrix element will always be multiplied by 0, so we don't bother writing it), where we have defined

$$M_w = \frac{g_2 v}{2}, \qquad \text{and} \qquad M_Z = \frac{M_w}{c_w} = \frac{g_2 v}{2c_w} = \frac{v}{2}\sqrt{g_1^2+g_2^2}. \quad (4.425)$$

Thus, we see that, through symmetry breaking, we have given mass to the W_μ^+, the W_μ^-, and the Z_μ fields, while the A_μ remains massless.

These massive gauge fields are the W and Z vector bosons, which are the force carrying particles of the **Weak Force**. Each of these particles have an extremely large mass ($M_W \approx$80.4 GeV, and $M_Z \approx$ 91.2 GeV), which explains why they only act over a very short range ($\approx 10^{-18}$ meters). It was when the renormalizability was proven that Weinberg's paper started to be cited.

Note that A_μ remains massless, and there is still a single $U(1)$ symmetry that remains unbroken. This $U(1)$ and A_μ are the gauge group and field of **Electromagnetism**, as discussed in Sect. 4.5.3.

The point of all of this is that at very high energies (above the breaking of the $SU(2) \times U(1)$), we have only a Higgs complex scalar field, along with four massless vector boson gauge fields ($W_\mu^1, W_\mu^2, W_\mu^3, B_\mu$), each of which behave basically like a photon. At low energies, however, the $SU(2) \times U(1)$ symmetry of the Higgs is broken, and the low energy effective theory consists of a linear combination of the original four fields. Three of those linear combinations have gained mass, and one remains massless, retaining the photon-like properties from before symmetry breaking. The theory above the symmetry breaking scale is called the **Electroweak Theory** (with four photon-like force carrying particles), whereas below the breaking scale they become two separate forces: the broken **Weak** and the unbroken **Electromagnetic**. This is the first and most basic example of unification we have in our universe. At low energies, the electromagnetic and weak forces are separate. At high energies, they unify into a single theory that is described by $SU(2)_L \times U(1)_Y$.

There are two big results here. The first is that we have a renormalizable theory that may describe the weak force. The second big result is the discovery of a new type of Weak force. Let us explain.

Is the short-range force found by breaking $SU(2)_L \times U(1)_Y$ the same short range force we see in the Weak force? How could we know? Remember the neutrino that collided with the electron in the Sect. 4.8.1? Neutrinos have no electromagnetic charge, so the only possible interaction would be with the Weak force. Most examples of the Weak force involve a charge transfer. Consider a neutron decaying to a proton, an electron, and an anti-electron-neutrino. There is a charge transfer when the neutron decays to a proton that is then transferred to the electron and anti-electron neutrino production. Because of this charge transfer, it is called a decay via a Weak charged current. However, the neutrino scattering an electron to an electron witnessed in 1973 did not transfer charge. It must have involved a Weak neutral current. Subsequent studies found a collection of Weak interactions that could only be explained by a Weak neutral current. The Z_μ vector boson is the mediator of that Weak neutral current! There was now both a renormalizable theory with the prediction of a Weak neutral current, and then the observation of a Weak neutral current that pushed the Standard Model into the leading hypothesis. Since then the properties of the massive Weak bosons have passed every test... except a direct observation of their coupling to the Higgs boson. As of 2011, we have yet to see a Higgs produced at a collider.

4.8.4 The Gauge Bosons and Their Coupling to the Higgs Boson

Before moving on to include leptons (and then quarks), we first write out the full Lagrangian for the effective field theory for $h(x)$ and the gauge fields.

We start with the complete Lagrangian term for $h(x)$. We have written the original field ϕ as in equation (4.410). Our potential in equation (4.407) is now

$$V(\phi^\dagger,\phi) = \frac{1}{4}\lambda\left(\phi^\dagger\phi - \frac{1}{2}v^2\right)^2 = \cdots = \frac{1}{4}\lambda v^2 h^2 + \frac{1}{4}\lambda v h^3 + \frac{1}{16}\lambda h^4. \quad (4.426)$$

The first term on the right-hand side is a mass term giving the mass of the Higgs $(=\sqrt{\frac{\lambda}{2}}v)$, and the second two terms are interaction vertices. The kinetic term for the Higgs will be the usual $-\frac{1}{2}\partial_\mu h \partial^\mu h$.

Now, following loosely what we did in Sect. 4.5.3, we want to find kinetic terms for the gauge fields. We start by finding them for the original gauge fields before symmetry breaking (W^1_μ, W^2_μ, W^3_μ and B_μ). Using (4.386), (4.369), and (4.380), and the $SU(2)$ structure constants given in equation (3.118), we have

$$\begin{aligned}
F^1_{\mu\nu} &= 2Tr(F_{\mu\nu}T^1)\\
&= 2Tr\big((\partial_\mu W_\nu - \partial_\nu W_\mu - ig_2[W_\mu, W_\nu])T^1\big)\\
&= 2Tr\big((\partial_\mu W^a_\nu T^a - \partial_\nu W^a_\mu T^a - ig_2 W^a_\mu W^b_\nu [T^a, T^b])T^1\big)
\end{aligned}$$

$$
\begin{aligned}
&= 2Tr(\partial_\mu W_\nu^a T^a T^1 - \partial_\nu W_\mu^a T^a T^1 - ig_2 W_\mu^a W_\nu^b if^{abd} T^c T^1) \\
&= \partial_\mu W_\nu^a \delta^{a1} - \partial_\nu W_\mu^a \delta^{a1} + g_2 W_\mu^a W_\nu^b f^{abc} \delta^{c1} \\
&= \partial_\mu W_\nu^1 - \partial_\nu W_\mu^1 + g_2 W_\mu^a W_\nu^b f^{ab1} \\
&= \partial_\mu W_\nu^1 - \partial_\nu W_\mu^1 + g_2 W_\mu^a W_\nu^b \epsilon^{ab1} \\
&= \partial_\mu W_\nu^1 - \partial_\nu W_\mu^1 + g_2 (W_\mu^2 W_\nu^3 - W_\nu^2 W_\mu^3).
\end{aligned} \tag{4.427}
$$

Similarly,

$$
\begin{aligned}
F_{\mu\nu}^2 &= \partial_\mu W_\nu^2 - \partial_\nu W_\mu^2 + g_2 (W_\mu^3 W_\nu^1 - W_\nu^3 W_\mu^1), \\
F_{\mu\nu}^3 &= \partial_\mu W_\nu^3 - \partial_\nu W_\mu^3 + g_2 (W_\mu^1 W_\nu^2 - W_\nu^1 W_\mu^2).
\end{aligned} \tag{4.428}
$$

The gauge field strength corresponding to the $U(1)$ will be defined as in (4.232):

$$
B_{\mu\nu} = \partial_\mu B_\nu - \partial_\nu B_\mu. \tag{4.429}
$$

We can now write the kinetic term for our fields according to equation (4.385):

$$
\mathcal{L}_{Kin} = -\frac{1}{4} F_a^{\mu\nu} F_{\mu\nu}^a - \frac{1}{4} B^{\mu\nu} B_{\mu\nu}. \tag{4.430}
$$

We can then use (4.420–4.423) to translate these kinetic terms into the new fields. We will spare the extremely tedious detail and skip right to the Lagrangian for the $SU(2)_L \times U(1)_Y$ gauge fields along with the Higgs field:

$$
\begin{aligned}
\mathcal{L}_{eff} = &-\frac{1}{4} F^{\mu\nu} F_{\mu\nu} - \frac{1}{4} Z^{\mu\nu} Z_{\mu\nu} - D^{\dagger\mu} W^{-\nu} D_\mu W_\nu^+ + D^{\dagger\mu} W^{-\nu} D_\nu W_\mu^+ \\
&+ie(F^{\mu\nu} + \cot\theta_w Z^{\mu\nu}) W_\mu^+ W_\nu^- \\
&-\frac{1}{2}\left(\frac{e^2}{\sin^2\theta_w}\right)(W^{+\mu} W_\mu^- W^{+\nu} W_\nu^- - W^{+\mu} W_\mu^+ W^{-\nu} W_\nu^-) \\
&-\left(M_W^2 W^{+\mu} W_\mu^- + \frac{1}{2} M_Z^2 Z^\mu Z_\mu\right)\left(1 + \frac{h}{v}\right)^2 \\
&-\frac{1}{2}\partial^\mu h \partial_\mu h - \frac{1}{2} m_h^2 h^2 - \frac{1}{2}\frac{m_h^2}{v} h^3 - \frac{1}{8}\frac{m_h^2}{v^2} h^4,
\end{aligned} \tag{4.431}
$$

where we have chosen the following definitions:

$$
\begin{aligned}
F_{\mu\nu} &= \partial_\mu A_\nu - \partial_\nu A_\mu, && \text{(Electromagnetic Field Strength)} \\
Z_{\mu\nu} &= \partial_\mu Z_\nu - \partial_\nu Z_\mu, && \text{(Kinetic term for } Z_\mu\text{)} \\
D_\mu &= \partial_\mu - ie(A_\mu + \cot\theta_w Z_\mu).
\end{aligned} \tag{4.432}
$$

and the rest of the terms were defined previously in this section.

The $\mathcal{L}_{eff}$ allows us to see what kinds of interactions will involve the Higgs and the vector bosons. The photon A_μ does not couple to h. However, the $W^\pm$ do. By expanding the second to last line of Eq. (4.431), we see that the interaction is proportional to W^+W^-h. This interaction is responsible for charged Weak gauge bosons fusing to form a Higgs. In a collider this interaction would have the second largest cross-section for Higgs production. If the Higgs has a mass of 150 GeV, then the LHC after shooting about 10^{21} protons across each other through a 15 micron beam diameter, should see on the order of 200000 Higgs produced by this mechanism (about 6 months to 1 year of running). Producing a Higgs is still a rare event even at 14 TeV of energy like at the LHC!

4.8.5 The Lepton Sector: The Origin of Mass

We now turn to the lepton sector, which is still in the $SU(2)_L \times U(1)_Y$ part of the Standard Model gauge group. A **Lepton** is a spin-$1/2$ particle that does *not* interact with the $SU(3)$ color group (the strong force). There are six **Flavors** of leptons arranged into three **Families**, or **Generations**. The table on page 50 shows this. The first generation consists of the electron (e) and the electron neutrino (ν_e), the second generation the muon (μ) and the muon neutrino (ν_μ), and the third the tau (τ) and tau neutrino (ν_τ). Each family behaves nearly exactly the same way, so we will only discuss one generation in this section (e and ν_e). To incorporate the physics of the other families we will make three copies of the below structure for the three generations and allow mixing between generations in the most general possible way.

The Standard Model needs to simplify to the $V - A$ model at low energies. One characteristic of the $V - A$ model is that only left-handed fields were included in the Weak interactions. Because the neutrino only participates in the Weak interactions (and gravity) there is only a left-handed neutrino needed for a Weak interactions theory. The electron exists in both a left-handed and right-handed state. How do we include this asymmetry?

The neutrino is added *as part of a left-handed* $SU(2)_L$ doublet *with* the left-handed electron,

$$L = \begin{pmatrix} \nu_e \\ e \end{pmatrix}. \tag{4.433}$$

This is why it is arranged as it is on page50 with the electron under the $(\mathbf{2}, -1/2)$ representation of $SU(2)_L \times U(1)_Y$. The right-handed electron is an $SU(2)_L$ singlet. It will be useful to remember that the right-handed electron is the same field as the left-handed positron under the relationship in equation (3.256).

This may seem confusing, but we hope the following will make it clear. We will proceed in what we believe is the cleanest way to see this (primarily following [86]). We start with two fields, L and $\bar{e}$. The $SU(2)$ doublet field L is defined in (4.433) and is a purely left-handed Weyl spinor. The $SU(2)$ singlet field $\bar{e}$ is also a purely left-handed Weyl spinor. (See Sect. 4.3.7 to review Weyl spinors.) As we have said, L is in the $(\mathbf{2}, -1/2)$ representation, $\bar{e}$ is in the $(1, 1)$ representation. The neutrino ν_e is part of the L doublet and has no representation of its own. A priori there is no link between the e that is part of L and $\bar{e}$. These fields are completely unrelated at this point in our story. The naming convention is foreshadowing.

Mimicking what we did in equation (4.403) in the previous section, we can write down the covariant derivative for each field,

$$(D_\mu L)_i = \partial_\mu L_i - ig_2 W^a_\mu (T^a)_{ij} L_j - ig_1 B_\mu Y_L L_i, \tag{4.434}$$

$$D_\mu \bar{e} = \partial_\mu \bar{e} - ig_1 B_\mu Y_{\bar{e}} \bar{e}. \tag{4.435}$$

The field $\bar{e}$ has no $SU(2)$ term in its covariant derivative because the 1 representation of $SU(2)$ is the trivial representation - this means it doesn't carry $SU(2)$ charge. Also, we know that

$$Y_L = -\frac{1}{2}\begin{pmatrix} 1 & 0 \\ 0 & 1 \end{pmatrix}, \tag{4.436}$$

and

$$Y_{\bar{e}} = (1)\begin{pmatrix} 1 & 0 \\ 0 & 1 \end{pmatrix}. \tag{4.437}$$

Following the Lagrangian for the spin-1/2 fields we wrote out in equation (4.112), we can write out the kinetic term for both (massless) fields:

$$\mathcal{L}_{Kin} = iL^{\dagger i} \bar{\sigma}^\mu (D_\mu L)_i + i\bar{e}^\dagger \bar{\sigma}^\mu D_\mu \bar{e}. \tag{4.438}$$

If we try to add mass terms for L and $\bar{e}$ fields, the gauge symmetry is broken. For example Lorentz invariant combination of $\bar{e}^a \bar{e}^b \epsilon_{ab} = (\bar{e}\bar{e})$ is $SU(2)_L$ invariant (both are singlets) but violates $U(1)_Y$. The term has a net hypercharge of $+2$. Likewise $L^{ai} L^{bj} \epsilon_{ab} \epsilon_{ij}$ is Lorentz invariant on the a, b indices and $SU(2)_L$ invariant on the i, j indices, but is not invariant under $U(1)_Y$ having a net hypercharge of -1. If we conjugate one of the terms, then we cannot form the Lorentz invariant combination and we lose Lorentz invariance. Therefore we cannot add a mass term. But, we know experimentally that electrons and neutrinos have mass, so obviously something is wrong. We must incorporate mass into the theory, but in a more subtle way than merely adding a mass term. It turns out that the Higgs mechanism comes to the rescue.

Adding a direct mass terms destroys gauge invariance, but we can add a Yukawa term (cf. equation (4.347)),

$$\mathcal{L}_{Yuk} = -y\epsilon^{ij} \phi_i (L_j \bar{e}) + \text{h.c.}, \tag{4.439}$$

where y is the Yukawa coupling constant, ϵ^{ij} is the totally antisymmetric tensor that combines the two $SU(2)$ doublets into a singlet, the $(L\,e)$ term indicates the Lorentz invariant combination of the spinor indices for L and $\bar{e}$ that are suppressed, and $h.c.$ is the Hermitian conjugate of the first term. We see that the term has a net hypercharge of zero, and the $SU(2)_L$ and Lorentz indices are all contracted to form singlets. At this point there is still no relationship between e and $\bar{e}$. They are unrelated fields in a model.

Now that we have added $\mathcal{L}_{Yuk}$ to the Lagrangian, we want to break the symmetry exactly as we did in the previous section. We continue to work in the unitary gauge where $\phi_2 = 0$. First, we replace ϕ_1 with $\frac{1}{\sqrt{2}}(v + h(x))$ and ϕ_2 with 0, exactly as we did in equation (4.410). So,

$$\begin{aligned}
\mathcal{L}_{Yuk} &= -y\epsilon^{ij}\phi_i L_j \bar{e} + \text{h.c.} \\
&= -y(\phi_1 L_2 - \phi_2 L_1)\bar{e} + \text{h.c.} \\
&= -\frac{1}{\sqrt{2}} y(v+h) L_2 \bar{e} + \text{h.c.} \\
&= -\frac{1}{\sqrt{2}} y(v+h)(e\bar{e}) - \frac{1}{\sqrt{2}} y(v+h)(\bar{e}^\dagger e^\dagger) \\
&= -\frac{1}{\sqrt{2}} y\, v\, \bar{\mathcal{E}}\mathcal{E} - \frac{1}{\sqrt{2}} y\, v\, h\, \bar{\mathcal{E}}\mathcal{E},
\end{aligned} \tag{4.440}$$

where

$$\mathcal{E} = \begin{pmatrix} e \\ i\sigma^2 \bar{e}^\dagger \end{pmatrix} \tag{4.441}$$

is the Dirac field for the electron (e is the left-handed electron and $\bar{e}^\dagger$ is the left-handed antielectron, or positron) and $\bar{\mathcal{E}} = \mathcal{E}^\dagger \gamma^0$. Comparing (4.440) with (4.112), we see that it is a mass term for the electron and positron where $m_e = yv/\sqrt{2}$. Therefore when the theory undergoes spontaneous symmetry breaking, the fields e and $i\sigma^2 \bar{e}^*$ which initially are completely unlinked join to form the left- and right-handed parts of the electron field $\mathcal{E}$. In the presence of the $SU(2)_L$ symmetry, ν_e and e can be rotated into each other without affecting the theory.

Now, we want a kinetic term for the neutrino. It is believed that neutrinos are described by Majorana fields (see Sect. 4.3.7), so we begin with the field

$$\mathcal{N}' = \begin{pmatrix} \nu_e \\ i\sigma^2 \nu_e^* \end{pmatrix}. \tag{4.442}$$

Now, we employ a trick. The kinetic term for Majorana fields has only one term (because Majorana fields have only one Weyl spinor), whereas the Dirac field sums over both Weyl spinors composing it. Hence, instead of working with the Majorana field $\mathcal{N}'$, we can instead work with the Dirac field

$$\mathcal{N} = \begin{pmatrix} \nu_e \\ 0 \end{pmatrix}. \tag{4.443}$$

So, the Dirac kinetic term $i\bar{\mathcal{N}}\gamma^\mu\partial_\mu\mathcal{N}$ will result in the correct kinetic term from (4.438), or $i\nu^\dagger\bar{\sigma}^\mu\partial_\mu\nu$.

Continuing with the symmetry breaking, we want to write the covariant derivative (4.434) and (4.435) in terms of our low energy gauge fields (4.415–4.418). We said in the previous section (which echoed our discussion in Sect. 3.2.16) that the gauge fields corresponding to Cartan generators (A_μ and Z_μ) act as force carrying particles, but do not change the charge of the particles they interact with. On the other hand, the non-Cartan generators' gauge fields ($W_\mu^\pm$) are force carrying particles which *do* change the charge of the particle they interact with. Therefore, to make calculations simpler, we will break the covariant derivative up into the non-Cartan part and the Cartan part.

The non-Cartan part of the covariant derivative (4.434) is

$$\begin{aligned} g_2(W_\mu^1 T^1 + W_\mu^2 T^2) &= \frac{1}{2}g_2\left(W_\mu^1\begin{pmatrix}0 & 1\\ 1 & 0\end{pmatrix} + W_\mu^2\begin{pmatrix}0 & -i\\ i & 0\end{pmatrix}\right)\\ &= \frac{1}{2}g_2\begin{pmatrix}0 & W_\mu^1 - iW_\mu^2\\ W_\mu^1 + iW_\mu^2 & 0\end{pmatrix}\\ &= \frac{g_2}{\sqrt{2}}\begin{pmatrix}0 & W_\mu^+\\ W_\mu^- & 0\end{pmatrix}, \end{aligned} \tag{4.444}$$

and the Cartan part is

$$\begin{aligned} g_2 W_\mu^3 T^3 + g_1 B_\mu Y &= \frac{e}{s_w}(s_w A_\mu + c_w Z_\mu)T^3 + \frac{e}{c_w}(c_w A_\mu - s_w Z_\mu)Y\\ &= e(A_\mu + \cot\theta_w Z_\mu)T^3 + e(A_\mu - \tan\theta_w Z_\mu)Y\\ &= e(T^3 + Y)A_\mu + e(\cot\theta_w T^3 - \tan\theta_w Y)Z_\mu. \end{aligned} \tag{4.445}$$

We have noted before that A_μ is the photon, or the electromagnetic field, and e is the electromagnetic charge. Therefore, the linear combination $T^3 + Y$ must be the generator of electric charge. Notice that the electromagnetic generator is a linear combination of the two Cartan generators of $SU(2) \times U(1)$.

We know that

$$T^3 = \frac{1}{2}\sigma^3, \tag{4.446}$$

and Y_L and $Y_{\bar{e}}$ are defined in equations (4.436) and (4.437), so we can write

$$\begin{aligned} T^3 L &= \frac{1}{2}\begin{pmatrix}1 & 0\\ 0 & -1\end{pmatrix}\begin{pmatrix}\nu_e\\ e\end{pmatrix} = \frac{1}{2}\begin{pmatrix}\nu_e\\ -e\end{pmatrix},\\ Y_L L &= -\frac{1}{2}\begin{pmatrix}1 & 0\\ 0 & 1\end{pmatrix}\begin{pmatrix}\nu_e\\ e\end{pmatrix} = -\frac{1}{2}\begin{pmatrix}\nu_e\\ e\end{pmatrix}. \end{aligned} \tag{4.447}$$

Further, $\bar{e}$ carries no T^3 charge, so its T^3 eigenvalue is 0, while $Y_{\bar{e}}$ is +1. Summarizing all of this,

$$\begin{aligned} T^3 \nu_e &= +\frac{1}{2}\nu_e, & T^3 e &= -\frac{1}{2}e, & T^3 \bar{e} &= 0, \\ Y\nu_e &= -\frac{1}{2}\nu_e, & Ye &= -\frac{1}{2}e, & Y\bar{e} &= +\bar{e}. \end{aligned} \tag{4.448}$$

Then defining the generator of electric charge to be $Q = T^3 + Y$, we have

$$Q\nu_e = 0, \qquad Qe = -e, \qquad Q\bar{e} = +\bar{e}. \tag{4.449}$$

Hence, the left-handed neutrino ν_e has no electric charge, the left-handed electron e has negative electric charge, and the left-handed antielectron, or positron, has plus one electric charge – all exactly what we would expect.

We can now take all of the terms we have discussed so far and write out a complete Lagrangian. However, doing so is both tedious and unnecessary for our purposes. The primary idea is that electrons/positrons and neutrinos all interact with the $SU(2)_L \times U(1)_Y$ gauge particles, the $W^\pm$, Z_μ, and A_μ. The Z_μ and A_μ (the Cartan gauge particles) interact but do not affect the charge. On the other hand, the $W^\pm$ act as $SU(2)$ raising and lowering operators (as can easily be seen by comparing (4.415) and (4.416) to equation (3.122)). The $SU(2)_L$ doublet state acted on by these raising and lowering operators is the doublet in equation (4.433). The W^+ interacts with a left-handed electron, raising its electric charge from minus one to zero, turning it into a neutrino. However W^+ does not interact with left-handed neutrinos. On the other hand, W^- will lower the electric charge of a neutrino, making it an electron. But W^- will not interact with an electron.[29] The exchange of $W^\pm$ will reduce to a charged current Weak interaction. The exchange of a Z will reduce to a neutral current Weak interaction.

At this point you are likely beginning to understand why most find the Standard Model to be "ugly". When first learning it, students often feel that the Standard Model is a tool to take a set of beautifully elegant mathematical tricks and turn them into something hideous. Keep in mind that while it is a mess, it is the most experimentally successful theory in physics.

---

[29]This does not mean that no vertex in the Feynman diagrams will include a W^- and an electron field, but rather that if you collide an electron and a W^-, there will be no interaction at lowest order. For example a Z boson exchange will mediate a scattering between W^- and $\bar{e}$ at next to lowest order.

4.8.6 The Quark Sector

A Quark is a spin-$1/2$ particle that interacts with the $SU(3)$ color force. Just as with leptons, there are six flavors of quarks, arranged in three families or generations (see the table on page 50).

Following very closely what we did with the leptons, we work with only one generation. Extending to the other generators is then trivial. To begin, define three fields: Q, $\bar{u}$, and $\bar{d}$, in the representations $(\mathbf{3}, \mathbf{2}, +1/6)$, $(\bar{\mathbf{3}}, 1, -2/3)$, and $(\bar{\mathbf{3}}, 1, +1/2)$ of $SU(3)_C \times SU(2)_L \times U(1)_Y$. The field Q will be the $SU(2)_L$ doublet

$$Q = \begin{pmatrix} u \\ d \end{pmatrix}. \tag{4.450}$$

This is exactly analogous to equation (4.433). Analogously, Q is only the left-handed part of u and d.

Again, following what we did with the leptons, we can write out the covariant derivative for all three fields:

$$(D_\mu Q)_{\alpha i} = \partial_\mu Q_{\alpha i} - i g_3 A^a_\mu (T^a_3)^\beta_\alpha Q_{\beta i} - i g_2 W^a_\mu (T^a_2)^j_i q_{\beta j} - i g_1 \left(\frac{1}{6}\right) B_\mu Q_{\alpha i}, \tag{4.451}$$

$$(D_\mu \bar{u})^\alpha = \partial_\mu \bar{u}^\alpha - i g_3 A^a_\mu (T^a_{\bar{3}})^\alpha_\beta \bar{u}^\beta - i g_1 \left(-\frac{2}{3}\right) B_\mu \bar{u}^\alpha, \tag{4.452}$$

$$(D_\mu \bar{d})^\alpha = \partial_\mu \bar{d}^\alpha - i g_3 A^a_\mu (T^\alpha_{\bar{3}})^\alpha_\beta \bar{d}^\beta - i g_1 \left(\frac{1}{3}\right) B_\mu \bar{d}^\alpha, \tag{4.453}$$

where i is an $SU(2)_L$ index and α is an $SU(3)_C$ index. The $SU(3)$ index is lowered for the $\mathbf{3}$ representation and raised for the $\bar{\mathbf{3}}$ representation. The 8 generators of $SU(3)_C$ acting on the $\mathbf{3}$ are given by T^a_3, and acting on the $\bar{\mathbf{3}}$ are given by $T^a_{\bar{3}} = -(T^a_3)^*$. The vector field A^a_μ is the gluon field. It is distinguished from the photon by the gauge index. Again $\bar{u}$ and $\bar{d}$ are simply foreshadowing their relationships to other left-handed fields after spontaneous symmetry breaking.

Just as with leptons. we cannot write down a gauge invariant mass term for these particles, but we can include a Yukawa term coupling these fields to the Higgs:

$$\mathcal{L}_{Yuk} = -y' \epsilon^{ij} \phi_i Q_{\alpha j} \bar{d}^\alpha - y'' \phi^{\dagger i} Q_{\alpha i} \bar{u}^\alpha + \text{h.c.}. \tag{4.454}$$

Likewise, we can break the symmetry according to equation (4.410), and writing out this Yukawa term, we get

$$\begin{aligned} \mathcal{L}_{Yuk} &= -\frac{1}{\sqrt{2}} y'(v+h) \left(d_\alpha \bar{d}^\alpha + \bar{d}^\dagger_\alpha d^{\dagger\alpha}\right) - \frac{1}{\sqrt{2}} y''(v+h) \left(u_\alpha \bar{u}^\alpha + \bar{u}^\dagger_\alpha u^{\dagger\alpha}\right) \\ &= -\frac{1}{\sqrt{2}} y'(v+h) \bar{\mathcal{D}}^\alpha \mathcal{D}_\alpha - \frac{1}{\sqrt{2}} y''(v+h) \bar{\mathcal{U}}^\alpha \mathcal{U}_\alpha, \end{aligned} \tag{4.455}$$

where we have defined the Dirac fields for the up and down quarks,

$$\mathcal{D}_\alpha = \begin{pmatrix} d_\alpha \\ \bar{\iota}\sigma^2 d_\alpha^* \end{pmatrix}, \qquad \mathcal{U}_\alpha = \begin{pmatrix} u_\alpha \\ \bar{\iota}\sigma^2 u_\alpha^* \end{pmatrix}. \tag{4.456}$$

Notice that, whereas both the up and down quarks were massless before breaking, they have now acquired masses

$$m_d = \frac{y' v}{\sqrt{2}}, \qquad m_u = \frac{y'' v}{\sqrt{2}}. \tag{4.457}$$

Again, the spontaneous symmetry breaking of ϕ_i links a term in the doublet with the singlet $\bar{u}$ or $\bar{d}$ to form a mass.

Writing out the non-Cartan and Cartan parts of the covariant derivatives in terms of the lower energy $SU(2) \times U(1)$ gauge fields, we get

$$g_2 W_\mu^1 T^1 + g_2 W_\mu^2 T^2 = \frac{g_2}{\sqrt{2}}, \begin{pmatrix} 0 & W_\mu^+ \\ W_\mu^- & 0 \end{pmatrix}.$$

$$g_2 W_\mu^3 T^3 + g_1 B_\mu Y = eQA_\mu + \frac{e}{s_w c_w}(T^3 - s_w^2 Q) Z_\mu, \tag{4.458}$$

where Q was defined in the previous section (above equation (4.449)).[30]

It is again straightforward to find the electric charge eigenvalue for each field:

$$Qu = +\frac{2}{3}u, \qquad Qd = -\frac{1}{3}d, \qquad Q\bar{u} = -\frac{2}{3}\bar{u}, \qquad Q\bar{d} = +\frac{1}{3}\bar{d}. \tag{4.459}$$

Again, we can collect all of these terms and write out a complete Lagrangian. However, doing so is extremely tedious and unnecessary for our purposes.

The primary idea to take away is that the $SU(2)_L$ doublet (4.450) behaves exactly as the lepton doublet in (4.433) when interacting with the "raising" and "lowering" gauge particles $W^\pm$. This is why the u and d are arranged in the $SU(2)$ doublet Q in (4.450), and why Q carries the $SU(2)$ index i in the covariant derivative (4.451), whereas $\bar{u}$ and $\bar{d}$ carry only the $SU(3)$ index.

The $SU(3)$ index runs from 1 to 3, and the three values are conventionally denoted *red, green*, and *blue* (r, g, b). These obviously are merely labels and have nothing to do with the colors in the visible spectrum.

The eight gauge fields associated with the eight $SU(3)$ generators are called **Gluons**, and they are represented by the matrices in (3.211), (3.214) and (3.215). We label each gluon as follows:

$$g_\alpha^\beta = \begin{pmatrix} r\bar{r} & r\bar{g} & r\bar{b} \\ g\bar{r} & g\bar{g} & g\bar{b} \\ b\bar{r} & b\bar{g} & b\bar{b} \end{pmatrix}. \tag{4.460}$$

[30] We apologize for the notational inconvenience of using Q for two different things. The distinction can be generally seen by whether or not it has any indices.

The upper index is the anti-color index, and denotes the column of the matrix, and the lower index is the color index denoting the row of the matrix.[31] Then, from (3.214), consider the gluon

$$g_r^{\bar{g}} \propto \begin{pmatrix} 0 & 1 & 0 \\ 0 & 0 & 0 \\ 0 & 0 & 0 \end{pmatrix}, \tag{4.461}$$

and the quarks

$$q_r = \begin{pmatrix} 1 \\ 0 \\ 0 \end{pmatrix}, \qquad q_g = \begin{pmatrix} 0 \\ 1 \\ 0 \end{pmatrix}, \qquad q_b = \begin{pmatrix} 0 \\ 0 \\ 1 \end{pmatrix}. \tag{4.462}$$

It is easy to see that this gluon will interact as

$$g_r^{\bar{g}} q_r = 0, \qquad g_r^{\bar{g}} q_g = q_r, \qquad g_r^{\bar{g}} q_b = 0. \tag{4.463}$$

In other words, the gluon with the anti-green index will only interact with a green quark. There will be no interaction with the other quarks. Multiplying this out, and looking more closely at the behavior of the $SU(3)$ generators and eigenstates as discussed in Sects. 3.2.12–3.2.15, you can work out all of the interaction rules between quarks and gluons. You will see that they behave exactly according to the root space of $SU(3)$.

4.8.7 Yukawa Couplings Among Generations

To represent the three generations we make three copies of the Standard Model structure; one for each generation. However there is no symmetry preventing terms from different generations from coupling to each other. Therefore the Yukawa couplings need to be generalized to allow a $\bar{c}$ to couple to Q from the first generation. Let us denote the generations with the indices A, B, C, etc. The Yukawa coupling now becomes

$$\mathcal{L}_{\text{Yuk}} = -Y_{AB}^{d}\phi Q^A \bar{d}^B - Y_{AB}^{u}\phi^\dagger Q^A \bar{u}^B - Y_{AB}^{e}\phi L^A \bar{e}^B + \text{h.c.} \tag{4.464}$$

where we have suppressed the $SU(3)_C$ and the $SU(2)_L$ and the Lorentz indices. Because neutrinos have mass, we should add several Yukawa terms to give the neutrino generations mass. For the moment, we'll stick to the original approximation where the neutrinos are massless.

[31] The generators are also required to be traceless.

The kinetic terms for the three generations $L^{A\dagger} i\sigma^\mu D_\mu L^A$ are invariant under global $SU(3)$ transformations of the fields $L'^A = R^A{}_B L^B$. The same is true for three generations of Q, $\bar{u}$, $\bar{d}$ and $\bar{e}$. Using the freedom to redefine L and $\bar{e}$ we can diagonalize Y^e. These entries will be proportional to the mass of the leptons. However if we diagonalize Y^u by redefining Q and $\bar{u}$ we are left without enough freedom to diagonalize Y^d. This observation predicts that the quarks will decay from heavier to lighter quarks directly. The remaining structure is called the CKM matrix. Verifying the unitarity of the CKM matrix is another significant test of the Standard Model that it has passed at ever experimental challenge.

4.9 References and Further Reading

The primary source for this chapter is [86], which is an exceptionally clear introduction to Quantum Field Theory. We also used a great deal of material from [18, 66, 78, 90, 103], all of which are outstanding QFT texts. The derivation of the Dirac equation came from [61], which is written mostly above the scope of this book, but is an excellent survey of some of the mathematical ideas of Non-Perturbative QFT and Gauge Theory.

The sections on the Standard Model come almost entirely from [86], in that Srednicki's exposition could hardly be improved upon for the scope of this book. The section on the geometrical interpretation of spin came from [65].

For further reading, we also recommend [8, 30, 63, 69, 70].

Chapter 5
Beyond the Standard Model of Particle Physics

We've discussed in some detail now the underlying mathematical and physical tools we currently use to understand our universe. With the exception of the Higgs boson, everything we've discussed so far has been experimentally observed, and it is reasonable to believe that we'll find the Higgs in the near future. However, when it comes to theory, what we've seen so far is not the end of the story (in some respects it's only the beginning). So, before concluding this book, we'll spend a chapter discussing where physics "goes" from here. For better or worse, we live in a time when theory has gone far beyond where experiment can currently go. While the empiricism that served as the foundation for science for centuries is not entirely there, theoretical research into what we currently can't study is an extremely active area. It is necessary for every particle physicist to have some feel for what is going on in these realms.

As we stated in the introduction, this chapter is meant to be a vast, mountaintop view of these ideas, and it will no doubt leave you wanting more. We'll be coming back to almost every topic here in significant detail in later volumes, so don't be alarmed if some of the ideas still a little unclear. This is very much a "drinking from the fire hydrant" chapter. For the (physical and mathematical) concepts in this chapter that we haven't introduced yet, we'll try to comment on them in the footnotes.

5.1 Overview of Physics Beyond the Standard Model

The Standard Model is a rather unsatisfying end to a quest for understanding the fundamental nature of 'everything' (typical example of physicist humility). First of all despite its wide success in many areas, it doesn't work in all cases. The basic model only has massless left-handed neutrinos, and we know from recent observations of neutrino oscillations that at least two of the neutrinos have some non-zero mass. Second, there are 19 free parameters in the Standard Model (13 from the Yukawa sector, 2 from the Higgs sector, 3 from the gauge sector,

M. Robinson, *Symmetry and the Standard Model: Mathematics and Particle Physics*, DOI 10.1007/978-1-4419-8267-4_5, © Springer Science+Business Media, LLC 2011

and 1 from θ_{QCD}). Neutrino masses introduce 8 new free parameters making the new total 27. Beyond these we also have Newton's constant. To the reductionist physicist, 28 unrelated measured numbers suggest a more fundamental theory may exist. There are many puzzles left unanswered in the SM. Physics beyond the Standard Model (BSM) searches for some possible explanations and reduction in free parameters.

BSM physics has taken many forms over the years. The main tracks currently popular are massive neutrinos, technicolor, supersymmetry, extra-dimensional models, grand unified theories, string theory, loop quantum gravity, and several other less developed quantum gravity theories. The observed neutrino mass has led to many models linking massive neutrinos to the SM: some where the neutrinos have a Dirac mass, others where the neutrinos have Majorana mass, and finally others with a mix. The more popular models link the neutrino's small mass to the large mass scales of grand unified theories discussed below through a so-called "see-saw" mechanism.

Because fundamental scalars, like the Higgs in the SM, have poor renormalization properties, technicolor and supersymmetry were suggested. Technicolor models give the W and Z mass from a 5th force (the technicolor force) and do not require a fundamental Higgs. Rather the Higgs is produced as a composite scalar formed from a condensate of spin-$1/2$ fermions that carry charges of an additional $SU(n)$, with $n > 3$, gauge symmetry. As the renormalization equations for coupling strengths show, the coupling strength of the $SU(n > 3)$ technicolor force would become strong before the $SU(3)_C$. Around the electroweak scale, two fermion states with technicolor charge would then be induced to combine into a scalar condensate, forming a non-fundamental Higgs. Thus, in technicolor the W and Z are given mass not by a fundamental Higgs scalar but by a composite pion-like particle of this fifth force.

A second alternative to a fundamental Higgs came from string theory and then from brane models. Models with extra dimensions might not (and need not) carry an actual Higgs scalar. Instead, the role could be played by the fifth component of a gauge field (that acts like a scalar in our four-dimensional spacetime).

Supersymmetry introduces a set of superpartner particles appropriate to compensate for the poor renormalization properties of the Higgs. For every standard scalar, a superpartner spin-$\frac{1}{2}$ fermion is introduced. For every spin-$\frac{1}{2}$ matter fermion, a superpartner scalar is introduced. Further, for every type of spin-1 gauge boson, there is a corresponding spin-$\frac{1}{2}$ "gaugino," while for the spin-2 graviton, there is an associated spin-$\frac{3}{2}$ "gravitino".

There is a special choice of couplings that allows fermions and bosons superpartners to be interchanged with each other, which yields much nicer renormalization properties for the scalars. This symmetry requires the scalar and fermion superpartners to have equal mass. However, no supersymmetric particles, such as the scalar electron, have yet been detected at a particle accelerator. Hence they must have a much higher mass than their standard particle counterparts. Thus, supersymmetry must be a broken symmetry.

Large extra dimensional models add dimensions to spacetime that are either coiled up or confine matter to a 4-D slice (e.g., a three-brane) of the larger dimensional space. Grand unified theories (GUTs) are efforts to combine the 3 gauge coupling parameters into one. Finally, loop quantum gravity and string theory each try to combine quantum physics with gravity and provide an explanation for some or all of the above puzzles (and for the origin and existence of the universe itself).

One of the most magnificent things about physics is that small observations can have so much meaning. The meaning is not from direct observation but rooted in the models for the universe (like supersymmetry, GUTs, string theory, etc.) that we are working with, testing, and falsifying. For example: after staring at tons and tons of ultra pure water surrounded by super sensitive photodetectors for years, 'scientists' have determined that nothing happened to the protons at the core of the hydrogen atoms on the H_2O. At first, this sounds like a commercial for wasted taxpayer dollars. However, we'll see the importance of these results as we follow our tank of water through these next few sections.

In the following sections, we will discuss first the GUT approach to BSM physics and then the quantum gravity proposals. The latter take more "top-down" approaches, which seek to resolve elementary particle physics questions by first resolving the quantum gravity paradoxes. The topics selected will allow us to make use of the algebraic toolbox you have studied thus far.

5.2 Grand Unified Theories

Grand unified theories build on the idea that the symmetries at long distances (low energies) may be broken-symmetries, and that at small-distances (high energies) the world may be more symmetric. Recall that the Standard Model (SM) is a spontaneously broken theory where the Higgs field acquires a vacuum expectation value (VEV) of order 100 GeV and gives the gauge bosons of the broken generators masses of order 100 GeV, thereby breaking the SM gauge group from $SU(3)_C \times SU(2)_L \times U(1)_Y$ down to $SU(3)_C \times U(1)_{EM}$. This means that at distances shorter than about 10^{-16} cm, the symmetry group is $SU(3)_C \times SU(2)_L \times U(1)_Y$ whereas at longer distances the symmetry group is only $SU(3)_C \times U(1)_{EM}$. The artifacts of the broken hidden symmetry can be found in gauge fields associated with the broken generators which now have masses of the scale of the VEV. As has been discussed earlier, the SM has so many "odd" hypercharge assignments and non-symmetric matter assignments that it feels quite contrived. Could the SM come from a more symmetric model who's enhanced (but hidden) symmetry is broken at a much larger energy scale known as the grand unified theory (GUT) scale (M_X)? Grand unified theories (GUTs) are models where the SM itself arises from a more fundamental theory with a simpler gauge group which, after a GUT-scale symmetry breaking, becomes the SM.

The SM's massive gauge fields become important to physical processes at interaction energies near or above the scale of their mass, which are about 100 GeV. At much lower energies than the VEV energy scale, these massive gauge fields manifest as effective interactions that govern the rate of rare processes. Radioactive decay is the low-energy manifestation of the massive gauge bosons of the SM. We will find that proton decay and other baryon or lepton number-violating processes (which as of this writing have never been observed) are low-energy manifestation of broken grand unified symmetry. The predictions for the decay channel and the decay rate are determined by the specific details of the GUT being considered.

As you continue your career in fundamental physics, you'll notice that every "anomaly" reported by experimentalists is followed by a flurry of theorists trying to explain it with a new proposed theory. This is largely because the study of beyond-the-Standard Model physics is really a study of rare processes and trying to align the rare processes that are observed and not observed with a compatible theory. Remember the tank of water? By watching it for years and verifying that 'nothing happened', we place a lower bound on the lifetime of the proton. This rare process will speak volumes about excluding huge classes of GUTs. Other rare processes are also predicted and studied with GUTs.

The task of specifying plausible GUTs for the SM involves a lot of guess work and understanding how the SM group can be embedded into larger groups. First we'll provide some background on how the strength of the forces tend toward a unified strength at a very small distance scale. Then, we'll do some explicit embeddings of the SM into GUT symmetry groups where we'll discover some of the simplifications and subsequent complications associated with GUT model building.

5.2.1 Unification of the Coupling Constants

The unification hypothesis is motivated by two things. First it is motivated by the promise of increased symmetry, which is itself a "holy-grail" type search. Second, the strengths of the three non-gravitational forces (based on experimentation) tend towards a common value at smaller distances. In this section, we'll focus on the second of these. The remainder of our GUT discussions will focus on the first.

To understand what we mean by the strength of the coupling constants tending to a common value at high energies you need to know a few more of the generic properties of renormalization. Renormalization is the process of hiding the infinities in QFT, and it manifests as changes in the apparent strength of a force at different distances. Precise QFT calculations of what will happen when two particles collide requires renormalization to deal with infinities and requires that the renormalization process introduce an arbitrary energy scale μ into the dynamics. Even if the theory you begin with did not have an energy scale, you will need to introduce an arbitrary renormalization energy scale μ to do the renormalization calculations. Renormalization was referred to in Sect. 4.6.11 and the related concept of asymptotic freedom of quarks in Sect. 2.5.3.

Renormalization scales are like choosing the zero of potential energy in classical mechanics: choosing a reference is required, the choice is arbitrary, but all the steps and values of parameters in the calculations depend on the choice. Just as for potential energy, some choices for the renormalization scale μ make for easier calculations than others. The results of perturbative calculations using renormalization end up depending on factors similar to $g(\mu)\log(\mu/E)$ where E is the center-of-mass energy of the particles participating in the collision and $g(\mu)$ is the strength of the gauge coupling for a particular choice of renormalization scale μ. When doing classical mechanics calculations the total energy depends on the choice of the potential energy reference; likewise in renormalization calculations all three coupling constant strengths $g_i(\mu)$ and even the mass parameters (like the electron mass m_e or up-quark mass m_u) depend on the choice of the renormalization scale. In classical mechanics the final answer does not depend on the choice of the potential energy reference, and likewise in renormalization calculations the final answers do not depend on the scale μ chosen if a complete calculation is done.

However, complete calculations are never done because we cannot exactly solve these QFT problems. We are always working to some order in perturbation theory or working with a lattice QCD approximation. There can be many simplifications in the perturbative calculations achieved by choosing the renormalization scale to be the same order of magnitude as the collision being studied. Keeping the size of $\log(\mu/E)$ small ensures perturbation theory converges faster and a more precise answer is achieved at lower-order calculations. If the log gets too big, the perturbation series may even diverge and low-order perturbation theory is meaningless.

A few more nuances with renormalization. There are many different ways to do renormalization: dimensional, mass-shell, minimal subtraction, $\overline{MS}$, $\overline{DR}$, etc; a description of these can be found in many standard QFT texts. Each way to do renormalization has its own coupling constant and mass parameters for every choice of renormalization scale μ; when doing calculations and using values in a book for mass or coupling constants, one has to be careful to choose values based on the same renormalization scale and the same renormalization technique.

In many respects the renormalization scale is similar to looking at the world with a particular 'map resolution'. The energy scale μ can switch to the distance scale $d = \hbar c/\mu$ of the 'map resolution'. This allows us to interpret the renormalization scale as a distance. Choosing μ to match the distance scale minimizes the order in perturbation theory to get approximately accurate answers. Therefore, we can think of the parameters $g(\mu)$ and the masses $m(\mu)$ as producing the picture that most closely resembles the view of the universe through a magnifying glass studying distances $d = \hbar c/\mu$.

The renormalization of the coupling constants is another way to say that you are eliminating the infinities and discovering the μ dependence of the coupling constants. This is done by calculating Feynmann diagram loops. The following differential equations result from a one-loop calculation for the renormalization group flow of the coupling constant g of an $SU(N)$ gauge theory coupled to fermions and scalars:

$$\frac{\partial}{\partial \log \mu} g(\mu) = -\frac{g^3}{16\pi^2}\left[\frac{11}{3}S(R_A) - \frac{2}{3}n_F S(R_F) - \frac{1}{3}n_S S(R_S)\right]. \tag{5.1}$$

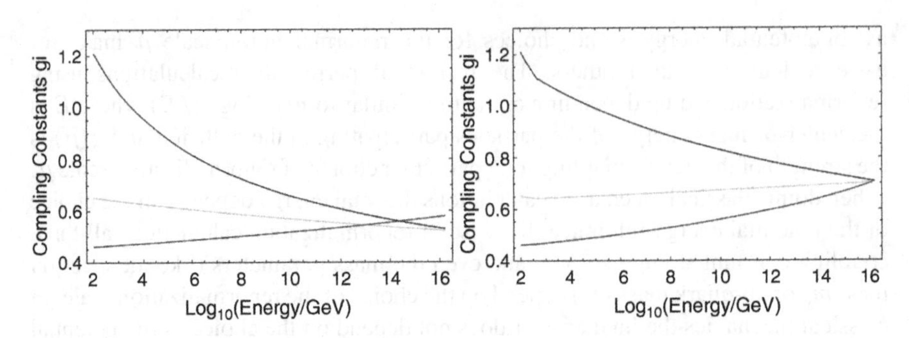

Fig. 5.1 Gauge couplings for the three non-gravitational forces as a function of log (base 10) of the energy scale for the (*left*) Standard Model and (*right*) Minimal Supersymmetric Standard Model

In this equation n_F is the number of 2-component fermions in that representation, n_S is the number of complex scalars, and $S(R_{A,F,S})$ is the Dynkin index for the representation of adjoint, the fermions, or scalars respectively. The Dynkin index is defined as $S(R)\delta_{ab} = Tr(T_a T_b)$ for the generators that transform that representation. For the fundamental representation of $SU(N)$ we have $S(R) = 1/2$, while for the adjoint representation $S(R) = N$. For a $U(1)$ gauge theory, $S(R) = \sum(Q)^2$.

Applying the renormalization group formula to the SM with one Higgs field leads to

$$\frac{\partial}{\partial \log \mu} g_i = \frac{g_i^3}{16\pi^2} b_i \tag{5.2}$$

$$\text{SM} \begin{cases} b_1 = n_G \, \frac{4}{3} + n_H \, \frac{1}{10} \\ b_2 = -22/3 + n_G \, \frac{4}{3} + n_H \, \frac{1}{6} \\ b_3 = -11 + n_G \, \frac{4}{3} \end{cases}$$

where g_1 is associated with $U(1)_Y$ and g_2 is associated with $SU(2)_L$ and g_3 is associated with $SU(3)$, $n_G = 3$ is the number of generations, and n_H is the number of Higgs doublets ($n_H = 1$). With a little foreshadowing to the GUT result, we have defined g_1 associated with $U(1)_Y$ hypercharged states such that $S(R) = 3/5 \sum(Y)^2$. There is a freedom in how to choose the definition of the charges Y and the coupling g as they can be arranged to always appear as the product of Yg.

The figure above shows the solution to the differential equation reflecting the coupling constants as a function of energy scale; the low-energy boundary condition is set by experiment. To foreshadow the importance of supersymmetry, we have also included the results of renormalizing the coupling constants for the minimal supersymmetric Standard Model (MSSM). Figure 5.1 shows the three forces approach a common strength at somewhere between 10^{14} and 10^{17} GeV depending on the assumptions (SM vs. MSSM) made.

Two lines with non-equal slopes will intersect. The intersection of the third line is what makes unification compelling. The unification in the SM is not perfect. When it was first proposed in the early 1980s, the strong coupling constant g_3 had enough uncertainty that unification was consistent within experimental bounds. However, during the 1990s the uncertainty became small enough that unification was no longer possible with the SM.

However, when supersymmetry is imposed unification is still consistent (to within about a 10% uncertainty) with our observed values of the three forces. There is a freedom in the supersymmetric case to choose the mass scale of the supersymmetric particles; this can be used to keep the unification successful but requires the supersymmetric particles to have masses under a few hundred TeV. There are also GUT scale threshold corrections which explain that we don't actually expect a perfect unification; the threshold corrections are beyond the scope of this introduction.[1]

Although the crossings of the three forces in the figure on page 274 look like they occur at a particular region, we pulled a quick one on you. Because the abelian gauge theory Lagrangian lacks products of gauge couplings $(A_\mu A^\mu)^2$, you can convince yourself that the $U(1)_Y$ charges initially all have an arbitrary scale which makes such a unification seem trivial. We can rescale the charges and absorb an extra factor into the coupling constant so that all three coupling constants unify by definition! Two lines with different slopes will always intersect, and if there is a free-parameter for $U(1)$ coupling to tune then a three-way intersection is almost guaranteed. When the $SU(3)_C \times SU(2)_L \times U(1)_Y$ is embedded in a larger GUT group like $SU(5)$, the $U(1)_Y$ coupling g' is rescaled to the normalization appropriate to the $SU(5)$ generator that becomes hypercharge (we'll show details below). This rescaling causes us to work with $g_1 = \sqrt{5/3}\, g'$. This rescaling is the origin of the strange $3/5$ in the definition of $C(R)$ for $U(1)_Y$.

However, if the scale of the $U(1)$ coupling is fixed by the GUT embedding, and the three forces still (nearly) unify, then this *is* surprising and suggests to us that GUTs may be part of the more fundamental description of nature. The right figure on page 274 shows that supersymmetry makes the unification of the coupling constants even more dramatic.

The observation that the three forces may have had the same coupling constant value at a smaller distance scale suggests that the three forces may be a single force at this small distance scale. If there is only a single force, then what distinguishes an electron from a down-quark? a neutrino from an up-quark? or an up-quark from a down-quark? GUT models explore the ways in which these seemingly different particles are at some small distance scale indistinguishable.

Here are some remaining puzzles associated with the SM to keep in mind as we explore what GUTs may contribute: Why are the hypercharge assignments so strange? Where does the weak mixing angle come from? Why is charge quantized?

[1] Though they will be dealt with extensively in a later volume.

Why are there exactly three generations of matter particles? Where do the Yukawa couplings for the mass come from? Do the SM forces have any connection with gravity?

Another wrinkle for the SM arose from cosmology several decades prior: essentially all spiral galaxies rotate too fast for their stellar limbs to remain attached due to center of mass gravitational attraction from a galaxy's SM matter. Rather, stability of galaxies requires inclusion of dark (a.k.a. hidden) matter not composed of any of the SM particles. Dark matter is so defined because it does not interact (or in the least, interacts only extremely weakly) with any of the SM particles through strong or electroweak forces, but interacts only through gravity. Galaxy and galaxy cluster stability predicts dark matter to be roughly eight times as common as ordinary SM matter.

The gauge theory part of the Standard Model is very elegant, and expanding it to include GUTs will answer the first three of these questions. However, a GUT in itself cannot provide an explanation for the number of generations. The Yukawa couplings also remain a mystery. Lastly, a GUT containing the SM offers no explanation for dark matter, nor does it provide for any connection with gravity. Concepts beyond GUTs will be required to address these latter issues and will be introduced following the GUTs discussions.

These issues (and additional ones not addressed here) regarding the SM might be summed up as follows: Although the basic features of the underlying forces of the universe specified by the SM are known, an explanation for the SM itself is not currently understood.[2] The goal of the rest of this chapter is to examine several concepts that, either individually, or in combination, may provide deeper meaning to, and (at least partial) explanation for the SM and a possible connection with gravity.

Remember the large vat of water which was watched while 'nothing happened'? They were used to place a lower limit on the lifetime of proton. The GUT remnants that have masses of order 10^{14} to 10^{17} GeV manifest themselves as rare decays like the decay of a proton and the structure of neutrino masses and mixings. The observations of these large vats of water for proton decay has placed limits on proton decay and has ruled out GUTs that have mass scales 10^{14} to 10^{15} GeV, thereby making the only viable GUTs supersymmetric. Quite an amazing discovery made simply by carefully watching protons *not* decaying.

GUTs provide a model to explain unification of forces and a more symmetric origin for the SM particle zoo. They also provide a structure to study the very, very small. Energy scales of 10^{16} GeV correspond to distance scales of about 10^{-30} cm. GUTs provide a microscope to this distance scale. Experiments may one day specify the rate and type of proton decay and measure the remaining neutrino masses and mixings. When they do, then those experiments will dramatically narrow down

---

[2]This is, of course, a fundamental problem in science. Whatever explains the SM will itself need to be explained, and so on. The real goal is simply to take the question as far as science can. And as of now it seems that there is quite a bit farther we can go.

Table 5.1 Standard Model field's transformation properties

Field	Lorentz	$SU(3)_C$	$SU(2)_L$	$U(1)_Y$
$(L)^i$	$(1/2, 0)$	1	2	$-1/2$
$(e^c)^i$	$(1/2, 0)$	1	1	1
$(\nu^c)^i$	$(1/2, 0)$	1	1	0
$(Q)^i$	$(1/2, 0)$	3	2	$1/6$
$(u^c)^i$	$(1/2, 0)$	$\bar{3}$	1	$-2/3$
$(d^c)^i$	$(1/2, 0)$	$\bar{3}$	1	$1/3$
H	1	1	2	$+1/2$
B_μ	$(1/2, 1/2)$	1	1	0[a]
W_μ	$(1/2, 1/2)$	1	3[a]	0
G_μ	$(1/2, 1/2)$	8[a]	1	0

[a]The field does not transform covariantly but rather transform as a gauge transformation

the number of viable GUTs. This combination of theory with experiment will give scientists the most powerful microscope ever achieved into the physics of the very small. GUTs may also help guide searches for a quantum theory of gravity.

We have now reviewed how the coupling constants for the three forces tend to a common value, and how that process leads to the exclusion of most non-supersymmetric GUTs. We will now turn to the second compelling reason to study GUTs, the algebraic elegance of embedding the SM into larger symmetry groups.

5.2.2 The Basic $SU(5)$

We begin with details of the prototypical GUT. To introduce the mechanics of working with GUTs, we'll go over the explicit embedding of the SM into a basic $SU(5)$ GUT. This model is not viable because it predicts proton decay faster than is experimentally observed.[3] We should have seen more than a few hundred hydrogens in the large vats of water decay by now. Although no longer viable, understanding the basic $SU(5)$ is good to learn because the concepts are used over and over again in other GUTs with larger symmetry groups. Afterward the basic $SU(5)$, we'll provide a glimpse at embedding the SM into larger GUT symmetry groups.

The first basic task is to place all the SM fermion particles into left-handed (this is an arbitrary choice) representations. All the particles in a given representation possess similar spacetime transformation properties, so that we can more easily see how to form more symmetric 'internal' symmetry groupings. The field content of the Standard Model with an extra neutrino state (for future use in models with neutrino mass) is given in the Table 5.1 on page 277: As was discussed in the section starting

[3]In contrast, some $SU(5)$ models derived from string theory in particular may contain additional mechanism for slowing down proton decay to acceptable levels.

on page 201, the gauge bosons do not transform covariantly (or linearly) under the gauge group, rather they transform with an extra term as in equation (4.223) or (4.364). The SM's left-handed leptons (electron and neutrino) form $SU(2)$ doublets L^i, the right-handed charged leptons which form $SU(2)$ singlets $(e^c)^i$, and the Higgs fields H which is an $SU(2)$ doublet. The basic $SU(5)$ has no place for a right-handed neutrino. We include it here for future use.

So what is the explicit embedding of the SM into $SU(5)$? We know from Sect. 3.2 that the rank of $SU(5)$ is 4 (cf. paragraph on page 111) and so there are 4 generators in the (all diagonal) Cartan subalgebra. Because at low energies, the $U(1)$, $SU(2)$ and $SU(3)$ symmetry groups are independent, the generators representing these three subgroups must commute with each other. For example, all the $SU(2)$ generators need to commute with all the $SU(3)$ generators. Likewise, the $U(1)$ subgroup needs to commute with the entire $SU(2)$ and $SU(3)$ subgroup. We will now search for the first three elements of the Cartan subalgebra belonging to the $SU(2)$ and $SU(3)$ subgroups, and see if we have the freedom to choose the fourth Cartan subalgebra generator of $SU(5)$ to be consistent with an embedding of the SM.

The generators of $SU(5)$ are in the adjoint of the group. There are $24 = 5^2 - 1$ (again cf. page 111) $SU(5)$ generators, each of which is realized as a 5×5 matrix in the adjoint representation. We can take the first 8 generators to be the Gell-Mann generators of $SU(3)$ (equation 3.208), where $j = 1 \ldots 8$, and for which we have two diagonal generators

$$T^j = \frac{1}{2}\begin{pmatrix} \lambda^j & 0 \\ 0 & 0 \end{pmatrix}. \tag{5.3}$$

Here the λ^i are 3×3 matrices and the zeros fill in the remainder of the 5×5 matrix. The next three $SU(5)$ generators can be taken as the three pauli Matrixes with $j = 1 \ldots 3$ where we have one more diagonal generator

$$T^{j+8} = \frac{1}{2}\begin{pmatrix} 0 & 0 \\ 0 & \sigma^j \end{pmatrix}. \tag{5.4}$$

Again the σ_j is a 2×2 matrix and the zeros pad the 5×5 matrix.

We need to find the fourth diagonal generator. We chose a basis by placing the $SU(3)$ subgroup to be the first three components of the vector space and the $SU(2)$ in the last two dimensions. The 12th generator is the fourth and final traceless diagonal generator that commutes with the basis we have already chosen. This generator is chosen so that it is diagonal in the subspace spanned by the $SU(3)$ generators and the subspace spanned by the $SU(2)$ generators. These constraints uniquely determine (up to a sign) the generator[4]

---

[4]You should confirm that this is indeed the only choice for this matrix.

$$T^{12} = \sqrt{\frac{3}{5}} \begin{pmatrix} -\frac{1}{3} & 0 & 0 & 0 & 0 \\ 0 & -\frac{1}{3} & 0 & 0 & 0 \\ 0 & 0 & -\frac{1}{3} & 0 & 0 \\ 0 & 0 & 0 & \frac{1}{2} & 0 \\ 0 & 0 & 0 & 0 & \frac{1}{2} \end{pmatrix}. \tag{5.5}$$

The normalization must satisfy $\mathrm{Tr}(T^a T^b) = \frac{1}{2}\delta^{ab}$ which is the same normalization as the other 23 generators. If this generator does indeed give $U(1)_Y$, then the fixed normalization this provides is one way to solve the charge quantization puzzle.[5]

The remaining 12 generators consist of the 12 ways in which we make Hermitian generators with only the non-diagonal elements that have been zero in all the preceding 12 generators:

$$T^{13} = \frac{1}{2} \begin{pmatrix} 0 & 0 & 0 & 1 & 0 \\ 0 & 0 & 0 & 0 & 0 \\ 0 & 0 & 0 & 0 & 0 \\ 1 & 0 & 0 & 0 & 0 \\ 0 & 0 & 0 & 0 & 0 \end{pmatrix} \quad T^{14} = \frac{1}{2} \begin{pmatrix} 0 & 0 & 0 & i & 0 \\ 0 & 0 & 0 & 0 & 0 \\ 0 & 0 & 0 & 0 & 0 \\ -i & 0 & 0 & 0 & 0 \\ 0 & 0 & 0 & 0 & 0 \end{pmatrix} \tag{5.6}$$

and continues on for all 12 of the off diagonal entries until

$$T^{23} = \frac{1}{2} \begin{pmatrix} 0 & 0 & 0 & 0 & 0 \\ 0 & 0 & 0 & 0 & 0 \\ 0 & 0 & 0 & 0 & 1 \\ 0 & 0 & 0 & 0 & 0 \\ 0 & 0 & 1 & 0 & 0 \end{pmatrix} \quad T^{24} = \frac{1}{2} \begin{pmatrix} 0 & 0 & 0 & 0 & 0 \\ 0 & 0 & 0 & 0 & 0 \\ 0 & 0 & 0 & 0 & i \\ 0 & 0 & 0 & 0 & 0 \\ 0 & 0 & -i & 0 & 0 \end{pmatrix}. \tag{5.7}$$

What possible physics has thus far been discovered by the embedding of the SM into $SU(5)$? The first eight generators show how we are embedding $SU(3)_C$ of the SM, the next three generators show how we are embedding the $SU(2)_L$, and the T^{12} shows how we are going to embed $U(1)_Y$. The generators 13 to 24 correspond to generators which mix $SU(3)_C$ and $SU(2)_L$. Leptons are only charged under $SU(2)_L$ and $U(1)_Y$, and quarks are additionally charged under $SU(3)_C$. Because the generator 13 through 24 would link the $SU(2)_L$ from the leptons to $SU(3)_C$ which couple to quarks, they represent a process that would convert a lepton into a quark or the reverse process. Because this process is very rare (as yet unobserved), these generators must be very massive; the more massive the more rare these events will be. Giving a very large mass to generators 13 through 24 is equivalent to breaking the $SU(5)$ symmetry (cf. Sect. 4.7).

[5]There are many other proposals. Some purely geometrical while others rely on anomaly cancelation conditions.

Just because we can embed $SU(3)$ and $SU(2)$ and $U(1)$ into $SU(5)$ doesn't mean that we will automatically have the fermions with the right charges. In the $SU(5)$ model the fermions are arranged into a $\bar{\mathbf{5}}$ (ψ_a^i) and a $\mathbf{10}$ ($\phi^{ab\,i}$) where $a, b, c, ..$ are the $SU(5)$ indexes and $i, j, ..$ are the family indexes. The $\mathbf{10}$ can be understood as the antisymmetric part of a two-index $SU(5)$ tensor (antisymmetric tensors have $n(n-1)/2$ independent terms). Which particles should go into the $\bar{\mathbf{5}}$? We can see that the $SU(3)$ generators in equation (5.3) act on the first three components and that the generators $SU(2)$ act on the last two. If we have a $\bar{\mathbf{5}}$, then we see that the first three components would transform as a $\bar{\mathbf{3}}$ of $SU(3)_C$ and a $\mathbf{1}$ of $SU(2)_L$.[6] The last two components of the $\bar{\mathbf{5}}$ would transform as a $\bar{\mathbf{2}}$ under $SU(2)_L$ and a singlet under $SU(3)_C$. To make the hypercharge generator consistent leads us to identify $Y = \sqrt{\frac{5}{3}}T^{12}$. The properties of these subgroups can be summarized by grouping the $SU(3)$ representation, the $SU(2)$ representation, and the $U(1)_Y$ representation into the combination: $(A, B)_Y$ where A is the $SU(3)$ representation, B is the $SU(2)$ representation and Y is the hypercharge. This notation is common when discussing GUTs. Now the first three components of $\bar{\mathbf{5}}$ transform as $(\bar{3}, 1)_{1/3}$ and the next two as $(1, \bar{2})_{-1/2}$ where we have been careful to use the generators that would act on the $\bar{\mathbf{5}}$. Looking back to Table 5.1 on page 277 we can see exactly which SM fermions transform in this way. This thought process identifies the components of the $\bar{\mathbf{5}}$ as the quintuplet of five fields given by the three d^c components and the two components $i\sigma^2 L$ which together are: $(d^c, i\sigma^2 L)^j$.

What about the $\bar{\mathbf{2}}$ and $\mathbf{2}$ and that extra $i\sigma^2$? Why are the signs opposite what is given in T^{12}? These details require a quick note on working with these representations. The generators given above act on a $\mathbf{5}$ of $SU(5)$. The $SU(2)$ group is special because it can form a scalar either by combining a $\mathbf{2}$ with a $\bar{\mathbf{2}}$ as in $\phi^a\bar{\phi}_a$ or by using the two component antisymmetric tensor $\epsilon_{ab}\phi^a\phi^b$. As a result we can couple $\mathbf{2}$ with $\mathbf{2}$ to make a scalar. The $(i\sigma^2)_{ab} = \epsilon_{ab}$ is used to convert L into a $\bar{\mathbf{2}}$. Because it is so easy to switch between $\mathbf{2}$ and $\bar{\mathbf{2}}$, most references don't fret over the distinction between $\mathbf{2}$ and $\bar{\mathbf{2}}$. This explains the $\bar{\mathbf{5}}$ fermions. To accommodate all the fermions we will need to include other representations. We have been discussing the $\bar{\mathbf{5}}$. Here are the basic toolbox elements for doing calculations with such representations. The generators $-T^*$ act on the complex conjugate representation $\bar{\mathbf{5}}$ (cf. Sect. 4.7.4).

If the representation has more than one index, like $\mathbf{10}$ where we have a and b antisymmetric as in ϕ^{ab}, then the generators will act on each index. For example an infinitesimal shift along the axis $\delta\theta_c$ gives:

$$\delta\phi^{ab} = i\,(\delta\theta)_c((T^c)^a_d\phi^{db} + (T^c)b_d\phi^{ad}). \tag{5.8}$$

[6]The term $\mathbf{1}$ means that the components are inert under that symmetry. They are singlets and do not transform.

If we have a lower index as in the adjoint representation this becomes

$$\delta\phi^a{}_b = i\,(\delta\theta)_c((T^c)^a{}_d\phi^d{}_b - (T^c)^d{}_b\phi^a{}_d). \tag{5.9}$$

Notice that because the generators are Hermitian, the conjugate of T equals its transpose.

What about the other 10 particles (we neglect the ν^c in basic $SU(5)$ GUTs)? As we said earlier they are going to fit into an antisymmetric product of two $SU(5)$ indices. Let's see how the different parts of the **10** transform under the Standard Model. Lets begin with ϕ^{ab} where a and b are limited to $1\ldots3$. This transforms as a $\bar{\mathbf{3}}$ of $SU(3)$ but does not transform at all under $SU(2)$. The $\bar{\mathbf{3}}$ can be understood because $\phi^{ab}\epsilon_{abc}$ could couple to a **3** and form a scalar. Using the rule given earlier for working with multi-index objects, the hypercharge for this group will come from adding the contributions from the diagonal generators on both indices giving $Y = -2/3$. Together this indicates that the subgroup for the first three rows and columns of the **10** transforms as $(\bar{\mathbf{3}}, \mathbf{1})_{-2/3}$ which matches u^c. What about the bottom right 2×2 block. Here we have an antisymmetric combination of **2**; this is an $SU(2)$ scalar giving it the transformation properties $(1, 1)_1$ which match the properties of e^c. Following the same steps, we can find that u and d also belong to the **10** and where in the matrix they end up. Another way to check the assignments is to calculate the electric charge generator on the state:

$$Q = \tau^3 + Y = T^{11} + \sqrt{\frac{5}{3}}T^{12} \tag{5.10}$$

where we have used our hypercharge assignment of

$$Y = \sqrt{\frac{5}{3}}T^{12}. \tag{5.11}$$

Using this we can confirm the electric charge of each of the fermions.

The final $SU(5)$ fermion assignments for the $\bar{\mathbf{5}}$ and the **10** are

$$\psi^i_a = \left(d^c_r\ d^c_g\ d^c_b\ e\ \nu\right)^i \tag{5.12}$$

$$\phi^{ab\,j} = \frac{1}{\sqrt{2}}\begin{pmatrix} 0 & u^c_b & -u^c_g & -u_r & -d_r \\ -u^c_b & 0 & u^c_r & -u_g & -d_g \\ u^c_g & -u^c_r & 0 & -u_b & -d_b \\ u_r & u_g & u_b & 0 & -e^c \\ d_r & d_g & d_b & e^c & 0 \end{pmatrix}^j. \tag{5.13}$$

where we have included with the three quarks the (r, g, b) labels. We have now discovered which components of the $\bar{\mathbf{5}}$ and **10** correspond to which fermions after the $SU(5)$ symmetry is broken. Before the symmetry is broken, all of these components are interchangeable; that is what the $SU(5)$ symmetry means. It was a pure miracle that the 'odd' hypercharge assignments of the SM turned out to be exactly how the T^{12} generator would couple. In this way we have explained another ugly feature of the SM by embedding it into an $SU(5)$ GUT.

We have also confirmed the embedding of the SM gauge group into $SU(5)$. We have found the gauge bosons associated with generators T^1 through T^8 correspond to the gluons of $SU(3)_C$, the gauge bosons associated with T^9-T^{11} correspond to weak $W^{\pm}_{\mu}$ and Z^o_{μ} bosons, and the gauge boson associated with T^{12} corresponds to the B_{μ} hypercharge gauge boson. Furthermore we have discovered an extra 12 gauge bosons that must be heavier and mediate very rare processes and are typically labeled X^a_{μ} and Y^a_{μ}. Traditionally they are categorized according to their electric charge Q with X^a_{μ} being the gauge boson's with charge $\pm 4/3$ and Y_{μ} being the gauge bosons with charge $\pm 1/3$. They carry both lepton number and color and therefore referred to as leptoquarks. We will study the mass of the X^a_{μ} and Y^a_{μ} bosons later in this chapter.

The embedding of the SM into $SU(5)$ specified a normalization for the $U(1)_Y$ coupling constant. This both solves a charge quantization puzzle, and it provides a prediction for the weak mixing angle. The weak mixing angle is given by $\sin\theta_W = g'/\sqrt{g^2 + g'^2}$. Because $g' = \sqrt{3/5}g_1$ and at the unification scale $g_1 = g_2 = g_3$, we see that at the unification scale $\sin^2\theta_W = 3/8$. This prediction is for the GUT scale, and to compare to experiment the parameters must be moved with the renormalization group to the scale where we are measuring the weak mixing angle in accelerators. Looking at Fig. 5.1, we see that the unification assumption underlying this prediction is poor for non-supersymmetric $SU(5)$, but quite good for supersymmetric $SU(5)$ unification.

Embedding the gauge group is not enough for a complete grand unified theory. Next we need to include the Higgs field so that mass terms can be formed. The SM Higgs transforms as $(1,2)_{\frac{1}{2}}$ which suggests it should transform as a **5** as $(H)^a$. The bottom two components of $(H)^a$ are the SM Higgs. The top three components are a color triplet which had better be very heavy if we expect the $SU(5)$ theory to simplify to the SM at low energies. To find the couplings we basically need to find the ways that we can take the given fields and contract all the indices. We will form two invariants that can be used to find the Yukawa couplings:

$$L_Y = \sqrt{2}(Y^d)_{ij}\phi^{ab\,i}(i\sigma^2)\psi_a{}^j(H^*)_b + \frac{1}{4}(Y^u)_{ij}\phi^{ab\,i}(i\sigma^2)\phi^{dc\,j}H^e\epsilon_{abcde} \quad (5.14)$$

where the $i\sigma^2$ is used to form a Lorenz scalar from the two left-handed Weyl fermions. To give the mass couplings we expect the Higgs VEV to be $H^a = (0,0,0,0,v_o)$, where v_o is of the order of the electroweak symmetry breaking scale.

The Yukawa coefficients give us the ability to make our next possible discovery with the GUT model. The coefficients are chosen so that Y^u and Y^d correspond to the SM Yukawa couplings. However, there is a surprise in these couplings. The Y^d matrix entries multiply the group $(d^c_r d_r + d^c_g d_g + d^c_b d_b + e^c e)^{ij}$. This means that we must give the down quark entries equal Yukawa couplings as the leptons!

The bottom quark and the tau lepton would have the same mass? Although this seems absurd, recall that the GUT symmetry is only a good symmetry near the unification scale. Also remember that the parameters must be renormalized and will be different near that GUT length scale. This 'prediction' that the tau and bottom have equal masses was actually experimentally consistent within the error bars of the 1980s. Today we have tighter measurements and this mismatch is being used to possibly provide insight to what other structure must be at the GUT scale.

What about the other two generations? The ratios of the lighter two generations remain largely constant in the renormalization process. This is because they share the same gauge couplings and we can neglect the contribution from the smaller Yukawa couplings of the first and second generations. The ratios of the respective generations are nowhere near equal. The fact that the ratios $m_s/m_d \approx 20$ is so different than $m_\mu/m_e \approx 200$ suggest that maybe there needs to be some other couplings for the different generations; some non-minimal secondary Higgs field like an $(H_{45})^a_{cd}$ coupling to the second generation.[7] Many, many GUT papers are written studying these alternative ways to construct the Yukawa couplings and make them consistent with experiment over all three generations. In short, although $m_b = m_\tau$ at the GUT scale is not unreasonable, in general GUTs do not seem to easily solve the Yukawa coupling problem.

We have now shown that the SM embeds into $SU(5)$ nicely. We still have to explain what breaks $SU(5)$ and what gives the Higgs color triplet mass. Our next question is: how do we break $SU(5)$ with the prescribed pattern? There are two parts to this question. First we need to ask what new field and VEV will give GUT-scale masses to the generators T^{13} to T^{24} GUT? Second we need to ask what potential for this new field will give the desired VEV? Afterwards we will return to the question of the Higgs color triplet. The gauge fields associated with the SM are all considered massless at this stage. We do not expect them to acquire mass until the much, much smaller electroweak scale.

Recall from the SM chapter that the masses of the gauge fields in a broken theory can be found by diagonalizing the kinetic term of the scalar sector evaluated at the scalar's VEV. This technique is known as the Higgs mechanism. By analogy to the low-energy Higgs mechanism of the SM which gives mass to the $W^\pm$ and Z, giving VEVs to the a new field in the **5** representation will give GUT scale masses to the $SU(3)_C$ bosons or the $SU(2)_L$ bosons and the $U(1)_Y$ bosons. We want them to have EW-scale masses, not GUT scale masses. To keep the SM gauge bosons light, we need to consider a more complex field where the VEV somehow leaves the SM generators without mass at the GUT-scale transition. The problem becomes easier after considering the adjoint transformation properties. Consider the kinetic term of the **24** field $\Sigma^a_{\ b}$ which transforms like the adjoint in $SU(5)$:

$$L_{\Sigma Kin} = (D_\mu \Sigma)^a_{\ b} (D^\mu \Sigma^*)^b_{\ a} \tag{5.15}$$

---

[7]The H_{45} is antisymmetric on the bottom two indices and traceless on any contraction of top and bottom indices.

where

$$(D_\mu \Sigma)^a{}_b = \partial_\mu \Sigma^a{}_b - igV^c_\mu (T^c)^a{}_d \Sigma^d{}_b + igV^c_\mu (T^c)^d{}_b \Sigma^a{}_d \tag{5.16}$$

$$= (\partial_\mu \Sigma - igV^c_\mu [T^c, \Sigma])^a{}_b. \tag{5.17}$$

Evaluating at Σ's VEV gives

$$L_{\Sigma Kin} = g^2 V^c_\mu V^{d\,\mu} [T^c, \langle \Sigma \rangle]^a{}_b [T^d, \langle \Sigma \rangle]^b{}_a. \tag{5.18}$$

From this equation we can read off the key property of the Σ's VEV needed to leave the SM gauge fields massless at the GUT scale breaking: the VEV $\langle \Sigma \rangle$ must commute with all the generators of the Standard Model. With this concept, one now realizes that the VEV $\langle \Sigma \rangle \propto T^{12}$ will commute with all the SM gauge fields. Because the hypercharge generator is proportional to T^{12}, we know that T^{12} commutes with T^{12} leaving hypercharge generator massless. We also recall that T^{12} was precisely the generator which commuted with both the $SU(2)_L$ subgroup and the $SU(3)_C$ subgroup. All that is left to check is that T^{12} does not commute with all of the remaining generators T^{13} through T^{24}. Identifying the $\langle \Sigma \rangle$ VEV as

$$\langle \Sigma \rangle^a{}_b = v(T^{12})^a{}_b \sqrt{15} = v \,\mathrm{diag}\left(1, 1, 1, -\frac{3}{2}, -\frac{3}{2}\right) \tag{5.19}$$

we find the mass of the X^a_μ and Y^a_μ gauge fields to be

$$L_{\Sigma Kin} = \frac{1}{2} M^2_{X,Y} (X^c_\mu X^{c\,\mu} + Y^c_\mu Y^{c\,\mu}) \tag{5.20}$$

where $M^2_{X,Y} = \frac{25}{8} g^2 v^2$. This means the leptoquark gauge bosons X^a_μ and Y^a_μ both have mass of the order of the vacuum expectation value of Σ. The breaking of the $SU(5)$ symmetry by $\langle \Sigma \rangle$ also triggers the splitting of $SU(5)$ into $SU(3)_C \times SU(2)_L \times U(1)_Y$. In a viable GUT theory, the value of v needs to be specified by the coupling constant unification and will also need to agree with the rate of proton decay if it is ever observed.

Now that we know what the VEV of Σ should be, we now need to create the potential that will result in $\Sigma \propto T^{12}$. We write down the potential for Σ consistent with the gauge symmetries, a discrete symmetry $\Sigma \to -\Sigma$, and with mass dimension less than or equal to 4:

$$V(\Sigma) = -\mu^2 \Sigma^a{}_b \Sigma^b{}_a + \frac{a}{4} \left(\Sigma^a{}_b \Sigma^b{}_a\right)^2 + \frac{b}{2} \Sigma^a{}_b \Sigma^b{}_c \Sigma^c{}_d \Sigma^d{}_a. \tag{5.21}$$

To find the direction and magnitude of the Σ VEV note that only the b coefficient affects the direction. Why not the first two? Because the **24** is the adjoint, the VEV can be parameterized in terms of the generators $\langle \Sigma^a_{\ b} \rangle = \sqrt{15} v_j (T^j)^a_{\ b}$ where we normalize $(v_j)(v_j) = v^2$ and the $\sqrt{15}$ is there to be compatible with equation (5.19). Each generator satisfies the same normalization $\text{tr}(T^a T^b) = \frac{1}{2}\delta^{ab}$. Therefore each direction contributes the same coefficient of $1/2$ to the sum v^2. The first two terms in $V(\Sigma)$ involve the trace of two generators which gives $v_j v_j = v^2$. The μ^2 term multiplies v^2, and the a term multiplies $(v^2)^2$. Therefore neither of these select out a direction. The b term selects a direction. The trace of the fourth power of any of the 20 off-diagonal generator contributes a factor of $\text{tr}(T^4_{OD}) = 1/8$. The diagonal generators of the $SU(2)$ and $SU(3)$ subgroups also have the same factor of $\text{tr}(T^4_{SU(2),SU(3)}) = 1/8$. Therefore any one of these 23 generators is equal to any other. Only the generator T^{12} has the distinct trace of $\text{tr}((T^{12})^4) = 7/120$. Thus to make the T^{12} direction a preferred direction, one only needs to make $b > 0$ because $7/120 < 1/8$. This ensures that the lower energy direction is that parallel to T^{12}. To ensure that the VEV is non-zero, we need $\mu^2 > 0$. Finally to ensure that the potential is bounded below, we need $a > -(7/15)\, b$.

A consistent theory requires that all the terms consistent with the given symmetries be present. If they are absent, then the process of renormalization will reintroduce them through loop diagrams. We have included the Σ self-interactions, and shown that for a set of parameters the VEV is given by equation (5.19). However there are other invariants that can be formed with Σ and fields already present in the theory; for example:

$$L_{int} = \alpha \Sigma^a_{\ b} \Sigma^b_{\ a} H^c (H^*)_c + \beta (H^*)_a \Sigma^a_{\ b} \Sigma^b_{\ c} H^c. \tag{5.22}$$

Looking at the coupling of the H fields to the $\langle \Sigma \rangle$, we discover that the Higgs now has a mass of order the GUT scale v. Although careful choice of α and β can arrange it so that only the Higgs color triplet gets a GUT-scale mass, this choice is unstable to small perturbations in the values induced by the renormalization group.

In summary, we start with an $SU(5)$ gauge symmetry and a new field Σ. With a special choice of potential for Σ the $SU(5)$ is broken leading to heavy gauge bosons and the SM gauge fields. The Higgs has a color triplet that is problematic to keep it sufficiently massive while keeping the SM Higgs light. What has all this complication bought us? First, we may be able to exchange three gauge couplings for one gauge coupling. It explains the odd hypercharge assignments of the SM. It explains charge quantization. It provides a possible source for the weak mixing angle. We have mass relations and coupling constant relations that, although esthetically attractive, disagree with experiments. We also have predictions for the rate of proton decay based on the derived energy scale of approximate unification which also disagree with experiment. There is no room for massive neutrinos which are now needed to agree with neutrino oscillation observations. The $SU(5)$ GUT requires some upgrades.

5.2.3 *Supersymmetry*

There are many upgrades to the original $SU(5)$ unification which have been extensively researched. The most significant upgrade is that of supersymmetry, which solves many of the issues highlighted above. First, supersymmetry changes the renormalization group resulting in the coupling constants unifying at a higher-energy GUT scale as seen in the right figure found on page 274. The unification is also now nearly perfect and in excellent agreement with experiment. This also adjusts and makes the prediction of the weak mixing angle in excellent agreement with experiment. Also, because unification occurs at about a 100 times higher energy scale, and because the rate of proton decay is suppressed with the fourth power of the unification scale, the rate of proton decay is now also in agreement. Each of these agreements involves some model dependent assumptions.

Supersymmetry also provides two more surprising bonus GUT predictions. The renormalization of the Higgs field's mass squared is driven negative in supersymmetry provided the top quark is sufficiently massive. The process is known as radiative electroweak symmetry breaking. This explains the odd Higgs potential; specifically the negative coefficient of the Higgs squared is what forces the Higgs to acquire a VEV and also provided an early prediction for the mass of the very massive top quark. Another surprise is that in supersymmetry the top quark's mass is very near a fixed point of the supersymmetric renormalization group. Further, the most often mentioned reason for supersymmetry is that it dramatically improves the renormalization properties of the Higgs mass squared term. Without supersymmetry, the mass of the Higgs should naturally run to the Planck scale via mass renormalization contributions from loop diagrams. To keep the physical mass of the Higgs on the 100 GeV scale, fine tuning of counter-terms to one part in 10^{16} is necessary. In contrast, exact supersymmetry eliminates all mass renormalization since contributions between loop diagrams of supersymmetry partners exactly cancel, because fermion loops pick up an extra factor of -1 compared to the corresponding loop containing a supersymmetric bosonic partner.

While complete cancellation between fermion and boson loop contributions does not occur after spontaneous supersymmetry breaking, the remaining contributions are a function of the fermion/boson mass differences. This results in very soft running of the Higgs mass, which is much more naturally cancelled by counter-terms, than is the SM fine tuning. Last, supersymmetry provides one possible source of stable dark matter (which accounts for approximately 24% of all mass-energy in the universe being approximately 8 to 9 times more common than baryonic matter). The lightest stable supersymmetric particle (LSP) may well be the glue that binds galaxies together, as well as galactic clusters.

Unfortunately this supersymmetry does not solve all the problems with GUTs. Supersymmetry for example does not help with the Higgs color triplet's mass. Nor does it do much to help explain the Yukawa couplings or the neutrino masses and mixings. Since supersymmetry doubles the number of all force and matter particles from 63 (including the Higgs doublet and the graviton) to 126, many more free parameters arise. Actually, in supersymmetric models all matter interactions must be expressed in terms of particles of one chirality. While the SM Higgs doublet h and its complex conjugate $h^\dagger$ provide mass for each component of a weak doublet matter particle, supersymmetry necessitates the existence of separate h_{up} and h_{down} doublets of the same chirality to provide mass to the each component of a weak doublet. Hence, the number of particles in the simplest supersymmetry version of the SM, referred to as the Minimal Supersymmetric Standard Model (MSSM), contains 130 particles–65 bosons and 65 fermions. After allowing for soft supersymmetry break, the simplest MSSM consistent with the SM possesses over 100 additional free parameters that can only be determined experimentally. Most theorists do not expect a fundamental, underlying theory to possess more than a couple of free parameters, let alone over 100.

5.2.3.1 Extended Supersymmetry

While standard supersymmetry may seem complicated enough, extended numbers of supersymmetries are also possible, up to $N=8$. The N corresponds to the number of particles of different spin that combine into a single multiplet with identical gauge transformation properties. For example, $N=2$ supersymmetry pairs up two bosons with two fermions. Associated with every left-handed spin-$\frac{1}{2}$ fermion are two complex spin-0 bosons and a right-handed spin-$\frac{1}{2}$ fermion (forming a so-called 'hyper-multiplet'), which guarantees that all matter representations are non-chiral. These $N = 2$ matter reps are then contained in all higher numbers of supersymmetry. Hence, $N = 1$ supersymmetry is the only supersymmetry consistent with a chiral model like the SM; all extended ($N \geq 2$) supersymmetries produce non-chiral models. For $N=2$ supersymmetry, the spin-2 graviton is paired with two spin-$\frac{3}{2}$ gravitinos and one spin-1 vector particle (called a gravi-photon). Spin-1 gauge particles in the adjoint representation are paired with 2 spin-$\frac{1}{2}$ gauginos and 1 spin-0 scalar.

The pattern of particle spin-content in a generic N supersymmetry is perhaps becoming obvious: Representations of N supersymmetry each contain 2^N particles. For a highest spin $s \geq N/2$ particle, the super-multiplet of N-supersymmetries contains $\binom{N}{0} = 1$ spin-s particle, $\binom{N}{1} = N$ different spin-$(s-\frac{1}{2})$ particles, $\binom{N}{2} = N$ different spin-$(s-1)$ particles down to $\binom{N}{N} = 1$ spin-$(s-N/2)$ particles. If $s < N/2$, then there can be a change in chirality within a super-multiplet and correspondingly the magnitude of spin increases again, as in the $N = 2$ matter hyper-multiplet given above. In that event spin-$(s-p<0)$ is replaced by spin-$(p-s)$.[8] $N+1$ supersymmetry reps may be formed by combining two N supersymmetry reps, one

[8]The number of particles in a super-multiplet is in some cases doubled by the demand of parity invariance, such as for a graviton super-multiple.

with lowest spin s and the other with lowest spin $s + \frac{1}{2}$, which follows from $\binom{N}{i} + \binom{N}{p+1} = \binom{N+1}{p+1}$. (These distributions may be recognized in connection with the well known Pascal's triangle.)

A representation of N supersymmetry is generated by N pairs of spin-$\frac{1}{2}$ raising and lowering operators $\bar{Q}_{\dot{\alpha}i}$ and Q^i_α, respectively, with $i = 1$ to N, and α denoting the spinor index. These spinors obey the anti-commutation algebra,

$$\{Q^i_\alpha, \bar{Q}_{\dot{\beta}j}\} = 2\sigma^m_{\alpha\dot{\beta}} P_m \delta^i_j \tag{5.23}$$

$$\{Q^i_\alpha, Q^j_\beta\} = \{\bar{Q}_{\dot{\alpha}i}, \bar{Q}_{\dot{\beta}j}\} = 0, \tag{5.24}$$

where the σ are the Pauli matrics. Combining these relations with the commutation rules

$$[P_m, Q^i_\alpha] = [P_m, \bar{Q}_{\dot{\beta}j}] = 0 \tag{5.25}$$

$$[[P_m, P_n]] = 0 \tag{5.26}$$

forms a superalgebra (a.k.a. a graded Lie algebra). The numbers of states of given spin $s + \frac{1}{2}p$ correspond to the number of distinct anti-commuting sets of p $\bar{Q}$'s that can be chosen from the complete set of N $\bar{Q}$'s. In general supersymmetry is a global symmetry. However, it is made local by the gravitino gauging this symmetry, i.e., transforming it into supergravity. N supersymmetries requires N distinct massless gravitinos. When the gravitino gains mass, this local supersymmetry is spontaneously broken.

As stated above, there is a new phenomenological difficulty with higher numbers of supersymmetry. That is, $N > 1$ supersymmetry cannot provide chiral matter, while matter in the universe is definitely chiral. Nevertheless, extended supersymmetries are interesting to study because of their connection with additional non-renormalization theorems and their association with the simplest geometric compactifications of string models from 10 to 4 large spacetime dimensions. Simple geometric compactification of a 10 spacetime dimensional $N = 1$ supersymmetric model down to 6 large spacetime dimensions increases the supersymmetry $N = 2$, and compactifying further down to 4 large spacetime dimensions increase the supersymmetry to $N = 4$. Only special classes of compactified space can reduce these extended supersymmetries to $N = 1$.

5.3 Higher-Rank GUT Unification

Another class of upgrades involves unifying the SM into GUT groups of the same or higher rank than $SU(5)$. First, a variation of $SU(5)$ has been proposed based on the fact that the SM matter (including now a left-handed anti-neutrino) can actually be divided among a set of **1**, $\mathbf{\bar{5}}$ and **10** $SU(5)$ reps in two ways. In the first case, the electromagnetic charge is completely embedded in $SU(5)$ (making this a true GUT); in the second, the electromagnetic charge is shared between the $SU(5)$ and an additional $U(1)$ (making this a semi-GUT). Models of the latter type are referred to as flipped $SU(5)$–the reason for this particular nomenclature will be obvious shortly. As mentioned prior, in the case of the standard $SU(5)$, elimination of a phenomenologically dangerous $U(1)$ anomaly requires the sum of the electromagnetic charges of all states in a given GUT rep must be zero, i.e., the trace of $Q_{\rm EM} = 0$. This constraint is strong enough to completely determine the SM particle embedding, as specified in (5.12) and (5.13).

In the alternate version of $SU(5)$, the up and down quarks, and also the neutrino and electron leptons, are exchanged (or "flipped") between reps. This variation is thus known as "flipped" $SU(5)$. In terms of left-handed states,

- the $(\mathbf{1},\mathbf{1})_{+1}$ anti-electron becomes a $\mathbf{1}$ rep;
- the $(\mathbf{1},\mathbf{1})_{-1/2}$ neutrino/electron doublet with net $Q_{\rm em} = -1$ and the $(\overline{\mathbf{3}},\mathbf{1})_{-2/3}$ anti-up with net $Q_{\rm em} = -2$ are combined into a $\overline{\mathbf{5}}$ rep; and
- the $(\mathbf{3},\mathbf{2})_{1/6}$ up/down quarks with net $Q_{\rm em} = +1$, the $(\overline{\mathbf{3}},\mathbf{1})_{+1/3}$ anti-down with net $Q_{\rm em} = +1$, and the $(\mathbf{1},\mathbf{1})_0$ anti-neutrino, with net $Q_{\rm em} = -1$ are combined into a $\mathbf{10}$.

Clearly, in each representation $Q_{\rm EM} \neq 0$ To enable the embedded component of the electromagnetic charge to be traceless, an additional electromagnetic charge component must be carried by the additional $U(1)$ factor, making the gauge group $SU(5) \times U(1)$, with the reps for each generation being $\mathbf{1}_{-5}$, $\overline{\mathbf{5}}_3$, and $\mathbf{10}_1$, where the subscript now denotes the additional $U(1)$ charge of the representation.

Why has flipped $SU(5)$ been studied alongside the standard $SU(5)$, when the latter better embeds the electromagnetic charge? One issue, as worked out prior, is that to break the standard $SU(5)$ to the SM gauge group, a spin-0 scalar in the adjoint (**24**) rep (or in some optional higher dimensional rep that could be substituted) must acquire a VEV to break the GUT into the SM, since lower dimension non-trivial reps do not contain any electromagnetic charge-free components. In contrast, in flipped $SU(5)$ a component of the **10** (in the new anti-neutrino position) is electromagnetic charge-free. Thus for flipped $SU(5)$ a **10** state can act as a GUT Higgs to break flipped $SU(5)$ to the SM. In quantum field theory this may not seem that significant. However, as later volumes in this series will discuss, it matters in string theory: string models with geometric interpretation of the compact dimensions and $N = 1$ supersymmetry do not contain any scalar particles in adjoint or higher reps. This poses a problem for generic GUT models in string theory–they cannot be broken to the SM gauge group via GUT adjoint Higgs. Flipped-$SU(5)$ is an exception.

Both standard and flipped $SU(5)$ still have a chirality because left- and right-handed states were in different representations of $SU(5)$. A possible solution is to unify into a left-right symmetric group. A common choice is the Pati-Salam gauge group with $SU(2)_L \times SU(2)_R \times SU(4)$. The Pati-Salam group is isomorphic to $SO(4) \times SO(6)$ and is the primary semi-GUT alternative embedding to (flipped) $SU(5)$. The Pati-Salam model contains a left-handed isospin gauge symmetry and a new right-handed isospin gauge symmetry. Lepton number is added as a fourth color making it $SU(4)$. The right-handed isospin symmetry incorporates a right-handed neutrino, therefore massive neutrinos are natural in this model. The lowest dimensional non-trivial reps of $SU(4)$ are the $\mathbf{4}$, $\overline{\mathbf{4}}$, and the $\mathbf{6}$. The adjoint is the $\mathbf{15}$ rep.

However, before considering this semi-GUT, let us first review a slightly simpler model and glean what we can from it first. This latter is known as the left-right symmetric model, with gauge group $SU(3)_C \times SU(2)_L \times SU(2)_R \times U(1)_{Y'}$. As the name implies, the gauge group contains a set of both left- and right-handed weak generators and interactions. The electromagnetic charge is embedded in both $SU(2)_L$ and $SU(2)_R$, but some of the charge is still carried by an extra $U(1)_{Y'}$ and is defined by $Q_{\text{em}} = T_{3L} + T_{3R} + Y'$. The left-handed SM states appear as:

- $(\mathbf{3}, \mathbf{2}, \mathbf{1})_{+1/6}$ for the up and down quarks,
- $(\overline{\mathbf{3}}, \mathbf{1}, \mathbf{2})_{-1/6}$ for the anti-up and anti-down quarks,
- $(\mathbf{1}, \mathbf{2}, \mathbf{1})_{-1/2}$ for the neutrino and electron, and
- $(\mathbf{1}, \mathbf{1}, \mathbf{2})_{+1/2}$ for the anti-electron and anti-neutrino.

The Higgs doublet now appears as a $(\mathbf{1}, \mathbf{2}, \mathbf{2})_0$ rep.

The remaining step to convert this to a Pati-Salam $SU(4)_C \times SU(2)_L \times SU(2)_R$ model is to form $SU(4)_C$ from $SU(3)_C$ and the Y'. In terms of these gauge groups, the adjoint of $SU(4)$ is expressed as

$$\mathbf{15} = (\mathbf{8})_0 + (\mathbf{1})_0 + (\mathbf{3})_{-4/3} + (\overline{\mathbf{3}})_{4/3}. \tag{5.27}$$

This embedding combines the $\mathbf{3}$ ($\overline{\mathbf{3}}$) and singlets of $SU(3)_C$ into $\mathbf{4}$ ($\overline{\mathbf{4}}$) reps of $SU(4)_C$ as:

- $(\mathbf{4}, \mathbf{2}, \mathbf{1})_{+1/6}$ for the up and down quarks, and the neutrino and electron, and
- $(\overline{\mathbf{4}}, \mathbf{1}, \mathbf{2})_{-1/6}$ for the anti-(up and-down quarks, electron and neutrino).

The Higgs doublet can remain as a $(\mathbf{1}, \mathbf{2}, \mathbf{2})_0$ rep since it was a singlet under both $SU(3)_C$ and Y'.

Another unification structure commonly used is $SO(10)$. The set of matter reps of a single generation of either the rank-4 (flipped) $SU(5)$ or the rank 5 Pati-Salam model can be combined into a single matter rep of dimension 16, which is the dimension of the smallest spinor rep of rank 5 $SO(10)$. Here the 16 particles of the Standard Model are arranged into a chiral spinor representation of $SO(10)$, the components of which can be defined through a set of 5 $\pm\frac{1}{2}$ charges, $(\pm\frac{1}{2}, \pm\frac{1}{2}, \pm\frac{1}{2}, \pm\frac{1}{2}, \pm\frac{1}{2})$. The $\mathbf{16}$ is the set of all combinations of the 5 charges that contain an even number of "−" signs; the $\overline{\mathbf{16}}$ is the set of all combinations of the five charges that contain an odd number of "−" signs. The first three charges specify the $SU(3)$ representation and the next three specify the $SU(2)$ representation of the SM: The hypercharge is defined as $\frac{1}{3}$ times the sum of the first three charges + $\frac{1}{2}$ times the sum of the last two charges. Thus, in terms of the signs of the five charges, the 16 states of each generation can be expressed as,

state	$SU(3)_C$	charges		$SU(2)_L$	charges
u_r	$-$	$+$	$+$	$+$	$-$
u_g	$+$	$-$	$+$	$+$	$-$
u_b	$+$	$+$	$-$	$+$	$-$
d_r	$-$	$+$	$+$	$-$	$+$
d_g	$+$	$-$	$+$	$-$	$+$
d_b	$+$	$+$	$-$	$-$	$+$
u_r^c	$+$	$-$	$-$	$-$	$+$
u_g^c	$-$	$+$	$-$	$-$	$+$
u_b^c	$-$	$-$	$+$	$-$	$+$
d_r^c	$+$	$-$	$-$	$+$	$-$
d_g^c	$-$	$+$	$-$	$+$	$-$
d_b^c	$-$	$-$	$+$	$+$	$-$
ν	$-$	$-$	$-$	$+$	$-$
e	$-$	$-$	$-$	$-$	$+$
ν^c	$+$	$+$	$+$	$+$	$+$
e^c	$+$	$+$	$+$	$-$	$-$

In $SO(10)$, two Higgs doublets, with two exotic Higgs triplets, appear in a vector **10** rep. For $SO(10)$, neutrino masses are natural and they predict a three-way mass unification $m_b = m_\tau = m_{top}$ at the GUT scale. In some supersymmetry models, this configuration is still experimentally viable after accounting for two-loop renormalization group running of the parameters. The $SO(10)$ unification group breaks down to either $SU(5)$, Pati-Salam $SU(2)_L \times SU(2)_R \times SU(4)$ or straight to the SM. In terms of $SU(5) \times U(1)$, the adjoint (**45**) of $SO(10)$ is expressed as

$$\mathbf{45} = (\mathbf{24})_0 + (\mathbf{1})_0 + (\mathbf{10})_4 + (\overline{\mathbf{10}})_{-4}. \tag{5.28}$$

Alternately, in terms of $SU(4) \times SU(2) \times SU(2)$, the adjoint (**45**) of $SO(10)$ is expressed as

$$\mathbf{45} = (\mathbf{15}, \mathbf{1}, \mathbf{1}) + (\mathbf{1}, \mathbf{3}, \mathbf{1}) + (\mathbf{1}, \mathbf{1}, \mathbf{3}) + (\mathbf{6}, \mathbf{2}, \mathbf{2}). \tag{5.29}$$

While no gauge group larger than $SO(10)$ is necessary to place each generation of matter states in a single gauge group representation,[9] extending $SO(10)$ into a gauge group of one higher rank, specifically E_6 (one of the 5 "exceptional" gauge groups), can also unify one pair of SM Higgs doublets with each generation, thereby guaranteeing the appearance of SM Higgs. When $SO(10)$ is enlarged to E_6, the generational matter representation in the **16** rep is combined with its own

[9]Note also that any larger gauge group will also contain additional exotic SM-charged states.

generational Higgs in a **10** rep, and with a singlet **1**, to form a **27** rep of E_6 (the smallest dimensional rep of E_6). The adjoint **78** rep of E_6 decomposes in $SO(10) \times U(1)$ reps as:

$$\mathbf{78} = \mathbf{45}_0 + \mathbf{1}_0 + \mathbf{16}_{-3} + \overline{\mathbf{16}}_{+3}. \tag{5.30}$$

There are many more gauge unification structures that are less common than those that we have presented. Our goal here, however, is to introduce you to the basics that are now accessible as a result of the algebraic toolbox built up in previous chapters. We leave examination of further GUTs to the reader.

5.3.1 A GUT Implication

Most GUT models have a interesting feature: they predict the existence of magnetic monopoles, theorized particles that carry a magnetic charge analogous to ordinary particles that carry electric charge. Magnetic monopoles were first hypothesized by Pierre Curie in 1884. The quantum version was developed by P.A.M. Dirac in 1931. In his seminal paper, Dirac proved that the existence of a single monopole anywhere in the universe invoked quantization of electric charge everywhere in the universe. A system consisting solely of an electric monopole–magnetic monopole pair was imagined. The electric monopole and magnetic monopoles were placed at a fixed distance away from each other. The total angular momentum of the electromagnetic field surrounding the two monopoles is proportional to the product of the respective electric and magnetic charges, $q_e q_m$, and is independent of the distance between particles. Since quantum mechanics requires angular momentum to be quantizated in integer units of $\hbar$, the charge product must be quantized in units proportional to $\hbar$. This relationship can also be derived from the Aharonov-Bohm effect (which will be considered in a later volume), when an electron is transported in a closed path around a magnetic monopole. Therefore, a large fundamental unit of electric charge implies an inversely small unit of magnetic charge throughout the universe and vice-versa.

Equivalently, if a single monopole exists anywhere in the universe, electric charge is quantized in integer multiples n of $\frac{1}{2}\hbar c/q_m$. For $n = 1$, the relationship between magnetic coupling strength $\alpha_m = q_m^2/\hbar c = 137$ and electric coupling strength $\alpha_e = q_e^2/\hbar c = 1/137$ is $\alpha_e \alpha_m = 1$. Hence, weak coupling to electric charge corresponds to strong coupling to magnetic charge, and vice-versa. This is an example of a naturally occurring duality. As we shall discuss in later volumes, string theory led to the prediction of a whole class of dualities like this.

Past monopoles searches have been based either on the change in flux a monopole produces when traversing a coil or by its ionization or excitation of atoms. Detection of a magnetic monopole has never been confirmed. However, magnetic monopoles are a natural outcome of GUTs. If the universe passed through a GUT stage soon after the big bang, magnetic monopoles would have been produced in abundance, with around one monopole per Planck volume. Further, magnetic monopoles are stable particles that would not have decayed. Lack of detection of any monopoles was one of the reasons that Alan Guth developed inflation theory. According to this, an observable universe that started out at Planck scale, should contain no more than a single monopole (which makes discovering the one in our universe–if it exists–extremely difficult, to say the least).

The discovery of monopoles has been hoped for by many. One reason is the symmetry their existence would bring to Maxwell's equations. Including a monopole density ρ_m and a monopole current $\mathbf{j}_m$, Maxwell's equations become,

$$\nabla \cdot \mathbf{E} = 4\pi\rho_e, \tag{5.31}$$

$$\nabla \cdot \mathbf{B} = 4\pi\rho_m, \tag{5.32}$$

$$-\nabla \times \mathbf{E} = \frac{1}{c}\frac{\partial B}{\partial t} + \frac{4\pi}{c}\mathbf{j}_m, \tag{5.33}$$

$$\nabla \times \mathbf{B} = \frac{1}{c}\frac{\partial E}{\partial t} + \frac{4\pi}{c}\mathbf{j}_e. \tag{5.34}$$

5.3.2 GUT Summary

The GUTs being studied in today's research are far from problem free; which means there is plenty of room for contributions by diligent graduate students and as-yet-undiscovered geniuses. GUTs introduce new puzzles like the origin of their symmetry breaking, the difference in scales between the GUT and Electroweak scale, and how to keep the SM Higgs field light while making the Higgs field components with $SU(3)_C$ charges super massive (the doublet-triplet splitting problem). Although it seems that GUTs would provide a unifying framework for flavor, they still don't provide a good toolbox for explaining the three generations or the Yukawa couplings.

5.4 Alternate Directions and Quantum Gravity

GUTs are not the end of the story for deeper origins–a GUT symmetry by itself cannot explain its own appearance in our universe (shades of Gödel!), nor does it have an explanation for gravity. For the latter one must turn to quantum gravity theories.[10] String theory (and its successor M/-theory), loop quantum gravity, and

[10]The meaning of theory can at times can be ambiguous. We wish to make it clear to the reader that we use the term here (and throughout the chapter) to mean a consistent (or believed to be consistent) mathematical structure, rather than a well-supported physical law.

more recent theories such as causal dynamical triangulation, causal sets, non-commutative geometry, twistor theory and Hořava-Lifshitz gravity are examples of quantum gravity theories that seek to resolve the issues of the symmetries of matter and of the non-gravitational forces by working backwards after first resolving the conundrums of gravity. Instead of starting with the physics of our universe (classical gravity, and quantum descriptions of strong, weak, electromagnetism, fermions, and Higgs) and then embedding these into a quantum theory of gravity, they start with proposals for a functioning valid quantum theory of gravity, often with no strong force, weak force, or fermion matter content.

From this "top-down" approach to quantum gravity, research is directed towards how it may be possible to reproduce the three other forces and the fermion and Higgs content. In order for any one of these theories to make contact with the reality of our universe, it must find either the SM/MSSM or a GUT inside of its structure, and then it must explain why the rest of the structure is not observed at low energy. The set of plausible viable GUT theories provide a window into the physics only a few orders of magnitude smaller than the Plank scale which is associated with quantum gravity. Therefore GUTs are an appealing place to bring these theories (string theory in particular) into contact with the physics of our world. Thus, as we review the advancements of quantum gravity alternatives, we will at times make connections with GUTs.

We start first with discussion of an aspect endemic to string theory that nevertheless originated long before it–extra spatial directions, which can have both classical and quantum gravitational aspects. Then we examine some generic features of quantum gravity, before presenting a brief summary of string/M theory. Since several more volumes in this series will be devoted to exploration of string/M theory, we limit our discussion of it at this time. We conclude this section by summarizing the fundamental features of loop quantum gravity, causal dynamical triangulation, causal sets, non-commutative geometries, twistor theory, and Hořava-Lifshitz gravity. We hope to pique the reader's interest, inspiring him or her to delve more deeply into these quantum gravity theories.

5.4.1 Extra Dimensions

The concerted drive to understand the forces of nature, including gravity, in a consistent, interrelated manner began long before the beginnings of the SM in the 1950s. Three centuries prior, Newton had worked out a classical understanding of the force of gravity; one century prior James Clerk Maxwell had derived the fundamental equations of electromagnetics, thereby proving that light and magnetism were manifestations of the same force. From Maxwell until the 1930s, gravity and electromagnetics were believed to be the only fundamental forces.

The first attempt at unification of these two fundamental forces was proposed in 1921 by the mathematician Theodor Kaluza. He attempted to unify electromagnetics with gravity by extending Einstein's general relativity (GR) to five-dimensional spacetime. Kaluza was able to separate the resulting equations into the 4-dimensional Einstein field equations (when both indices correspond to 4-dimensional spacetime), Maxwell's equations for electromagnetism (when one index corresponds to 4-dimensional spacetime and the other index to the new spatial direction), and the equation of motion for an extra scalar field (when both indices correspond to the new spatial direction).[11] Five years after Kaluza's proposal, Oskar Klein mathematically compactified the additional spatial direction by curling it up into a circle with very small radius. This compactification of a spatial direction was a pre-cursor to both modern field theory and to string theory. This compactification corresponded geometrically to the symmetry of the $U(1)$ group we've discussed extensively so far. Recall that $U(1)$ implies a rotation in the complex plane – a circle. So by curling the additional spatial direction into a circle, Klein was actually forming a geometrical, spacetime version of the electromagnetism forming $U(1)$.

In spite of its initial success in unifying gravity and electromagnetics (and gaining the eventual support of Albert Einstein), Kaluza-Klein theory lost favor in the 1930s with the realization that gravity and electromagnetism were not the only fundamental forces: The newly discovered strong and weak nuclear forces could not be explained as compact forms of gravity (yet)! Nevertheless sixty years later, the ideas of Kaluza-Klein returned with a vengeance, as the founders of string theory argued for not just 1, but a total of 6 extra spatial dimensions (all likely of Planck-scale length)! While point-like gravitons vibrating in a single compact direction could not produce the nuclear forces, string-like gravitons vibrating in more than one compact dimension with certain duality properties unique to string theory could produce these forces.

In the 1990s the number of proposed compact dimensions rose from 6 to 7 as string theory was subsumed into M-theory, but one or more of the 7 no longer needed to be of Planck length. Alternately, F-theory supporters then requested one further compact dimension, but a temporal direction rather than spatial. Then in the spirit of string theory, brane world theory also assumed the existence of extra dimensions, but ones for which string/M theory related constraints need not be imposed. In these current theories the compact dimensions may be (i) very small (e.g. 10^{-33} cm and closed (e.g., without ends) for string theory, (ii) very small to sub-millimeter and with endpoint boundaries for heterotic M-theory & the first Randall-Sundrum brane model, (iii) very small to sub-millimeter and compact for Type II M-theory, and (iv) very large and non-compact for Randall-Sundrum brane model II and other braneworld models.

With all of the current focus on extra (compact) dimensions, you might ask, "Has experimental evidence for any been found? If not, in what possible ways might they be detected in the near future?" The answers are "No, but experiments are currently underway that could detect one or more of these extra dimensions." Evidence of

[11]The first to examine the possible splitting of 5-dimensional field equations into 4-dimensional Einstein equations and 4-dimensional Maxwell equations was actually Gunnar Mordström in 1914 at the Helsinki University, but little attention was paid to his work then.

extra dimensions could appear as deviations of Newton's law of gravity or in particle interactions at accelerator centers such as Fermilab or CERN.

One of the world leaders in this test has been a research group at the University of Washington in Seattle. For around a decade the research team at U. of Washington has been performing a series of extended Eötvös experiments to place upper bounds on the size of the largest compact dimension, given that it exists. The experimental set-ups of the Eöt-Wash Group have been simultaneously relatively low cost and brilliant. Their first experiment involved measuring with extremely high precision the gravitational torque between two small flat, uncharged discs [1]. The discs were aligned on a common axis and were of equal radius. After asymmetric holes were drilled in the discs, the gravitational torque one disk exerted on the other was measured as a function of rotation angle and disc separation distance. Over a range of angles, the torque was determined to be consistent with a $1/r^2$ force, down to a minimum disc separation distance of ≈ 0.1 mm. Variation of the $1/r^2$ form of the gravity at sub-millimeter scales would provide firm evidence of extra dimensions, strongly inferring M-theory. These results imply that any extra dimensions that do exist must be significantly smaller than 0.1 mm, given that gravity is free to move along these directions. The latter is true for all string and M models, and most string-inspired brane models, but not for the extra non-compact dimensions of Randall-Sundrum type II models. Later, more-complicated Eöt-Wash experiments have provided more precise measurements and further reduced the upper limit on compact dimension size by an order of magnitude.

Another means of extra dimension detection is through time delays in photon emissions from sources in space. The existence of extra dimensions can cause the speed of light to become frequency dependent if light can oscillate in these dimensions. The time delay is least for Kaluza-Klein Planck/string-scale compactified directions and most for models with large extra directions. The SWIFT and GLAST satellite-based detectors, or the VERITAS ground-based TEV gamma-ray detector will be able to distinguish between the different extra-dimensions compactifications [39].

Constraints beyond the Eötvös limits can be placed on the number and size of extra dimensions. For example, if we assume we exist on a 3-brane embedded in a higher dimensional bulk wherein only gravity is allowed to propagate freely (without warping effects), stringent constraints are found. If the bulk (total) spacetime contains an additional n spatial dimensions compactified into an n-sphere of radius R, then the Planck scale for the full $(n+4)$-dimensional space, $M_{\text{Pl},n+4}$, is related to the standard 4-dimensional Planck scale $M_{\text{Pl},4}$ by

$$M_{\text{Pl},4}^2 = R^n M_{\text{Pl},n+4}^{n+2}. \tag{5.35}$$

A large radius R can reduce our effective Planck scale $M_{\text{Pl},n+4}$ to be much lower than $M_{\text{Pl},n+4}$. Therefore, the hierarchy problem can be solved by reducing $M_{\text{Pl},n+4}$ to near the electroweak scale, within the range of 10 to 100 TeV. In this case $n = 1$ is immediately excluded, because then $R \sim 10^8$ cm. $n = 2$ places R right at the current experimental limits of ~ 0.1 mm.

The possibility of sub-millimeter compact dimensions leading to creation of massive Kaluze-Klein gravitons at Fermilab and CERN was also investigated. Appearance of these particles would have led to production of a photon and missing energy at LEP, and production of either a photon or a jet and missing transverse energy at the Tevatron. None of these were seen, placing upper bounds on R at 200 μm to μm for two to five compact dimensions from CERN and upper limits for R at 40 fm for six compact dimensions [59].

Neutrino emission of SN1987A complements these constraints: Interference effects of Kaluza-Klein states could significantly alter the neutrino emission from a Type II supernova. This has led to the bound of $R \lesssim 0.66\ \mu\text{m}$ ($M_{\text{PL},n+4} \gtrsim 31$) TeV for $n = 2$ and $R \lesssim 0.8$ nm ($M_{\text{PL},n+4} \gtrsim 2.7$) TeV for $n = 3$ [38].

Additional tests for extra dimensions are scheduled for the Linear Hadron Collider (LHC) at CERN. Evidence for extra large directions forming the higher dimensional M theory bulk space would be the creation of mini black holes with very short half-lives. The black holes would be generated through particle annihilations that release energy equal to or greater than the energy scale corresponding to the length of an extra spatial dimension. This energy could escape briefly into the bulk, thereby producing an extremely short-lived mini black hole with a very distinct decay scheme [28].

Beginning in 2011, Fermilab also plans tests of a different manner. The lab will set up a detector to investigate the possible existence of a fourth kind of a neutrino that may be bouncing in and out of extra dimensions. This proposal is based on suggestions of the existence of this new type of neutrino from the past MiniBooNE experiment.

5.4.2 *What About (Quantum) Gravity?*

One question the reader may be asking is "Why wasn't gravity included in the SM with the electroweak and strong forces?" Surely Kaluza and Klein were not the last ones to attempt to unify gravity with the three non-gravitational forces! The answer is that gravity is not discussed in the context of the SM because as a quantum field theory, gravity is a drastically different beast compared to electromagnetic and the strong and weak forces [100]. We have discussed in prior chapters how, as quantum field theories, the three non-gravitational forces are exerted via exchange of spin-1 (or vector) particles. In contrast, quantum gravity is exerted by spin-2 (or 'tensor') particles. This might seem like a trivial difference, but it is actually very profound. Recall that in the high energy (equivalently short wavelength) limit of these spin-1 or spin-2 particle exchanges, infinities (a.k.a., anomalies) appear in computations of various amplitudes and probabilities. The significant difference is that for vector forces, the number of infinities can be made finite, whereas for tensor forces, the number of infinities is infinite.

As discussed prior, mathematical techniques known as "renormalization group equations" were developed during the 1950s through the 1970s. Through these equations, renormalization provides physically meaningful methods for eliminating the finite numbers of anomalies that can appear from exchange of spin-1 particles. Renormalization allows finite theoretical values of quantities to be compared (and in some cases set to match) experimentally measured values in a sensible manner. Nonetheless, parallel techniques could not be constructed to completely remove, in a physically meaningful manner, the infinite set of anomalies of a tensor force like gravity. To do so requires specification of an infinite number of parameters, which greatly reduces the usefulness of the theory (or often makes it useless). Such theories are referred to as non-renormalizable [86].

An approach to quantum mechanics that partially avoided the issue of gravitational anomalies was to assume that quantum gravity is, in fact, an effective low energy field theory of a more fundamental theory. In that case, an energy cut-off scale can be used in quantum gravity calculations. Above the cut-off, the effective field theory is not valid. At the cut-off, the anomalies are actually finite, but extremely large. They increase (usually exponentially) with cut-off, but can be absorbed into the infinite parameters of the theory. Then, at energies significantly below the cut-off, only a small finite number of the parameters remain and need to be set to physically consistent values. After this, the effective field theory can provide predictions for other independent physically measurable quantities.

Alternately, Steven Weinberg suggested the possibility of what he termed the **asymptotic safety scenario** [95].[12] Weinberg defined a theory as asymptotically safe if its parameters approach a fixed point as the renormalization momentum scale goes to infinity. Weinberg hypothesized there may exist a non-trivial ultraviolet fixed point at which one can define a consistent theory of quantum gravity, which therefore does not contain any anomalies. Quantum gravity theories that appear to follow the asymptotic safety scenario have appeared recently and will be discussed in some of the following sections.

The classical gravity of Einstein's general relativity and the quantum theory of the SM were seen as the underlying forces in the universe for nearly a century. However, there was a tension between the two nearly from the start. The nature of the underlying tension between the two is the difference between the fundamental natures of each. In general relativity, gravity is the curvature of spacetime and is, therefore, not constant–it changes locally as matter particles and energy sources move. General relativity describes the geometry of spacetime as dynamic. In contrast, quantum mechanics describes particles interacting within a fixed background geometry. Thus, for most of a century classical gravity and quantum theory could not be successfully merged into a unified theory that reduced to quantum mechanics in the low gravity limit, and to Einstein's general relativity in the large mass limit, while also providing a consistent description of interaction forces in the limit of both strong gravity and short distance, i.e., near Planck scale. The difficulties of

[12]The name was motivated by the analogy to the asymptotic freedom properties of non-Abelian gauge theories, indicating that physical quantities are safe from divergencies as the cutoff is removed.

combining general relativity and quantum mechanics can be observed even in the simpler case of calculating SM matter interactions on curved spacetime.

To improve the poor behavior of gravity at short distances (at and below the Planck scale), it has been understood for some time that the nature of spacetime must be significantly altered there. Several theories of quantum gravity have proposed that the number of effective spacetime dimensions must change as small (Planck scale) distances are approached [43].

Possible reconciliation of general relativity and quantum mechanics did not appear until the mid 1980s, with the announcement of a consistent quantum gravity born out of the quantum mechanics of a string-like fundamental particle [32]. From the string theory view point there was no gravitational anomaly–the infinities in standard quantum gravity appear only for quantum gravity as an effective field theory, which has application only significantly below the Planck (string) scale. The implication of string theory is that gravity and quantum mechanics should not be viewed as equals. Rather, quantum mechanics becomes the underlying law. In contrast, general relativity is the inevitable outcome of quantum mechanics and the existence of fundamental particles of finite extension (whether string-like or higher dimensional). String theory implies that extra dimensions open up, while M-theory (its enhancement) suggests our universe may be the boundary of higher-dimensional space. The four dimensions of our large spacetime becomes a course graining effect. The basic features of superstring theory are reviewed in Sect. 5.4.3.

In contrast to theories with dimensions opening up at short scale, alternate theories have been developed that suggest the reverse, that is, the (effective) reduction of spatial dimensions at short distances or the breaking of 4-dimensional spacetime into smaller dimensional units. The latter is referred to as a **Foliation**, which is essentially a decomposition of a manifold into a union of submanifolds of smaller dimension [43]. At sufficiently small scale, the apparent smooth and continuous geometry of the large scale universe may be replaced by discrete structures (submanifolds) or become non-geometric or 'foamy-like'. These proposals suggest the continuity of large-scale spacetime becomes an emerging concept as the distance scale grows.

A semi-recently proposal for a consistent version of quantum gravity fitting in this category is loop quantum gravity. However, in contrast to string theory, loop quantum gravity does not attempt to unify quantum gravity with the non-gravitational forces. Instead, it provides a means of quantizing gravity while keeping it distinct from the other forces. Loop quantum gravity seeks to resolve the fundamental issue of the drastically different natures of gravity and quantum mechanics through background-independent quantum gravity. To do so, it develops a form of quantum mechanics that does not necessitate a background. The equations of loop quantum gravity are not embedded into spacetime, except for its invariant topology. The existence of spacetime is not presupposed either. Loop quantum gravity is examined in more detail in Sect. 5.4.4.

A large number of additional methods for quantizing gravity have been proposed. These theories vary in terms of which properties of general relativity and quantum theory are kept and which are altered. Examples include causal dynamical triangulation (5.4.5), causal sets (5.4.6), non-commutative geometry (5.4.7), and twistor theory (5.4.8).

The infinities of quantum gravity can also be better controlled if higher-order spatial derivatives, but not higher-order time derivatives, appear in the Lagrangian. This process breaks the Lorentz symmetry of general relativity. Nevertheless a recent proposal, entitled Hořava-Lipshitz (HR) gravity, follows this approach. HR gravity has, in fact, been very well received because it argues this particular type of Lorentz symmetry breaking can be limited to the short distance scale and the symmetry is recovered at the large scale. HR gravity is reviewed in (5.4.9).

5.4.3 String Theory

During the 1980s, a primary motivating force of particle physicists and cosmologists was to better understand the unresolved issues of the SM (many of which we have discussed prior) and find solutions to them. They searched for answers in terms of something more fundamental. A consistent string theory resolution to the SM issues first appeared in the mid 1980s. String theory unifies the strong and electroweak forces of the SM with gravity, and also connects matter with these forces. As a "top-down" quantum gravity approach, many believe that string/M theory (or some additional future enhancement) holds the potential to provide complete explanation for the SM/MSSM and all higher energy effective field theories in nature, that it is a viable contender for a "Theory of Everything." Others believe it is far too early for this conclusion.

String theory reduces the number of basic particles from either 62 (SM) or 128 (MSSM) to 1. The fundamental particle proposed by string theory is essentially a closed string of compact pure energy of Planck-scale length, 10^{-33} cm. This string of energy can act analogous to a violin string. Just as vibrations travel up and down on a violin string, vibrations on the energy string can travel around both clockwise and counter-clockwise. Just as different vibrations on a violin string produce different notes, different combinations of clockwise and counter-clockwise vibrations on the energy string can produce the different "basic" particles of the MSSM (and a vast array more).

In 1984 research revealed a quite odd, but very valuable, constraint that quantum mechanical consistency (i.e., elimination of anomalies) placed on superstring theory: that spacetime could not be just $1 + 3$ dimensional – instead $9 + 1$ dimensions were required! However, since only three spatial dimensions (height, width, and depth) are observed, the founders of string theory theorized that these extra dimensions must be very small. Further, to unify the SM forces with gravity in a consistent way, the lengths of six compact spatial dimensions need to be on par with length as the string itself, 10^{-33} cm. (This means that in moving along one of these directions, the reader would return to his or her starting point after traversing only 10^{-33} cm.)

The existence of these compact directions provided a means by which all forces and matter can be related. In the language of Type II string theory,[13] a spin-two (tensor) particle with the properties of a graviton is produced by a string of energy for which both the clockwise and counterclockwise vibrations are in the everyday large-scale spatial directions (of height, width, and depth). The spin-one (vector) non-gravitational SM (and additional non-SM) forces can be produced when only one of the clockwise or counter-clockwise vibrations is in the large spatial directions and the alternate vibration is in the compact directions. Thus, the non-gravitational forces are non-fundamental. They result from the existence of general relativity and compact dimensions. Without compact dimensions, the only force would be gravity. That is, the non-gravitational forces are in essence special modes of the stringy graviton whereby some oscillations are along the compact dimensions, rather than along the non-compact directions.

Similarly, spin-zero (scalar) matter, (including the Higgs), can be produced when both clockwise and counter clockwise vibrations are in compact directions. What is identified as the spin of a particle in quantum field theory corresponds to the number of a string's vibrations in our large-scale spatial directions. Hence, all scalar matter is also a new mode of gravity. Just as water can come in different forms, so too does the string-derived gravitational force. The specific types and numbers of non-gravitational forces and of matter particles depends on both the size and shape of the compactified dimensions.

What about spin-$\frac{1}{2}$ matter particles though? Recall that superstring theory is, as the name implies, a theory with supersymmetry. Thus, accompanying all integer spin string modes are half-integer spin modes: the spin-two graviton is accompanied by the spin-$\frac{3}{2}$ gravitino; spin-1 gauge particles by spin-$\frac{1}{2}$ gauginos, and (critically importantly for us) spin-0 scalar matter by spin-$\frac{1}{2}$ matter (including the SM particles).

In addition to providing a unified explanation for forces and particles, string theory also offers an explanation for why the MSSM forces and matter are as they are: the specific types and classes of non-gravitational forces and matter particles produced by the vibrations of the energy string depend on the shape of the six compactified directions, just as the sounds emitted by a musical instrument depend on the shape of the instrument. The MSSM forces and matter particles correspond to a specific class of shapes of the compact dimensions! Further, the range of dark matter forces and particles correspond to variations within this class of shapes. In total, our universe would correspond to one particular shape of the compact dimensions within a set of around 100 trillion known as Calabi-Yau manifolds.[14]

Thus, for many years following 1984 the primary effort of string theorists was to determine which of the 100 trillion or so Calabi-Yau manifolds (or their

---

[13]Don't worry if you don't know what "Type II" means. There are different (but physically equivalent) ways of formulating string theory and this is just one of them.

[14]More on what a "Calabi-Yau manifold" is in a much later volume. For now just think of it as a particular way of wrapping up the extra dimensions.

mathematical equivalent) could be the shape of the extra six compact directions of our universe. If the correct compact shape could be found, string theory had a much strengthened argument to likely be the actual theory of everything (ToE). This search continued full scale for over a decade with very significant progress made. First, models containing the MSSM states along with several dozen unwanted exotic particles carrying MSSM charge were constructed. Eventually a few models were found that did not contain these exotic particles (which if they existed and had masses at scales around 100 GeV or lower would already have been discovered). Yet these models still had various phenomenological problems, such as incorrect masses for the SM particles or wrong particle couplings.

An even more underlying, troublesome issue of string theory was that it wasn't actually one single theory, but five alternative theories, with the odd nomenclatures[15] of "Type I," "Type IIA," "Type IIB," "Heterotic $E_8 \times E_8$," and "Heterotic $SO(32)$." In each theory the fundamental string appeared to possess slightly different properties in 10 dimensions. First, Type I and the Heterotic theories had $N = 1$ supersymmetry, while the Type II theories had $N = 2$ supersymmetry (recall our prior discussion of extended supersymmetries). Second, Type I and Heterotic $SO(32)$ have an $SO(32)$ gauge group in 10 dimensions, Heterotic $E_8 \times E_8$ has an $E_8 \times E_8$ gauge group, and the Type II theories had no gauge group in 10 dimensions. Was one theory better than the other four? String theorists worked ardently, but nonetheless unsuccessfully, for a decade to the answer to this question. Simultaneously they were investigating string models in all five theories. However, very soon Type II models compactified from 10 to 4 large directions appeared unable to ever reproduce all of the SM matter particles using the Calabi-Yau's manifolds that provided the correct SM gauge groups. So most of the focus was on the Type I and Heterotic theories.

That is, until 1994 when a few string theorists began to accumulate mathematical arguments that all five string theories were actually the same theory, with equivalent physics expressed by different mathematics. An analogy is like a book that has been written in five different languages. If a person knowing only one of these five languages receives a copy of all five language versions (and the book contains no pictures or artwork), that person would likely assume all five books to be different stories. But one who understands all five languages would immediately recognize that the five books tell the same story. And so it was with the five "different" string theories. During 1994–1995, a mathematical "Rosetta Stone" was developed that translated between the theories. This Rosetta Stone also revealed that the fundamental particle wasn't actually string-like, but was actually torus-like.

Replacement of a 1-dimensional string with a 2-dimensional torus (a.k.a. membrane) as the shape of the fundamental particle shape had profound effects. First, per quantum mechanical consistency, the number of spatial directions could not be

[15]The notation E_8 refers to a Lie group we didn't talk about in this book. It will play a major role in later volumes, but for now it is only necessary to understand that it is simply another Lie group, like $SU(n)$ or $SO(n)$.

"just" 9, but had to be increased to 10. Even more surprising, going from 9 to 10 spatial dimensions had more profound implications that the initial jump from 3 to 9! First, the number of possible shapes for six compact dimensions, a mere 100 trillion or so, increased unfathomably for seven compact dimensions to somewhere between 10^{100} and 10^{1000} (at least)! Second, while string theory was consistent with the existence of a single universe (ours), the new membrane theory more likely implies the existence of not just one, but (at least) a number of universes that is on part with the number of 7-dimensional compact shapes! In other words, not only may our universe not be alone – there could be more than we ever would have guessed! Thus began the current era of Membrane-theory (or M-theory for short).

With the advent of this vast "landscape" of M-theory models (possible universes), the phenomenological goal shifted from studying individual string models to better statistical understanding of the characteristics of the landscape as a whole. Some of the related questions that arose are: how probable is our universe to be found on the landscape? (The answer to which gives evidence for or against the validity of M-theory.) Are the properties of our universe consistent with the landscape? To answer these questions, global properties of the landscape must be better understood. Attention thus focused on mathematically obtaining statistical distributions of various independent and interdependent phenomenological properties. The distributions of physical features, such as observable and hidden sector gauge groups, numbers of matter generations, scales of supersymmetry breaking, etc., and possible correlations between these, began. Initial studies have suggested, for instance, that the SM gauge group may appear in roughly one out of a trillion M-theory models.

Within these ongoing studies, one physical feature receiving much attention is the cosmological constant. The range of cosmological constants found in M-theory models covers the full spectrum from -1 to $+1$, in units of M_{Plank}^4. Further, the models appear to be roughly equally distributed. If our universe is without supersymmetry, its cosmological constant is naively expected to be of order 1. If our universe has supersymmetry, but it broken as expected near the electroweak scale, its cosmological constant should instead be around 10^{-60}. In contrast, based on measured expansion rates of the universe since 1997, it has been discovered to be on the order of 10^{-120}. Why such a small value of the cosmological constant is unexplainable from quantum field theory. So far, the only light shed on the subject has come from Steven Weinberg. In 1984 Weinberg showed that a universe with cosmological constant greater than 10^{-119} could not support life–it would have expanded far too quickly for structures such as galaxies to form. However, many seek an explanation other than from such an anthropic argument. Thus, the goal of M-theory is not to just locate our universe on the landscape of M models, but to also understand why our universe is at that location, *i.e.*, to understand the physical reason for its small cosmological constant. With these thoughts we end our introductory discussion of string/M-theory. Far more detailed discussions will be found in upcoming volumes in this text series. From here we wish to briefly introduce the reader to a more recent (and thus less developed) proposed alternative to string/M-theory, loop quantum gravity.

5.4.4 Loop Quantum Gravity

Many of the features of general relativity carry over to loop quantum gravity (LQG) [76]. However, LQG goes beyond these features by quantizing both space and time at the Planck scale. The origin of LQG goes back to 1986, when Abhay Ashtekar of Penn State University recast general relativity in a form analogous to Yang-Mills field theory [2]. In Ashtekar's approach, Einstein's gravitational field is replaced by a field that acts much like an electromagnetic potential, called the Ashtekar potential. Use of the Ashtekar potential made the appearance of loop variables very natural. Ashtekar's variables describing general relativity were then used in a series of papers by Carlo Rovelli, Lee Smolin, and Ashtekar to construct a basis of states of the related quantum geometry [3, 76, 77]. They showed that the quantum operators for area and volume have a discrete spectrum. The size of these elementary quanta of space can be computed within the theory once a parameter is set.

This theory was constructed as a non-perturbative quantization of diffeomorphism – invariant group theory.[16] Diffeomorphism invariance is essentially general covariance – the invariance of physical laws under arbitrary coordinate transformations. This makes the theory effectively background independent. Diffeomorphism invariance is maintained by the requirement that physical states remain invariant under its generators. In contrast to LQG, perturbative string theory does not have a manifestly background independent formulation.

In LQG, space is represented by a network structure which evolves in discrete steps through time. The network is composed of finite quantized loops of excited gravitational fields, also known as spin networks. When evolved through time, these spin networks develop into a so-called spin foam. These spin networks can be associated with group representation theory. Relatedly, the fundamental objects in loop quantum gravity which are assumed as the basis for a non-perturbative quantization of gravity are a connection (what we have been calling gauge fields), a vierbein, and Wilson loops. The former two will be discussed in the next volume in this series (in addition to explaining why the first is called 'connections') while the latter will be discussed in a later volume of this series. The regularization of gravity is controlled by the diffeomorphism invariance of the vacuum state.

A strength of LQG is that is does not require any currently unobserved phenomenology such as supersymmetry or extra spatial dimensions, and as mentioned, it is a background independent theory.

On the other hand, several critical issues regarding LQG exist. First, while it was shown early on that LQG could produce a massless spin 2 state [4], with the correct propagators in the low energy limit for a graviton [12], an issue under continued debate has been whether LQG actually has a low energy limit that yields general

[16]We'll talk about "diffeomorphisms" in more detail in a later volume.

relativity and then finally Newtonian gravity. Arguments has been constructed very recently by Smolin that general relativity and Newtonian mechanics do indeed result in low energy limits [84].

Smolin's arguments make use of the discovery by T. Jacobson that Einstein's equations of general relativity can be derived directly from the laws of thermodynamics in tandem with the Bekenstein-Hawking proportionality between entropy within a volume of space and the surrounding area (known as a holographic principle) [50] and on E. Verlinde's analogous arguments for similar derivation of non-relativistic Newtonian gravity [91].

Connected with both Jacobson's and Verlinde's proofs is the close connection between topological field theories (which will be explored in upcoming volumes) and the dynamics of general relativity. Topological field theories are to diffeomorphism invariant quantum field theories what harmonic oscillators are to standard quantum field theories. Both respectively provide the Hilbert states on which the full non-linear dynamics act. Topological field theories are also directly connected with LQG and underlie Ashtekar's reformulation of gravity in [2] and the spin-foam concept that followed [84].

Interestingly, in his derivation of Newtonian gravity, Verlinde supported the view that gravitation is an emergent phenomena and, relatedly, suggested that there are no fundamental degrees of freedom associated with the geometry of spacetime. Instead, the fundamental degrees of freedom would come from the quantization of general relativity or another diffeomorphism invariant theory.

It should be emphasized that Smolin did not claim that classical spacetime emerges from LQG. Rather, he acknowledged he had to assume the existence of a classical spacetime on the large scale. Thus, what he showed is, if a classical spacetime emerges from LQG, then it will necessarily exhibit Newton's law of gravity.

Another issue with LQG is there are several varieties of its key dynamic equation, the Wheeler-DeWitt equation, and it is not known which, if any, is the correct one. (Unlike the five original perturbative theories of string theory, these varieties are not equivalent.) Further, while the mathematics of LQG may be well defined, a systematic way of computing scattering amplitudes and cross-sections has not been developed.

LQG also contains an undetermined variable known as the Barbero-Immirzi parameter [10, 48]. It is interpreted by some as a renormalization of Newton's constant and specifies the size of the quantum of area. Its numeric value was fixed phenomenologically by matching the number of microstates in LQG with the semiclassical blackhole entropy S (in Planck unites) which was determined by Stephen Hawking to be $S = A/4$, where A is the area of the event horizon of a black hole. The LQG prediction for the entropy only matches S for a specific value of the Barbero-Immirzi parameter. Requiring this parameter to take on an alternate value for other phenomenological reasons, will be considered proof that LQG cannot reproduce general relativity at long distance.

Pure LQG does not include any matter. Nevertheless, there are arguments suggesting that matter will necessarily appear in any quantization of gravity in

(3+1)-dimensions. Thus, while LQG quantized gravity apart from the other forces, matter carrying gauge group charges will likely eventually need to be fit into the theory.

At present, LQG does not make any experimental predictions beyond those of general relativity that can be currently verified or falsified (an unfortunate common trait of most quantum gravity proposals). It does make one prediction that might be verified or falsified in the future though–violation of the constancy of the speed of light. LQG's granular structure of space means that different wavelengths of light could travel at different speeds, therefore violating Lorentz invariance. A similar effect is produced when light travels through crystals.

5.4.5 Causal Dynamical Triangulation

Causal dynamical triangulation (CDT) [58] is another background independent theory that proposes how spacetime itself evolves. The approach of the theory is to divide spacetime into minimal 2-dimensional Planck-size triangular sections (a foliation) and then uses a dynamical triangulation process to map out how spacetime evolves into a 4-dimensional spacetime. The triangular sections interact to form 4-dimensional simplices (plural of simplex), which are also referred to as pentachorons. These are the generalized forms of triangles in 4-dimensions. These simplices become the building blocks of spacetime on the large scale.

This means that near the Planck scale spacetime is 2-dimensional (formed of triangles), while at large-scale it becomes (1+3)-dimensional (formed of simplices). Hence, the theory appears to possess a good semi-classical limit. Specifically, computer simulations have shown that, while in the short distance limit (before lattice effects play in), the **spectral dimension** d_s of CDT is 1.80 ± 0.25, in the long distance limit the spectral dimension undergoes a smooth transition to 4.02 ± 0.1 [5].

As discussed in [43], spectral dimension of a geometric object M is the effective dimension of M as seen by an appropriately defined diffusion process or random walk on M. It is characterized by the probability density $\rho(\mathbf{w}, \mathbf{w}'; t)$ of diffusion from point $\mathbf{w}$ to point $\mathbf{w}'$ in time t, with the initial condition $\rho(\mathbf{w}, \mathbf{w}'; 0) = \delta(\mathbf{w}-\mathbf{w}')$. The average return probability $P(t)$ is obtained by evaluating $\rho(\mathbf{w}, \mathbf{w}', t)$ at $\mathbf{w} = \mathbf{w}'$ and averaging over all points $\mathbf{w}$ in M. In terms of these, the spectral dimension d_s of M is defined as,

$$d_s \equiv -2\frac{d \ln P(t)}{d \ln t}. \tag{5.36}$$

CDT expands on an earlier dynamical triangulation approach by adding a causal constraint: the time-like edge of each 4-simplex was assigned an arrow on it indicating the forward direction in time. Then the only configurations of simplices that are allowed to form are those for which the time arrows of all joined edges are consistent (i.e., pointing in the same direction). This can be shown to preserve causality. By constraining the triangulations to respect a preferred foliation of spacetime (by slices of constant time), CDT avoids the pathological continuum phases of DT (especially by suppressing the appearance of baby universes) and leads to a different continuum limit than does DT, one with much more realistic physical properties [45]. CDT therefore allows a path integral to be calculated non-perturbatively, by summation over all possible allowed configurations of the simplices, yielding all possible spatial geometries.

One strength of CDT is considered to be its ability to construct 4-dimensional spacetime from a minimal set of assumptions that have no adjustable parameters. Another is that CDT offers a nice lattice framework to study non-perturbative field theoretical aspects of quantum gravity, including predictions from other candidate theories [7] (as we shall discuss below).

5.4.6 Causal Sets

An approach related to causal dynamic triangulation is called causal sets [71, 72], which also attempts to model spacetime through a discrete causal structure. However the causal set approach is more general and much more mathematical. It assumes that space-time events are related by a partial order. While space-time is fundamentally discrete, causal structure allows it to retain Lorentz invariance.

Mathematically speaking, a causal set is a set C with a partial-order relation $\preceq$ which is:

(1) Reflexive: If $x \in C$, then $x \preceq x$.
(2) Antisymmetric: If $x, y \in C$, then $x \preceq y \preceq x$ implies x = y.
(3) Transitive: If $x, y, z \in C$, then $x \preceq y \preceq z$ implies $x \preceq z$.
(4) Locally finite: If $x, z \in C$, then the number of elements in the set $\{y \in C \mid x \preceq y \preceq z\}$ is finite.

Causal sets of this form are embedded into a Lorentzian manifold, wherein the elements of the set are mapped into points in the manifold such that the order relation of the causal set matches the causal ordering of the manifold. The causal set is deemed "manifold-like" if the mapping is faithful, i.e., a mapping for which the number of causal set elements mapped into a region of the manifold is proportional to the volume of that region of the manifold. An underlying conjecture of the causal set community is that the same causal set cannot be faithfully mapped into two spacetimes that are not similar on large scales. A principle effort of causal set research is estimating the manifold dimension of a causal set and developing the correct dynamics for causal sets. It is believed this will provide a set of rules for determining which causal sets correspond to realistic spacetimes. A sum-over-histories of viable causal sets can then be performed. Because of its more general format, the casual set approach is much less developed than causal dynamical triangulation.

5.4.7 Non-commutative Geometries

Another growing realm of research that overlaps many different approaches to quantum gravity is non-commutative geometries. It has become a large field that has followed many different paths. However, a fundamental feature of non-commutative geometries common to most all approaches is that spatial translations are equivalent to gauge transformations [73]. In such theories, the distinction between internal and geometrical degrees of freedom fades [26].

Non-commutative algebraic structures like $xy \neq yx$ offers a geometric structure that modifies the standard concepts of points, distances, and differentials, which fit nicely with non-commutativity within the set of quantum mechanic's observables. Just as quantum theory replaces the classical phase-space coordinates of position and momentum with non-commutating operators, classical spacetime coordinates of general relativity can be replaced with generators of a non-commutative algebra. This yields a spacetime not composed of infinitesimal points, but of smeared-out unit cells at a fundamental length scale. This aspect is in common with string/M theory, since the string/membrane is also an indivisible fundamental object like a unit cell. The natural setting for non-commutative geometry is the non-commutative space resulting from the algebra of holonomy loops[17] on spaces of spin-connection. There is a similarity between this framework and the spin-network formalism of loop quantum gravity.

5.4.8 Twistor Theory

Twistor theory maps the geometric objects of 4-dimensional (3, 1) Minkowski space into geometric objects in a 4-dimensional space with metric signature (2, 2). Thus, coordinates in twistor space become complex-valued. Twistor theory was originally proposed by Roger Penrose [64] in his investigations of quantum gravity. However, connections were eventually seen by Witten [98] between twistor space and string theory compactified to 4-dimensions. This resulted in Witten's twistor string theory. Witten showed that several unexpected properties of perturbative scattering amplitudes are more clearly identified after Fourier transforming from momentum space to twistor space. He demonstrated this was a consequence of an equivalence between the perturbative expansion of $N = 4$ super Yang-Mills theory and an instanton[18] expansion of a certain topological string theory [98, 99]. F. Cachazo, P. Svrček, and Witten went on to explore the twistor space decomposition structure of a class of one-loop amplitudes in 4-dimensional gauge theory [17]. While

[17] Holonomy is yet another concept we'll talk about in the next volume.

[18] Instantons are non-perturbative solutions to the equations of motion of a classical field theory on a Euclideanized spacetime. They are localized in spacetime and appear in path integrals as leading-order quantum corrections to classical systems. Instantons will be studied in later volumes.

twistor theory cannot be generalized to other dimensions, it does suggest a way of studying quantum gravity via string theory without having to go beyond 4 spacetime dimensions.[19]

5.4.9 Hořava-Lifshitz Gravity

Hořava-Lifshitz (HR) gravity, as a possible completion of general relativity in the UV limit, has attracted much attention since Petr Hořava's first related publication appeared in 2009 [42]. (At the time of this writing, Hořava's HR gravity publications have received a total of nearly 700 citations.) It formulates quantum gravity as a quantum field theory, with the spacetime metric as the elementary field [45]. Like string theory, HR gravity treats quantum mechanics as more fundamental than general relativity. Like CDT, HR assumes a global time-foliation and enforces unitarity by construction.

However, unlike CDT, HR gravity specifically breaks the four-dimensional diffeomorphism symmetry of general relativity. Time and space are treated differently (anisotropically) to restore good (renormalizable) behavior to gravity at high energy/short distance. Second-order derivatives in time appear in the Lagrangian, in contrast to higher-order (specifically, $2d$ in d spatial dimensions) spatial derivatives appearing. Keeping time derivatives at second order renders the quantum theory unitary – higher-order time derivatives result in ghosts, which destroy unitarity. On the other hand, $2d$ or more spatial derivatives provide power counting renormalizability [87]. They remove (one class of) infinities of the graviton propagator at short distances.

Since time and space are treated differently, the Lorentz invariance of general relativity is clearly destroyed. Nevertheless, while Lorentz invariance is sacrificed in HL gravity as a fundamental symmetry at short distances (in the UV limit), the claim is that this symmetry will emerge at long distances (in the IR limit).

The anisotropic treatment of space and time dimensions appears in their chosen scaling properties:

$$\mathbf{x} \to b\mathbf{x}, \quad t \to b^z t, \tag{5.37}$$

with $z = d$.

z is referred to as the **Dynamical Critical Exponent** and is associated with a given fixed point of the renormalization group. Dynamical critical phenomena correspond to systems with many values of z and often appear in condensed matter physics. It was Lifshitz' early twentieth century work in phase transitions of condensed matter that inspired Hořava's proposal (hence the name) [57]. In condensed matter, a critical Lifshitz point in $(d + 1)$ spacetime dimensions corresponds

[19] We saw the first ideas of twister theory in Sect. 4.3.2.

to a dispersion relation between energy ω (in units of $\hbar$) and (magnitude of) momentum k,

$$\omega = \sqrt{m^2 + k^2 + \frac{k^4}{K^2} + \dots}\,, \tag{5.38}$$

which satisfies

$$\omega \to k^d \quad \text{as} \quad k \to \infty. \tag{5.39}$$

Lorentz invariance in the low momentum (large distance/IR) limit then requires

$$\omega \to \sqrt{m^2 + k^2} \quad \text{as} \quad k \to 0. \tag{5.40}$$

In Lagrangians for quantum field theory, terms appearing in the square root of (5.38) with powers of k^{2z} for $z > 1$ are regulators that help remove the infinities (anomalies) from the theory. They simultaneously produce a Lifshitz point of order z in $(d + 1)$ dimensions with a dispersion relation,

$$\omega = \sqrt{m^2 + k^2 + \sum_{i=2}^{z} \frac{k^{2i}}{K^{2i-2}}}\,. \tag{5.41}$$

In [42] Horava showed (very unexpectedly) that in $(d + 1)$ dimensions a quantum field theory containing polynomial self-interactions of the scalar field to any power is completely free of all infinities in its Feynman diagrams when $z = d$. While theories with gravity are significantly more complicated than scalar theories, Horava was also able to show that for $z = d$, gravitational infinities also disappear by power-counting.

The second underlying assumption of HL gravity is that the spacetime manifold M carries a co-dimension 1 foliation F, formed of $(d = 3)$-spatial dimensional leaves Σ of constant time. This is realized in the Arnowitt-Deser-Misner (ADM) decomposition of spacetime [85],

$$ds^2 = -N^2c^2dt^2 + g_{ij}(dx^i + N^i dt)(dx^j + N^j dt), \tag{5.42}$$

where N is called the **Lapse Function**, and N_i the **Shift Vector**.

To write down an action, this metric requires one to pick a preferred foliation of spacetime. Thus, unlike for general relativity, the corresponding action cannot be invariant under standard diffeomorphisms. Invariance is instead limited to foliation-preserving diffeomorphisms: space-independent time reparametrization and time-dependent spatial diffeomorphisms,

$$\delta t = f(t), \quad \delta x^i = x^i(t, x). \tag{5.43}$$

The surviving symmetry of HL gravity is invariance of the spatial metric $g_{ij}(\mathbf{x}, t)$ under time-independent spatial diffeomorphism. The lack of time-dependent gauge invariance means that not only do the tensor polarizations of the graviton propagate, but also the undesirable vector and scalar. Note that the model can be made gauge invariant under the foliation-preserving diffeomorphisms (5.43), in which case the metric also contains the lapse function and the shift vector. The latter play the role of the gauge fields of the surviving diffeomorphisms.

If N_i and N are assigned the same spacetime dependence as the generators, that is $N_i(\mathbf{x}, t)$ is a spacetime field and $N(t)$ is constant along Σ, the theory is said to be **Projectable**. The action for the projectable theory is

$$S = \frac{2}{\kappa^2} \int dtd^D x \sqrt{g} N \left[K_{ij} K^{ij} - \lambda K^2 - V \right], \tag{5.44}$$

where

$$K_{ij} \equiv \frac{1}{2N}(g_{ij} - \delta_i N_j - \delta_j N_i) \tag{5.45}$$

is the extrinsic curvature of Σ, $K = g^{ij} K_{ij}$, λ is a dimensionless running coupling (in analogy to the Barbero-Immirzi parameter of LQG), and V is an arbitrary remaining diffeomorphism-invariant local scalar functional build out of g_{ij}, its Riemann tensor, and the spatial covariant derivatives, but no time derivatives. The external curvature is the only covariant quantity under spatial diffeomorphisms that contains a time derivative.

While there is no fundamental principle behind projectability, it provides enough gauge freedom to allow one to set $N = 1$, as in general relativity. It also significantly reduces the number of invariants V can include.

If, on the other hand, the lapse N is promoted to be a spacetime field $N(\mathbf{x}, t)$, the theory is said to be **Nonprojectable**. Generic nonprojectable actions will produce additional terms for (5.44), constructed from the new field $a_i \equiv \partial_i N/N$.

HR gravity may also be chosen (but is not required) to possess **Detailed Balance**, which requires that the potential V has a special form. Specifically, V must be a square of the equations of motion of a D-dimensional theory of an action. Equivalently, this means V should be derivable from a prepotential W,

$$V = E^{ij} \mathcal{G}_{ijkl} E^{kl}, \tag{5.46}$$

where

$$E^{ij} = \frac{1}{\sqrt{g}} \frac{\delta W}{\delta g_{ij}}, \tag{5.47}$$

and

$$\mathcal{G}_{ijkl} = \frac{1}{2}(g_{ik} g_{jl} + g_{il} g_{jk}) - \lambda g_{ij} g_{kl}. \tag{5.48}$$

The demand for detailed balance significantly reduces the number of terms in V needed to be considered [85]. Nevertheless, detailed balance appears to brings with it serious physical issues, including a non-zero cosmological constant of the wrong sign and intrinsic gravitational parity violation [92]. The first issue has resulted in detailed balance being discarded by many working in the field.

Independent of whether or not a theory is projectable and/or has detailed balance, the extra scalar polarization of the graviton remains unless the HL action is further modified [85]. This unwanted scalar mode can be removed if a local $U(1)$ symmetry-an extension of the foliation-preserving diffeomorphisms-is imposed. The Abelian symmetry is a nonrelativistic form of general covariance. With its appearance, the number of independent spacetime symmetries per spacetime point matches that of general relativity. (Nevertheless, the preferred spacetime foliation is preserved.)

The needed local symmetry is provided by first noting that in a linearized approximation around flat spacetime, the theory exhibits an enhanced symmetry of the shift sector,

$$\delta N_i = \partial_i \alpha, \tag{5.49}$$

where $\alpha(\mathbf{x})$ is a time-independent local symmetry generator. The sought-after nonrelativistic general covariance arises when this symmetry is promoted to a spacetime-dependent gauge symmetry [44]. This requires the introduction of an additional field A that transforms as,

$$\delta A = \dot{\alpha} - N^i \partial_i \alpha. \tag{5.50}$$

For $d > 2$, one additional field is needed that transforms as a Goldstone field,

$$\partial \nu = \alpha. \tag{5.51}$$

All fields but ν can be assigned geometrical interpretations. Thus, the extra $U(1)$ symmetry associated with it does not seem to be related to any diffeomorphism and its physical interpretation remains undetermined so far.

The gauged HL action then appears as

$$S = \frac{2}{\kappa^2} \int dt d^D x \sqrt{g} \{ N \left[(K_{ij} K^{ij} - \lambda K^2 - V(g_{ij}, N) + \nu \Theta^{ij} (2K_{ij} + \nabla_i \nabla_j \nu) \right] \\ -A(R - 2\Omega) \}, \tag{5.52}$$

where

$$\Theta^{ij} \equiv R^{ij} - \frac{1}{2} g^{ij} R + \Omega g^{ij}, \tag{5.53}$$

and Ω is a cosmological constant-like coupling.

According to [44], in the long-distance limit, the propagating gravitons of HR gravity match those of general relativity. Whether this holds for the theory in the long-distance limit of the full non-linear theory remains to be seen. However, that the Schwarzschild spacetime appears to be an exact solution in the same limit is suggestive. In further support of good IR limit properties of HL gravity, [44] also claims the $U(1)$ symmetry forces λ to be equal to 1, as required for general relativity. However, this latter claim is contested in [19].

Lorentz invariance for HR gravity in the infrared also remains to be proven. It is not yet known whether, in the HL gravity models, different objects would experience different limiting speeds of propagation, not all equal to c. It does appear though that global Lorentz symmetries can be embedded into the Abelian gauged version [85]. A general opinion is that significantly more research needs to be performed before the fundamentals of HL models will truly be understood and that without this understanding detailed phenomenological studies may be premature [92].

Intriguingly, both Ambjørn et al. [7], and Hořava [45] have suggested that since both CDT and HL gravity have a preferred foliation structure of spacetime and enforce unitarity by construction, a connection may exist between them. Horařava conjectures that the CDT formulation of quantum gravity represents a lattice version of HR gravity with anisotropic scaling. From this Hořava offers an explanation how the effective dimension of spacetime can change continuously from 2 to 4 in CDT. He argues that as a well-defined geometric quantity, a spectral dimension can be calculated in the continuum approach to quantum gravity with anisotropic scaling. His result is

$$d_s = 1 + \frac{d}{z}, \tag{5.54}$$

where d is the number of large spacetime dimensions. Gravity flowing from a $z = 3$ UV fixed point to a $z = 1$ IR fixed point reproduces the qualitative crossover of d_s. Hořava interprets this to mean that the topological dimension of spacetime is always four, while the spectral dimension changes because of the anisotropic scaling at short distances [45].

Further evidence for the CDT/HL gravity connection comes from the similarity of the respective phase diagrams. Each offers three phases and has similar phase transition lines that meet at a tri-critical point. One phase in each describes a global de Sitter-like spacetime. For HL gravity, a second phase describes a recollapsing cosmology with a big bang and a big crunch, while the third phase breaks time reversal spontaneously, with an expanding big-bang cosmology or a contracting cosmology with a big crunch. In CDT there is a another phase where a string of small universes (uneven in size) are connected by small necks that appear arranged along the time direction. These mini universes can split and merge. In this phase, the overall geometry of a universe appears to oscillate in time. The remaining CDT phase is characterized by the apparent vanishing of the time direction, with a single large 3-dimensional spatial surface, but an immense number of neighboring minimally-sized 3-dimensional surface. This translates into any lack of classical geometry for this phase.

An especially interesting difference between CDT and HL is that the phase structure of the former allows potentially for both an anistropic and an isotropic UV fixed point [6, 7]. For HL, the boundaries between each of the three phases meet at the tri-critical point and are project outward without boundary or additional intersection. In contrast, for CDT, the boundary separating the de Sitter phase and the non-geometric phase appears to have an endpoint. This is believed to be the UV ultraviolet isotropic fixed point at which space and time coordinates scale isotropically rather than anisotropically, while still producing a de Sitter universe. Thus, the existence of a CDT de Sitter UV isotropic fixed point also implies a connection between CDT and the GR asymptotical safety scenario of Steven Weinberg [55,95]. Together CDT, HL, and asymptotic safety all suggest the UV completion of gravity is a quantum field theory of the fluctuating spacetime metric, without additional degrees of freedom or significant variation from standard QFT [45].

5.4.10 Quantum Gravity Summary

As the reader probably now realizes, quantum gravity is a multifaceted realm involving numerous research arenas. Numerous research opportunities abound in quantum gravity–the unanswered questions are too numerous to mention. The crux of the matter can perhaps be summarized as "What happens to spacetime at the Planck scale?" Does the number of spatial dimensions increase as in string/M-theory or (effectively) decrease as in loop quantum gravity, causal dynamical triangulation and Horava-Lifshitz gravity? Is Lorentz symmetry a true symmetry of nature at all scales or just a low-energy occurrence?

As we have pointed out, there are varying conceptual and phenomenological overlaps between the differing approaches to quantum gravity. Non-perturbative string theory likely has connections with non-commuting geometries. Loop quantum gravity and causal dynamical triangulation have much in common, as do causal dynamical triangulation, causal sets, and Hořava-Lifshitz gravity. Many more connections will likely be discovered. Perhaps most importantly, as a collection, these theories indicate that the nature of spacetime must radically change at the Planck scale of 10^{-33} cm. Consistency of quantum field theory demands it. Spatial dimensions may appear or disappear or become murky and undefinable. No one yet knows, but whatever happens to spacetime must be profound. Our hope is that our reviews of quantum gravity proposals may inspire readers to pursue new research paths in quantum gravity. Perhaps this reader will be the one to determine the true form of quantum gravity.

5.5 References and Further Reading

The GUT notes were strongly influenced by the presentation found in [75]. Additional GUT reviews can be found in [68, 82]. A nice introduction to modern Kaluza-Klein theory is [96]. A good first encounter for a string textbooks are [9, 104]. The fundamentals of M-theory are amazingly summarized in the very succinct [88], while a nice collection of M-theory lectures appear in [14]. Smolin reviews various approaches to quantum gravity, including loop quantum gravity, in

[83]. Some of the discussion of loop quantum gravity in Sect. 5.4.4, of the casual sets in Sect. 5.4.6, of non-commutative geometries in Sect. 5.4.7, and of twistor theory in Sect. 5.4.8 follow the presentations in [97], which does indeed provide an excellent and accurate account of these approaches. The review of Hořava-Lifshitz gravity closely follows the combination of discussions in [42–45, 85, 92]. For further reading, outstanding overviews of quantum gravity and the range of approaches to it may be found in [16, 100].

For further readings we suggest [11, 13, 21, 41, 52–54, 60, 67].

References

1. E. Adelberger, et al., Prog. Part. Nucl. Phys. 62 (2009) 102
2. A. Ashtekar, Physical Review Letters 57 (1986) 2244
3. A. Ashtekar, C. Rovelli, and L. Smolin, Journal of Geometry and Physics 8 (1992) 7
4. A. Ashtekar, C. Rovelli, and L. Smolin, Physical Review D44 (1991) 1740
5. J. Ambjørn, J. Jurkiewicz, and R. Loll, Physical Review Letters 95 (2005) 171301, [hep-th/0505113]
6. J. Ambjørn, A. Görlich, J. Jurkiewicz, R. Loll, J. Gizbert-Studnicki, and T. Trześniewski, The Semiclassical Limit of Causal Dynamical Triangulations, arXiv:1102.3929 [hep-th]
7. J. Ambjørn, A. Görlich, S. Jordan, J. Jurkiewicz, and R. Loll, J. Gizbert-Studnicki, and T. Trześniewski, CDT meets Hořava-Lifshitz gravity, arXiv:1102.3298 [hep-th]
8. D. Bailin and A. Love, "Introduction to Gauge Field Theory", Taylor and Francis (1993)
9. D. Bailin and A. Love, "Supersymmetric Gauge Field Theory and String Theory", Taylor and Francis (1996)
10. F. Barbero, Physical Review D51 (1995) 5507
11. D. Benedetti and J. Henson, Physical Review D80 (2009) 124036
12. E. Bianchi, L. Modesto, C. Rovelli, and S. Speziale, Classical and Quantum Gravity 23 (2006) 6989
13. D. Blas, O. Pujolas, and S. Sibiryakov, Journal of High Energy Physics 0910 (2009) 029
14. U. Bruzzo et al., editors, "Geometry and Physics of Branes", Institute of Physics (2003)
15. R. Cahn, "Semi-Simple Lie Groups and Their Representations", Dover (2006)
16. C. Callender and N. Huggett, Physics Meets Philosophy at the Planck Scale: Contemporary Theories in Quantum Gravity, Cambridge University Press, 2001
17. F. Cachazo, P. Svrček, and E. Witten, Journal of High Energy Physics 0410 (2004) 074
18. W. N. Cottingham and D. A. Greenwood, "Introduction to the Standard Model", Cambridge University Press (2007)
19. A. da Silca, arXiv:1009.4885 [hep-th]
20. F. Englert and R. Brout, Physical Review Letters 13 (1964) 321-323
21. V. Ezhela and B. Armstrong, "Particle Physics: One Hundred Years of Discoveries: An Annotated Chronological Bibliography", Springer (1996)
22. R. P. Feynman, R. B. Leighton, and M. Sands, "The Feynman Lectures Vols. I-III", Addison Wesley (2005)
23. J. B. Fraleigh, "A First Course in Abstract Algebra", Addison Wesley (2002)
24. A. French, "Special Relativity", CRC (1968)
25. S. Glashow, "Interactions: A Journey Through the Mind of a Particle Physicist and the Matter of This World", Grand Central Publishing (1988)
26. J. Gracia-Bondia, in "Proceedings of the Summer School 'New Paths Towards Quantum Gravity", pp. 3-58. Booss-Bavnbek, et al. editors, Spring (2010)

M. Robinson, *Symmetry and the Standard Model: Mathematics and Particle Physics*,
DOI 10.1007/978-1-4419-8267-4,

27. H. Georgi, "Lie Algebras and Particle Physics", Westview Press (1999)
28. S. Giddings and S. Thomas, Phys. Rev. D65 (2002) 056010
29. R. Gilmore, "Lie Groups, Lie Algebras, and Some of Their Applications", Dover (2006)
30. R. Gilmore, "Lie Groups, Physics, and Geometry", Cambridge University Press (2008)
31. H. Goldstein, C. P. Poole, and J. L. Safko, "Classical Mechanics", Addison Wesley (2001)
32. M. Greene and J. Schwarz, Physics Letters B149 (1984) 117
33. D. J. Griffiths, "Introduction to Classical Electrodynamics", Benjamin Cummings (1999)
34. G. S. Guralnik, C. R. Hagen, and T. W. B. Kibble, Physical Review Letters 13 (1964) 585-587
35. J. Hakim, "The Story of Science: Aristotle Leads the Way", Smithsonian Books (2004)
36. B. C. Hall, "Lie Groups, Lie Algebras, and Representations", Springer (2004)
37. F. Halzen and A. D. Martin, "Quarks and Leptons: An Introductory Course in Modern Particle Physics", John Wiley & Sons, Inc. (1984)
38. S. Hannestad, Physical Review D64 (2001) 023515
39. T. Harko and K.S. Cheng, Astrophysics and Space Science 297 (2005) 319-326
40. P. Higgs, Physical Review Letters 13 (1964) 508-509
41. C. Hoyle, U. Schmidt, R. Heckel, E. Adelberger, J. Gundlach, D. Kapner, and H. Swanson, Physical Review Letters 86 (2001) 1418-1421
42. P. Hořava, Physical Review D79 (2009) 084008
43. P. Hořava, Physical Review Letters 102 (2009) 161301
44. P. Hořava and C. M. Melby-Thompson, Physical Review D82 (2010) 064027
45. P. Hořava, General Covariance in Gravity at a Lifshitz Point, arXiv:1101:1081 [hep-th]
46. J. E. Humphreys, "Introduction to Lie Algebras and Representation Theory", Springer (1994)
47. T. W. Hungerford, "Algebra", Springer (2003)
48. G. Immirzi, Nuclear Physics Proceeding Supplements 57 (1997) 65
49. J. D. Jackson, "Classical Electrodynamics", Wiley (1998)
50. T. Jacobson, Physical Review Letters 75 (1995) 1260
51. J. V. Jose and E. J. Saletan, "Classical Dynamics: A Contemporary Approach", Cambridge University Press (1998)
52. T. Kaluza, Sitzungsber. Preuss. Akad. Wiss. (1921) 966-972
53. O. Klein, Zeitschrift für Physik a Hadrons und Nuclei 37 (1926) 895-906
54. K. Koyama and F. Arroja, Journal of High Energy Physics 1003 (2010) 061
55. O. Lauscher and M. Reuter, Journal of High Energy Physics 0510 (2005) 050, [hep-th/0508202]
56. R. Barate, et al., Physics Letters B565 (2003) 61-75
57. E. Lifshitz, Zh. Eksp. Teor. Fiz. 11 (1941) 255
58. R. Loll, J. Ambjorn, and J. Jurkiewicz, Contemporary Physics 47 (2006) 103
59. S. Mele, European Physics Journal C33 (2004) 919-923
60. J. Mehra and H. Rechenberg, "The Historical Development of Quantum Theory; v.6", Springer (2001)
61. G. Naber, "Topology, Geometry, and Gauge Fields: Interactions", Springer (2000)
62. G. Naber, "The Geometry of Minkowski Spacetime", Dover (2003)
63. M. Nakahara, "Geometry, Topology, and Physics", Taylor and Francis (2003)
64. R. Penrose, Journal of Mathematical Physics 8 (1967) 345
65. R. Penrose and W. Rindler, "Spinors and spacetime, Volume I", Cambridge University Press (1987)
66. M. E. Peskin and D. V. Schroeder, "An Introduction to Quantum Field Theory", Westview Press (1995)
67. A. Pickering, "Constructing Quarks: A Sociological History of Particle Physics", University of Chicago Press (1984)
68. S. Raby, arXiv:hep-ph/0608183
69. P. Ramond, "Field Theory: A Modern Primer", Westview Press (2001)
70. P. Ramond, "Journeys Beyond the Standard Model", Westview Press (2003)
71. D. Reid, Physical Review D67 (2003) 024034
72. D. Rideout and R. Sorkin, Physical Review D61 (2000) 024002

73. V. Rivelles, Physics Letters B558 (2002) 191
74. J. Rotman, "An Introduction to the Theory of Groups", Springer (1999)
75. G. G. Ross, *Reading, Usa: Benjamin/cummings (1984) 497 P. (Frontiers In Physics, 60)*
76. C. Rovelli and L. Smolin, Nuclear Physics B331 (1990) 80
77. C. Rovelli and L. Smolin, Nuclear Physics B442 (1995) 593
78. L. H. Ryder, "Quantum Field Theory", Cambridge University Press (1996)
79. B. Sagan, "The Symmetric Group", Springer (2001)
80. J. J. Sakurai, "Modern Quantum Mechanics", Addison Wesley (1993)
81. J. Schwinger, Classical Electromagnetics, Westview Press (1998)
82. Slansky, Physics Reports 79 (1981) 1
83. L. Smolin, Three Roads to Quantum Gravity, Basic Books (2001)
84. L. Smolin, Newtonian gravity in loop quantum gravity, arXiv:1001.3668.v2
85. T. Sotiriou, Hořava-Lifshitz gravity: a status report, arXiv:1010.3218v2 [hep-ph]
86. M. Srednicki, "Quantum Field Theory", Cambridge University Press (2007)
87. K. Stelle, Physical Review D16 (1977) 953
88. R. Szabo, "An introduction to String Theory and D-Brane Dynamics", Imperial College Press (2004)
89. TEVNPH Working Group, "Combined CDF and D0 Upper Limits on Standard Model Higgs-Boson Production with up to 6.7fb-1 of Data, arXiv:1007.4587v1
90. R. Ricciati, "Quantum Field Theory for Mathematicians", Cambridge University Press (1999)
91. E. Verlinde, Journal of High Energy Physics 1104 (2011) 029
92. M. Viusser, Status of Hořava gravity: A personal perspective, arXiv:1103.5587v2 [hep-th]
93. S. Weinburg, Physical Review Letters 19 (1967) 1264-1266
94. S. Weinberg, "The quantum theory of fields. Vol. 2: Modern applications," *Cambridge, UK: Univ. Pr. (1996) 489 p*
95. S. Weinberg, Ultraviolet divergences in quantum theories of gravitation, in General Relativity: Einstein centenary survey, eds. S.W. Hawking and W. Israel, Cambridge University Press, Cambridge, UK (1979) 790-831
96. P. Wesson, "Five-Dimensional Physics: Classical and Quantum Consequences of Kaluza-Klein Cosmology", World Scientific (2006)
97. www.wikipedia.org
98. E. Witten, Communications in Mathematical Physics 252 (2004) 189
99. E. Witten, Advances in Theoretical Mathematical Physics 8 (2004) 779
100. R. Woodard, "How Far Are We from the Quantum Theory of Gravity?", arXiv:0907.4238 [gr-qc]
101. N. M. J. Woodhouse, "Special Relativity", Springer (2007)
102. C. N. Yang and R. Mills, Physical Review 96 (1954) 191-195
103. A. Zee, "Quantum Field Theory in a Nutshell", Princeton University Press (2003)
104. B. Zwiebach, "A First Course in String Theory," Cambridge (2004)

Index

M. Robinson, *Symmetry and the Standard Model: Mathematics and Particle Physics*,
DOI 10.1007/978-1-4419-8267-4, © Springer Science+Business Media, LLC 2011